Handbook
Soil mix walls

Design and execution

bbri.be
Researches • Develops • Informs

CRC Press
Taylor & Francis Group
Boca Raton London New York

CRC Press is an imprint of the
Taylor & Francis Group, an **informa** business
A BALKEMA BOOK

Handbook Soil mix walls

Published by:
CRC Press/Balkema
Schipholweg 107C, 2316 XC Leiden, The Netherlands

First issued in paperback 2022

ISBN 13: 978-0-367-73396-4 (pbk)
ISBN 13: 978-90-5367-641-7 (hbk)

It is allowed, in accordance with article 15a Netherlands Copyright Act 1912, to quote data from this publication in order to be used in articles, essays and books, provided that the source of the quotation, and, insofar as this has been published, the name of the author, are clearly mentioned.

Editors	Nicolas Denies
	Noël Huybrechts
Publisher	Joshua Tourrich
Book design by	Lydia Slappendel
Translation	Textwerk, Amsterdam, Nicolas Denies & Noël Huybrechts

Liability
SBRCURnet and all contributors to this publication have taken every possible care by the preparation of this publication. However, it cannot be guaranteed that this publication is complete and/or free of faults. The use of this publication and data from this publication is entirely for the user's own risk and SBRCURnet hereby excludes any and all liability for any and all damage which may result from the use of this publication or data from this publication, except insofar as this damage is a result of intentional fault or gross negligence of SBRCURnet and/or the contributors.

CONTENTS

Preface

In recent years, the soil mix wall has become increasingly popular in both Belgium and the Netherlands. In practice, however, each supplier of these walls has more or less its own approach to design and execution (including in terms of safety philosophy and risk consideration). There is a lack of a generally accepted and widely used handbook which includes requirements and criteria regarding design, execution, management and maintenance. This makes it inconvenient for all concerned, because there is no direct reference serving as basic instructions for these type of walls.

For this reason, designers and contractors have been asked to pool knowledge about and experience with soil mix walls. The result is the handbook you are currently reading, which was created through a joint initiative between the BBRI and SBRCURnet.

There is a certain difference in the design approach between Belgium and the Netherlands. These differences will be extensively discussed in this handbook.

SBRCURnet/BBRI committee 1794 'Soil Mix Walls' has closely supervised the creation of this handbook.

The following people were involved in the publication of this handbook:
- Geerhard Hannink, chairman, Municipality of Rotterdam
- Noël Huybrechts, editor, BBRI/KU Leuven
- Nicolas Denies, editor, BBRI
- Alain Barthélemy, SECO
- Maurice Bottiau, Franki Geotechnics
- Richard Boogaarts, Municipality of Breda/COBc
- Flor De Cock, Geotechnical Expert Office Geo.be
- Koen Duyck, ABEF/Soiltech
- Marc Everaars, Grontmij Nederland B.V.
- Rijk Gerritsen, Witteveen+Bos
- Edward Grünewald, Engineering Agency Inpijn Blokpoel
- Flip Hoefsloot, Fugro GeoServices B.V.
- Erwin de Jong, Geobest B.V.
- Johan de Jongh, Heijmans
- Martin de Jonker, SGS Intron
- Oskar de Kok, Winmix
- Bart Lameire, ABEF/Lameire Foundation Technologies
- Eric Leemans, ABEF/Soiltech

- Jeroen de Leeuw, ConGeo b.v.
- Richard Looij, Smet Keller Foundation Technologies V.O.F.
- Jan Maertens, Jan Maertens BVBA
- Remmelt Mastebroek, Fugro GeoServices, formerly Wiertsema & Partners b.v.
- Harald Muller, Aduco Nederland b.v. - Lareco
- Peter Maas, Winmix
- Bas Snijders, CRUX Engineering B.V.
- Eelco van der Velde, Bauer Foundation Technologies B.V.
- Fred Jonker, coordinator, SBRCURnet

In addition, valuable contributions were made by Henri Havinga (Deltares), Erik Geurtjens (Engineering Agency Inpijn-Blokpoel), Dick-Peter Heikoop (ConGeo b.v.), Kees-Jan van der Made (Wiertsema & Partners b.v.), Hilda Pool-de Boer (Municipality of The Hague), Ad Verweij (ARCADIS Nederland BV, formerly Deltares), Aad van der Horst (BAM) and Aida Jusufagic (Municipality of Emmen). For the creation of this handbook, financial contributions were made by:
- ABEF
- Bauer Foundation Technologies B.V.
- Congeo
- FCK-CT
- Franki Geotechnics
- Geobest B.V.
- Heijmans
- Engineering Agency Inpijn Blokpoel
- Lareco Nederland BV (formerly Aduco Nederland b.v. - Lareco)
- Rijkswaterstaat GPO
- Winmix
- Wiertsema & Partners b.v.
- BBRI

SBRCURnet and the BBRI want to thank these institutions, as well as all members of the committee, who with great enthusiasm and effort collaborated on the creation of this handbook.

Finally, the SBRCURnet/BBRI committee would like to emphasise its conviction that this handbook meets a real need in the market and that it provides a solid, scientific foundation for the implementation of soil mix technology in the execution of earth-retaining, water-retaining and cut-off walls. This handbook will also serve as a basis for the further development and potential new applications of this promising and sustainable technology.

On behalf of the SBRCURnet/BBRI committee:

Geerhard Hannink - Chairman of the SBRCURnet/BBRI committee 1794 'Soil mix walls'
Noël Huybrechts - Head of the Geotechnical Division BBRI
Fred Jonker- Programme Manager Geotechnology and Soil - SBRCURnet

Summary

This manual on the design and execution of soil mix walls has been drawn up under the responsibility of a joint committee of SBRCURnet (the Netherlands) and the Belgian Building Research Institute (BBRI, Belgium).

For several decades now, the deep mixing method has been used for ground improvement works. A more recent application is the use of soil mix as structural elements for the construction of earth-water retaining structures and cut-of walls. Since 2000, due to the economic and environmental advantages of the method, these particular applications have shown an amazing growth in Belgium and in the Netherlands. This publication was initiated because no pragmatic standards or guidelines were available for the execution and the design of this kind of applications.

The publication is based on existing literature and the knowledge and experiences of committee members, and includes an extensive description of the design and execution processes. It also establishes the link between the conditions of use (functional requirements), the design and the quality control of the final soil mix structure that is especially important in the construction of soil mix walls.

Based on a large test campaign, a methodology is proposed for the design of the soil mix walls for which the interaction between steel and soil mix can possibly be taken into account dependent upon the application. Each potential function of the soil mix wall is described (e.g. earth retaining wall, cut-of wall, etc.) and the temporary or permanent character of the application (its lifetime) is always considered. Furthermore, the design methodology presented in this handbook is in agreement with the Eurocodes.

The publication includes aspects such as the hydro-mechanical characterisation and the durability of the soil mix material, the interaction between steel and soil mix and the monitoring and quality control of soil mix structures. With the application of the knowledge and methods included in this publication it is aimed to contribute to the realisation of soil mix walls of high quality and to minimize the risk of calamities or damage.

Notations

Below is a list of symbols and definitions used. In terms of the symbols and definitions that are related to the geometry of the wall (dimensions, position steel beam, cover, …), these are also clarified in the following image.

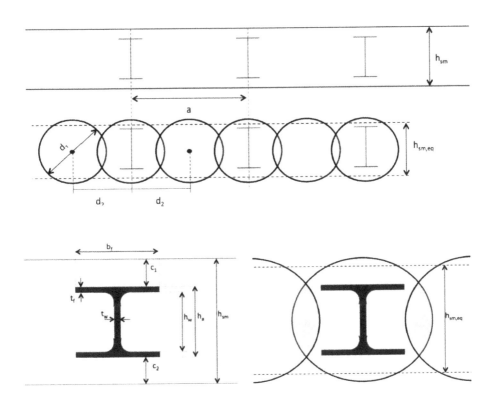

Definitions and symbols

Wall geometry

h_{sm}	thickness of a panel wall	m
$h_{sm,eq}$	equivalent thickness of a column wall	m
d_1 / d_2	column diameter resp. axis-to-axis distance of the columns of a wall	m
c_1 / c_2	concrete cover on the steel beam (c_1 side pressure zone, c_2 side tensile zone)	m
a	axis-to-axis spacing of the steel beam	m
a_{eq}	average axis-to-axis spacing of the steel beam	m
e	eccentricity of the steel beam	m
b_{e1}	effective width for the bending stiffness of the wall	m
b_{e2}	effective width for the moment capacity of the wall	m

Steel beam (values per beam)

b_f	flange width of the steel beam	mm
h_a resp. h_w	theoretical height of the beam resp. height of the web	mm
t_f resp. t_w	thickness of the flange resp. thickness of the web	mm
A_a	total surface area of the cross-section of the steel beam	mm^2
A_f	surface area of the cross-section of the flange of the steel beam = thickness t_f x effective width b_f (may also be separated into bottom flange A_{f1} and upper flange A_{f2})	mm^2
A_w	surface area of the cross-section of the web of the steel beam = thickness t_w x height h_w	mm^2
W_{el}	elastic section modulus	mm^3
W_{pl}	plastic section modulus	mm^3
f_{yk}	Yield strength of the steel	N/mm^2
γ_{M0}	partial safety coefficient on the steel stresses (e.g. 1.0 or 1.1)	-
E_a	design value of the modulus of elasticity of the steel	N/mm^2
l_a	moment of inertia of the beam compared to the strong axis y-y	mm^4

Soil mix material

$f_{sm,m}$	average cylinder compressive strength of the soil mix material	MPa
$f_{sm,k}$	characteristic cylinder compressive strength of the soil mix material after 28 days	MPa
$f_{sm,tk}$	characteristic cylinder tensile strength of the soil mix material after 28 days	MPa
$f_{sm,d}$	design value of the compressive strength $= \beta \times \alpha_{sm} \times f_{smk} / (\gamma_{SM} \times k_f)$	MPa
$f_{s,td}$	design value of the axial tensile strength of the soil mix material	MPa
$\tau_{Rsm,d}$	design value of the shear resistance of the soil mix material	MPa
f_{bd}	design value of the adhesion between the soil mix material and the reinforcement	MPa
β	time-correlation factor compared to the 28-day compressive strength for young material	-
α_{sm}	reduction factor on the compressive strength for long-term effects (e.g. 0.85)	-
γ_{SM}	partial safety coefficient on the compressive stresses (e.g. 1.50)	-
k_f	uncertainty factor for in situ realisation of the soil mix material (e.g. 1.1)	-
$E_{sm,d}$	design value of the secant elasticity modulus	MPa
Ψ	reduction factor on the modulus of elasticity for creep behaviour (e.g. 0 or 1)	-
I_{sm}	moment of inertia of the uncracked soil mix cross-section	mm^4
UCS	uniaxial compressive strength (Unconfined Compressive Strength) of the soil mix material	MPa
k	permeability coefficient of the soil mix material	m/s

Wall parameters of the composite section (soil mix + reinforcement)
(values applying to each average axis-to-axis distance a_{eq})

EI-1	bending stiffness of the uncracked composite section	Nmm^2 / axis-to-axis
EI-2	bending stiffness of the cracked composite section	Nmm^2 / axis-to-axis
EI-eff	effective bending stiffness of the composite section	Nmm^2 / axis-to-axis
EI-eff,k	characteristic value	Nmm^2 / axis-to-axis
EI,eff,d	design value	Nmm^2 / axis-to-axis
W_d	design value of the section modulus of the composite section	Nmm^3 / axis-to-axis

Soil parameters

γ	unit weight of the soil	kN/m^3
φ / φ'	undrained/drained friction angle	°
c / c'	undrained/drained cohesion	kPa
k	permeability coefficient	m/s
k_h	Modulus of sub-grade reaction	kN/m^3
δ	wall friction angle	°

Part 1
Chapter 1 Introduction

The soil mix technology, also called deep mixing method, is created using a mixing auger or milling cutter to blend soil up to a great depth (sometimes even up to 25 m) with a binding agent, allowing for the creation of soil mix columns or panels after curing.

Whereas soil mixing (or deep mixing) was originally used mainly for the overall stabilisation of weak soil massifs, since the early 2000s the technique has been increasingly used in our regions for the realisation of soil and water retaining walls. This has become possible due to new developments in the field of mixing equipement and the composition of the injected binding agents, allowing one to achieve typical soil mix material compressive strengths of 1 to 12 MPa.

The use of soil and water retaining soil mix walls has seen a significant increase over the past 10 years. As a result, the need for guidelines for the design and execution of these walls became increasingly important. This is partly because the existing guidelines and standards for deep mixing provide few solutions for these new applications.

In response to this need, the Belgian Building Research Institute (BBRI) – in collaboration with the Belgian Association for Contractors of Foundation Works (ABEF) and the University of Leuven – have conducted extensive research into the characterisation of the soil mix material. This research was funded by the agency for Innovation by Science and Technology (IWT Flanders), the Federal Public Service Economy (FOD Economy) and the Bureau for Standardisation (NBN), and included extensive experimental tests.

Together with the input from numerous experts from Belgium and the Netherlands, the results of this soil mix study have now been included in the SBRCURnet-BBRI 'Handbook Soil mix walls'. This handbook is the result of more than two years of hard work by a joint Belgian-Dutch committee.

The committee has decided to split the handbook into two parts. *Part 1* is a practical document that includes guidelines for execution, designing and control temporary and permanent earth-retaining, water-retaining and cut-off soil mix walls. It is based on the current international knowledge regarding soil mix technology. A detailed summary of this international expertise is included in a separate *Part 2* of the handbook.

Part 1 is divided into ten chapters, which will be explained briefly below. Chapter 2 (part 1) provides a general overview of the execution techniques that are common in Belgium and the Netherlands. The scope of application of the soil mix technology is clearly defined, as well as the conditions under which a soil mix wall can serve specific functions. It also focuses on the discussion with regards to definitive applications (use phase) of soil mix walls. Moving on to Chapter 3, this will translate into a list of requirements for the various possible functions a soil mix wall may serve.

In the form of a matrix, Chapter 4 provides a summary of the various risks that must be taken into account when using soil mix walls. Specific aspects that are related to preparatory activities and the necessary research are included in Chapter 5.

Chapter 6 presents an overview of the design strategy to be followed as well as a detailed description of the new design method that has been created for soil and water-retaining soil mix walls. This design method is inspired by the Eurocode 4 (EN 1994-1-1) approach and, in certain circumstances, although to a limited extent, allows one to take into account the interaction of soil mix and reinforcement steel, which may result in significant savings. Other aspects, such as fire, durability etc., are also discussed in Chapter 6. The new design method for soil mix walls is additionally illustrated by means of two fully-developed calculation examples, which are included in Appendix 4 of part 1. Specific execution requirements and tolerances in the realisation of the different types of soil mix walls as well as the requirements related to quality control of the soil mix material and/or soil mix walls are included in both Chapter 7 and Chapter 8. It is important to mention here that the type and number of control tests that must be conducted is fully dependent on the risk class of the work, wall function, environmental factors and experience of the executor.

Chapter 9 more closely examines the degradation mechanisms that may occur in respect of soil mix walls and how one may approach management and maintenance of soil mix walls.

Finally, Chapter 10 concludes with the main findings, recommendations and any aspects that require additional research.

It is the firm belief of the committee that this handbook meets a real need in the market and that it provides a solid, scientific foundation for the application of soil mix technology in the realisation of earth-water retaining walls and cut-off walls. It will also serve as the basis for future developments and new applications of this innovative and sustainable technology, which by now has proven its broad economic feasibility.

Chapter 2 Soil mix walls:
general observations and area of application

In the soil mix process, a specially-developed machine (auger or cutter) is used to mix local soil with a binding agent, often on the basis of cement, in order to improve the properties of the soil. The soil mix process was first introduced in Japan and Scandinavian countries in the 1970s. For many years now, the soil mix process has proven to be a ground improvement technology. Numerous articles that describe different soil mix methods are referenced in part 2 of this handbook. In the past, many (inter)national articles and research reports were published regarding this technology. In addition, in 2005, the European standard 'Execution of special geotechnical works - Deep Mixing' (EN 14679) was issued. Most of these research projects mainly focused on the general stabilisation of more soft, cohesive soils such as clay, silt, peat and gyttja.

Recently, the soil mix process has increasingly been used for the realisation of earth and water retaining walls in the execution of construction pits. A typical example of a construction pit equipped with such soil mix elements can be seen in I - Fig. 2.1. The increasing interest in soil mix walls can be explained by the fact that it is an economically interesting alternative to concrete secant pile walls, and in many situations, even to the soldier pile walls. By overlapping the soil mix columns or rectangular soil mix panels, one can create a conti-nuous wall, as illustrated in I - Fig. 2.2. Steel H or I beams are applied in the fresh soil mix material to provide resistance against shear forces and bending moments.

I - Fig. 2.1. Soil mix wall with earth- and water-retaining functions: on the left a soil mix wall made with column-shaped elements; on the right a soil mix wall made with panel-shaped elements.

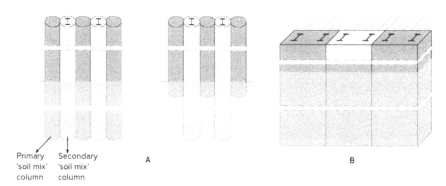

Primary | Secondary
'soil mix' | 'soil mix'
column | column

A

B

1 - Fig. 2.2. Schematic representation of the secant version of (A) soil mix columns and (B) rectangular soil mix panels.

The soil mix material thereby bridges the space between the steel beams (similar to the panels in soldier pile walls). Moreover, through arch effects, it transfers the forces to the reinforcement beams. The most important structural difference between a soil mix wall and the classic secant panel wall is the material used, which consists of a mixture of cement and soil (instead of concrete).

The soil mix technology is also regularly used for the realisation of walls with only an "impervious" function, also referred to as cut-off walls. Such walls are often installed without reinforcement beams.

2.1 Construction principles and execution processes

During the soil mix process, the soil is mechanically (and hydraulically or pneumatically) mixed in place, while the binding agent on the basis of cement or lime (more rarely in Belgium and the Netherlands) is injected using a specially designed machine. The soil mix technology can be classified in accordance with the execution process. In general, one can distinguish two installation methods on the basis of the way in which the binding agent is injected into the soil, namely with or without the addition of water, also referred to as the wet and dry methods.

In the wet method, which is almost exclusively used in Belgium and the Netherlands, a mixture of binding agent and water (if desired, with sand or additives) is injected and mixed into the soil. Depending on the type of soil and binding agent, a mortar-like mixture is formed which cures during the hydration process. For earth and water retaining walls, the amount of binding agent is usually 250 to 500 kg per m^3 column/panel. The injected grout includes a water-binder ratio between 0.6 en 1.2.

In most cases, a certain amount of bentonite (20 to 50 kg per m³) is added to the water-binder mixture to improve its processing properties. Moreover, bentonite in sandy soils can also contribute to promotion of the "impervious" character of the wall.

For applications with solely a cut-off function, the binding composition and wall thickness must be tuned to the project-specific requirements and soil conditions.

If the requirements for permeability are very strict, geomembranes may be installed during execution. Generally speaking, on the basis of the mixing equipment, one can make a distinction between three types of soil mix walls, namely: column walls, panel walls and walls realised by means of a chain cutter (type "chainsaw"device), such as illustrated in I - Fig. 2.3 and 2.4, respectively.

Column walls are installed by drilling one or multiple continuous or discontinuous augers vertically into the soil while panel walls are applied by inserting a cutter into the soil. Simultaneously with the drilling or milling, this equipment is used to inject the water-binder mixture and mix it with the soil.

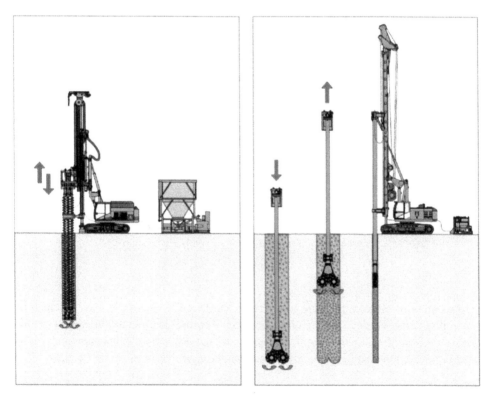

I - Fig. 2.3. Applying the column walls with a multiple shaft system (left) and panel walls (right).

Below is an overview of the characteristic dimensions of the soil mix walls realised in Belgium and the Netherlands by means of these three processes:

- the diameter of a soil mix column is often between 0.4 to 0.6 m (in certain cases this may be up to 1.2 m);
- the typical length for soil mix panels is between 2.2 to 2.8 m;
- on average, the thickness of a soil mix panel is 0.55 m. However, other thicknesses are also possible;
- generally speaking, the thickness of the soil mix trench realised with the chain cutters varies between 0.35 and 0.50 m;
- the execution depth of soil mix columns and panels is usually less than 20 m, but greater depths are also possible. The execution depth of the chain cutter is often limited to approx. 12 m.

Until now, the most common systems, available in Belgium and the Netherlands for the execution of soil mix walls are the CVR C-mix®, TSM by Smet F&C, MIP and SMW methods by Bauer, CSM systems by Bauer and Soletanche-Bachy-Tec System and the Lareco trench mixing. These systems are described in detail in part 2 of the handbook (§2.2.1 to §2.2.3).

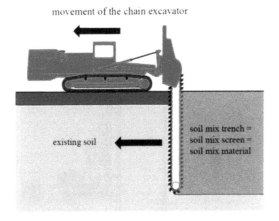

1 - Fig. 2.4. Trench mixing system.

The execution parameters (binding agent type, w/c factor, ...) and the way in which the mixing procedure is carried out are often determined by the base and the amount of binding agent one wishes to mix into the soil. In most cases, but depending on the experience of the contractor, these parameters are optimised on-site during the early stages of the work.

Generally speaking, the more mixing energy per unit volume soil mix material, the more homogeneous the end result will be. In the international literature, this mixing energy is described in terms of the 'blade rotation number' (or BRN). Depending on the type of soil, the literature talks about a minimum value of 350 to 430 rotations per metre in order to obtain a sufficiently homogeneous mixture. However, the definition and criterium are dependent on the type of mixing equipment, its configuration and the type of soil. They can thus not be unambiguously determined for all types of augers and cutters that are used for the soil mix process in Belgium and the Netherlands.

Depending on the soil type and the execution process, it is very likely that a certain volume of mixed material will flow back to the surface. In unsaturated and highly permeable soils, this volume will remain limited. In saturated soil, and especially in cohesive soils (clay), this volume may be up to 30% or more of the volume of realised soil mix element.

The execution of soil mix columns or panels is carried out in phases. In most cases, a series of primary and secondary columns (or panels) are executed in succession. Steel reinforcement beams are then placed in the fresh soil mix elements. The fact that the soil and water pressures applying on the wall must be transferred from the the soil mix to the reinforcement using a pressure arch (arch force) partly determines the distance between the beams. §6.4 includes a method of verifying the arching effect. For the potential application of reinforcement cages instead of reinforcement beams, it is recommended to perform additional research.

If the soil mix wall has a permanent water-retaining or cut-off function, it is recommended to create a wall with either panels or multiple column systems with sufficient overlap between individual panels/columns, or with a continuous trenchmixing system.

Chapter 7 elaborates in more detail on the phasing and requirements with regards to the execution and how to realise soil mix walls using these different methods of installation (tolerances, execution monitoring, etc.). With cut-off walls, generally speaking, no reinforcement is applied in the soil mix wall. If desired, geomembranes may be used. For more details with regards to geomembrane technology, please refer to §2.2.6 in part 2 of the handbook.

If anchors or struts are applied, the transfer of the forces of this horizontal support on the wall must occur with the aid of purlins (supporting beams). A local transmission of force with, for example, an anchor seat is also possible. In doing so, the anchors can be placed in between or at the location of the beams.

2.2   Characteristics of the soil mix material

Several parameters influence the (evolution of) characteristics of the produced soil mix material. After all, the quality of the soil mix material is dependent on the type of binder, binder content, any admixtures or additives, existing soil and execution process.

The main hydro-mechanical characteristics of soil mix material used in the design of soil mix walls are:
- the uniaxial compressive strength (unconfined compressive strength)
- the modulus of elasticity
- the tensile splitting strength
- the shear strength
- the adhesion between the reinforcement beam and soil mix
- the porosity
- the permeability

These characteristics will be discussed in more detail in part 2 of this handbook, including the typical ranges of the values of these variables. Chapter 6 of part 1 (design) lists the target values and correlations that may be used in the various design phases of soil mix walls. This is quite important, especially because the actual properties of soil mix can only be determined once the wall has been executed.

In addition, the conditions to which the soil mix elements are exposed during their life cycle will also have a certain influence on the strength development or any potential degradation of the soil mix material in the long term. The durability of the material is an especially important aspect to consider with regards to long term applications of soil mix walls. On the one hand, 'durability' concerns all aspects that could be related to the evolution and/or degradation of the properties (strength, permeability, pH, ...) of the soil mix material itself. On the other, there is also the issue of durability of soil mix walls realised in contaminated soils. In both cases that can have an influence on the corrosion or the increased rate of corrosion of the reinforcement steel. Unique to the soil mix process is that the impurities (or other components, such as, for example chlorides from brackish water) are mixed with the injected binding agent and the soil, and are therefore fully integrated into the soil mix matrix. The potential impact of impurities or other components is therefore greater than with, e.g., cast in place or precast concrete elements that are only exposed to contaminations at the contact zone with the soil.

For temporary structural soil mix elements there is especially the issue of the influence of impurities on the setting process and strength development of the soil mix material over time.

Depending on the concept, for permanent structural applications it is important that the soil mix can continue to serve its long-term function (arching effect to transfer forces onto the reinforcement beams, long-term permeability, ...) and that the risk of corrosion of the reinforcement elements as a result of various factors is taken into account.

Guidelines on how to practically approach the issue of soil mix durability are provided in §6.14 and Chapter 9.

2.3   Field of applications of soil mix walls

The field of applications of soil mix walls that this handbook covers has been divided into eight functions (A to H), summarised below:

- FUNCTION A: temporary earth retaining soil mix wall,
- FUNCTION B: permanent earth retaining soil mix wall,
- FUNCTION C: temporary earth-water retaining soil mix wall,
- FUNCTION D: earth-water retaining soil mix wall, at least one of the functions permanent,
- FUNCTION E: soil mix wall with a temporary bearing capacity function,
- FUNCTION F: soil mix wall with a permanent bearing capacity function,
- FUNCTION G: temporary cut-off soil mix wall,
- FUNCTION H: permanent cut-off soil mix wall.

In practice, a temporary function means that the lifetime of the wall is less than 2 years. If the lifetime of the wall is greater than 2 years, we speak of a permanent (or final) soil mix wall or a soil mix wall in the "use phase".

It should also be mentioned that a single wall may serve multiple functions. For example, a soil mix wall may have a temporary earth retaining function (A) at first. If the same wall is required to bear vertical loads in the use phase, it will take on function F. If in addition to vertical loads the wall is also required to retain horizontal earth pressure, then, in the use phase, the wall will take on both function B and F. It is self-evident that the requirements imposed on a soil mix wall are dependent on its function expected life cycle For example, for permanent applications special demands are placed on durability aspects and, depending on the approach to execution and application, additional provisions and/or measures must be taken to allow the wall to fulfil its function throughout its entire lifetime. This is reflected in a number of functional requirements for functions A to H, summarised in Chapter 3. In addition, the requirements related to quality control of the realised soil mix material are adjusted to the function the wall will serve. This aspect is discussed in more detail in Chapter 8.

Additionally, when realising soil mix walls the following matters must also be taken into account:

- the applicability of the technology is dependent on the local soil conditions. I - Table 2.1 provides an overview of the applicability of soil mix process depending on the type of soil;
- very limited vibrations occur when applying soil mix walls;
- underground obstacles may cause major problems. This is an important issue to be addressed during preparatory research.
- no lowering of the water level is required. In case of the presence of significant ground-water flows, the risk of soil mix corrosion must be studied extensively;
- execution within short-distance of existing structures is possible, provided that necessary preparatory measures (study) are taken, the execution phasing and dimensions of the soil mix elements are adjusted and the influence on nearby structures is properly evaluated.

I - Table 2.1. Applicability of soil mix process depending on the local soil conditions (V means applicable in certain conditions, VV means almost always applicable and VVV means always applicable).

SAND	LOAM	SOFT CLAY	SOLID CLAY
VVV	VV	VV	V
Note: special attention must be paid to the execution of soil mix in coarse gravels, peat (soils) and contaminated soils.			

Chapter 3 Functional requirements of the soil mix walls

Chapter 2 (part 1 of this handbook) provides an overview of the application area of soil mix walls. For this purpose, a division has been made in eight functions (A to H) that soil mix walls may serve. Table 3.1 shows which functional requirements apply per function .

It is important to recognise that the function, expected lifetime the soil mix wall and preconditions as established in §6.3.6 determine:

- whether long-term reduction factors must be applied on the strength and stiffness characteristics of the soil mix material;
- whether the soil mix wall must be protected with regards to exposure to the ambient air or freeze-thaw cycles;
- whether one can expect an interaction between the steel beam and the soil mix material;
- what type and amount of control tests must be conducted on the realised soil mix material.

I - Table 3.1. Focus points and minimum characteristics of the soil mix material.

A. Temporary earth retaining soil mix wall	
Preconditions, focus points and minimum characteristics	
A1	Minimum characteristic value of the compressive strength (UCS) of …* MPa- (§6.3 and §6.14)
A2	Minimum modulus of elasticity of …* GPa - (§6.3)
A3	Maximum vol% inclusions of 20%
A4	Consideration of the influence of the soil, the binder content, the execution parameters on the compressive strength and the stiffness of the soil mix material
A5	Verification of the arch effect/pressure arch (§6.4)
A6	Calculation of the bending stiffness of the soil mix wall – composite bending stiffness is permitted (§6.6)
A7	Verification of the internal forces in the wall: interaction soil mix – reinforcement may be taken into account (§6.7), provided there is a limit of τ_{Rd} in accordance with the criteria in §6.3.6. • moments • shear forces • normal forces
A8	Verification of the punching force and design of the anchor plates or purlins
A9	Verification of the horizontal displacement of the soil mix wall
A10	Consideration of the curing effect: evolution of the compressive strength over time
A11	Consideration of environmental influence: influence of polluted or brackish water and contaminated soil on the curing time and on the strength and stiffness
A12	Consideration of the risk of leaching of cement in the fresh soil mix material in important groundwater flows

*to be defined by designer depending on the requirements and conditions of the project

B. Permanent earth retaining soil mix walls

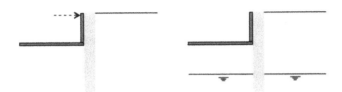

Preconditions, focus points and minimum characteristics

B1	Minimum characteristic value of the compressive strength (UCS) of …* MPa, including long-term reduction factor- (§6.3 and §6.14)
B2	Minimum modulus of elasticity of …* GPa, including long-term reduction factor- (§6.3)
B3	Maximum vol% inclusions of 20%
B4	Consideration of the influence of the soil, the binder content, the execution parameters on the compressive strength and the stiffness of the soil mix material
B5	Verification of the arch effect/pressure arch (§6.4)
B6	Calculation of the bending stiffness of the soil mix wall – composite bending stiffness is permitted (§6.6)
B7	Verification of the internal forces in the wall – only the reinforcement may be taken into account for the final function (§6.7), unless certain pre-conditions are met (cfr. 6.3.6) • moments • shear forces • normal forces
B8	Verification of the punching force and design of the anchor plates or purlins
B9	Verification of the horizontal displacement of the soil mix wall
B10	Consideration of the curing effect: evolution of the compressive strength over time
B11	Consideration of environmental influence: influence of polluted or brackish water and contaminated soil on the curing time and on the strength and stiffness
B12	Consideration of the risk of leaching of cement in the fresh soil mix material in important groundwater flows
B13	Consideration of potential creep behaviour of the soil mix material
B14	Calculation of the settlement of the soil mix wall

B15	Consideration of durability aspects (§6.14): • Protection of the surface of the soil mix wall against exposure to air using measures that allow the wall to serve its function during its entire lifetime. • Protection of (parts of) the wall against exposure to freeze-thaw cycles using measures that allow the wall to serve its function during its entire lifetime. • In contaminated soil, the risk of chemical reactions with side-effects must be properly investigated. • Corrosion of the steel beams
B16	Potential seismic or dynamic loads
B17	Accidental load (e.g. fire)

*to be defined by designer depending on the requirements and conditions of the project

C. Temporary earth-water retaining soil mix wall

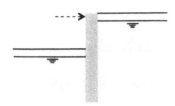

Preconditions, focus points and minimum characteristics	
C1	Minimum characteristic value of the compressive strength (UCS) of ...* MPa- (§6.3 and §6.14)
C2	Minimum modulus of elasticity of ...* GPa - (§6.3)
C3	Maximum vol% inclusions of 20%
C4	Maximum permeability of the wall 10^{-7} m/s, and of the soil mix material of 10^{-8} m/s
C5	Consideration of the influence of the soil, the binder content, the execution parameters on the compressive strength and the stiffness of the soil mix material
C6	Verification of the arch effect/pressure arch (§6.4)
C7	Calculation of the bending stiffness of the soil mix wall – composite bending stiffness is permitted (§6.6)
C8	Verification of the internal forces in the wall: interaction soil mix – reinforcement may be taken into account (§6.7), provided there is a limit of τ_{Rd} in accordance with the criteria in §6.3.6. • moments • shear forces • normal forces
C9	Verification of the punching force and design of the anchor plates or purlins
C10	Verification of the horizontal displacement of the soil mix wall
C11	Consideration of the curing effect: evolution of the compressive strength over time
C12	Consideration of environmental influence: influence of polluted or brackish water and contaminated soil on the curing time and on the strength and stiffness
C13	Consideration of the risk of leaching of cement in the fresh soil mix material in important groundwater flows

*to be defined by designer depending on the requirements and conditions of the project

D. Earth-water retaining soil mix wall, at least one of the functions permanent

D₁ permanent ER & temporary WR   D₂ permanent ER & permanent WR   D₃ perm./temporary ER & permanent WR
(polder construction)

	Preconditions, focus points and minimum characteristics
D1	Minimum characteristic value of the compressive strength (UCS) of …* MPa, including long-term reduction factor- (§6.3 and §6.14)
D2	Minimum modulus of elasticity of …* GPa, including long-term reduction factor- (§6.3)
D3	Maximum vol% inclusions of 20%
D4	Maximum permeability of the wall 10^{-7} m/s, and of the soil mix material of 10^{-8} m/s
D5	Consideration of the influence of the soil, the binder content, the execution parameters on the compressive strength and the stiffness of the soil mix material
D6	Verification of the arch effect/pressure arch (§6.4)
D7	Calculation of the bending stiffness of the soil mix wall – composite bending stiffness is permitted (§6.6)
D8	Verification of the internal forces in the wall – only the reinforcement may be taken into account for the final function (§6.7), unless certain pre-conditions are met (cfr. 6.3.6) • moments • shear forces • normal forces
D9	Verification of the punching force and design of the anchor plates or purlins
D10	Verification of the horizontal displacement of the soil mix wall
D11	Consideration of the curing effect: evolution of the compressive strength over time
D12	Consideration of environmental influence: influence of polluted or brackish water and contaminated soil on the curing time and on the strength and stiffness
D13	Consideration of the risk of leaching of cement in the fresh soil mix material in important groundwater flows
D14	Consideration of potential creep behaviour of the soil mix material
D15	Calculation of the settlement of the soil mix wall

D16   Consideration of durability aspects (§6.14):
        • Protecting the surface of the soil mix wall against exposure to air using measures that allow the wall to serve its function during its entire lifetime.
        • Protection of (parts of) the wall against exposure to freeze-thaw cycles using measures that allow the wall to serve its function during its entire lifetime.
        • In contaminated soil, the risk of chemical reactions with side-effects must be properly investigated.
        • Corrosion of the steel beams
        • If important groundwater flows are present and if the characteristic UCS value < 0.5 MPa, then the risk of leaching of the fresh soil mix material must be investigated for definitive water retaining functions (D2 and D3).

D17   Potential seismic or dynamic loads

D18   Accidental load (e.g. fire)

*to be defined by designer depending on the requirements and conditions of the project

E. Soil mix wall with temporary bearing capacity function

Preconditions, focus points and minimum characteristics	
E1	Minimum characteristic value of the compressive strength (UCS) of ...* MPa- (§6.3 and §6.14)
E2	Minimum modulus of elasticity of ...* GPa - (§6.3)
E3	Maximum vol% inclusions of 20%
E4	Consideration of the influence of the soil, the binder content, the execution parameters on the compressive strength and the stiffness of the soil mix material
E5	Consideration of the curing effect: evolution of the compressive strength over time
E6	Consideration of environmental influence: influence of polluted or brackish water and contaminated soil on the curing time and on the strength and stiffness
E7	Consideration of the risk of leaching of cement in the fresh soil mix material in important groundwater flows
E8	The superstructure forces are uniformly transferred in the soil mix wall. The vertical forces of the underground floors are transferred via connecting elements between the floor and steel beams: • One may assume a composite section, but the adhesion τ_{Rd} between the steel reinforcements and soil mix is limited according to the criteria in §6.3.6.
E9	Calculation of the geotechnical bearing capacity (§6.9) of a soil mix wall: • no positive wall friction is taken into account above the excavation level • below the excavation level, one may consider the positive shaft friction and base resistance; the nominal dimensions of the soil mix elements may be taken into account • the shape factor, group effect (for the soil mix wall consisting of columns) and the influence of the excavation must be taken into account
E10	Consideration of the vertical settlement of the soil mix wall
E11	The soil mix elements are preferably reinforced over their entire length

*to be defined by designer depending on the requirements and conditions of the project

F. Soil mix wall with permanent bearing capacity function

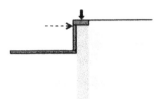

Preconditions, focus points and minimum characteristics	
F1	Minimum characteristic value of the compressive strength (UCS) of 3 MPa, including long-term reduction factor - (§6.3 and §6.14).
F2	Minimum modulus of elasticity of...* GPa, including long-term reduction factor - (§6.3).
F3	Maximum vol% inclusions of 20%.
F4	Consideration of the influence of the soil, the binder content, the execution parameters on the compressive strength and the stiffness of the soil mix material.
F5	Consideration of the curing effect: evolution of the compressive strength over time.
F6	Consideration of environmental influence: influence of polluted or brackish water and contaminated soil on the curing time and on the strength and stiffness.
F7	Consideration of the risk of leaching of cement in the fresh soil mix material in important groundwater flows.
F8	The superstructure forces are uniformly transferred in the soil mix wall. The vertical forces of the underground floors are transferred via connecting elements between the floor and steel beams: • One may expect a composite section, as long as the conditions as mentioned in §6.3.6 are met. • The adhesion τ_{Rd} between steel and soil mix is limited according to the criteria in §6.3.6.
F9	Calculation of the geotechnical bearing capacity (§6.9) of a soil mix wall: • No positive wall friction is taken into account above the excavation level. • Below the excavation level, one may consider the positive shaft friction and base resistance; the nominal dimensions of the soil mix elements may be taken into account. • The shape factor, group effect (for the soil mix wall consisting of columns) and the influence of the excavation must be taken into account.
F10	Consideration of the vertical settlement of the soil mix wall
F11	The soil mix elements are preferably reinforced over their entire length
F12	Consideration of potential creep behaviour of the soil mix material

F13	Consideration of durability aspects (§6.14): • Protecting the surface of the soil mix wall against exposure to air using measures that allow the wall to serve its function during its entire lifetime. • Protection of (parts of) the wall against exposure to freeze-thaw cycles using measures that allow the wall to serve its function during its entire lifetime. • In contaminated soil, the risk of chemical reactions with side-effects must be properly invesigated. • Corrosion of the steel beams.
F14	Potential seismic or dynamic loads
F15	Accidental load (e.g. fire)

*to be defined by designer depending on the requirements and conditions of the project

G. Temporary cut-off soil mix wall

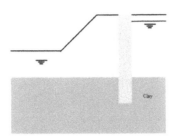

Preconditions, focus points and minimum characteristics

G1	The permeability of the wall as a whole and the soil mix material in particular must comply with the design principles that were applied during the preparation of the design. Arithmetically, one must be able to demonstrate that these principles meet the requirements regarding the expected leakage discharge. Typical values for these types of walls have a maximum permeability of the wall of 1.10^{-7} m/s, which for a wall with a thickness of approx. 0.55 m results in a minimum permeability of the soil mix material of approx. 5.10^{-8} m/s.
G2	Minimum characteristic value of the compressive strength UCS of 0.3 MPa
G3	Maximum vol% inclusions of 20%
G4	Consideration of the influence of the soil, the binder content, the execution parameters on the compressive strength and the stiffness of the soil mix material
G5	Consideration of the curing effect: evolution of the compressive strength over time
G6	Consideration of environmental influence: influence of polluted or brackish water and contaminated soil on the curing time and on the strength and stiffness
G7	Consideration of the risk of leaching of cement in the fresh soil mix material in important groundwater flows

*to be defined by designer depending on the requirements and conditions of the project

H. Permanent cut-off soil mix wall		

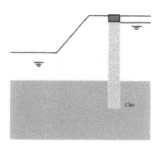

Preconditions, focus points and minimum characteristics	
H1	The permeability of the wall as a whole and the soil mix material in particular must comply with the design principles that were applied during the preparation of the design. Arithmetically, one must be able to demonstrate that these principles meet the requirements regarding the expected leakage discharge. Typical values for these types of walls have a maximum permeability of the wall of 1.10^{-7} m/s, which for a wall with a thickness of approx. 0.55 m results in a minimum permeability of the soil mix material of approx. 5.10^{-8} m/s.
H2	Minimum characteristic value of the compressive strength UCS of 0.5 MPa
H3	Maximum vol% inclusions of 20%
H4	Consideration of the influence of the soil, the binder content, the execution parameters on the compressive strength and the stiffness of the soil mix material
H5	Consideration of the curing effect: evolution of the compressive strength over time
H6	Consideration of environmental influence: influence of polluted or brackish water and contaminated soil on the curing time and on the strength and stiffness
H7	Consideration of the risk of leaching of cement in the fresh soil mix material in important groundwater flows
H8	Consideration of durability aspects (§6.14): • Protection of (parts of) the wall against exposure to freeze-thaw cycles using measures that allow the wall to serve its function during its entire lifetime. • In contaminated soil, the risk of chemical reactions with side-effects must be properly investigated. • In the presence of important ground water flows, and if the characteristic value of the compressive strength UCS value < 0.5 MPa, then the risk of leaching of the hardened soil mix material must be properly researched (affect on permeability).
H9	Potential seismic or dynamic loads

*to be defined by designer depending on the requirements and conditions of the project

40

Chapter 4 Risk management

The current chapter provides a risk assessment with regards to soil mix walls.
The information is based on a matrix in which the columns are divided into:

- Risk
- Consequence
- How to monitor (the risk)
- Preventive / corrective measures
- Reference to parts of this handbook that are related to the risk in question.

The mentioned risks are mainly related to the soil mix material or soil mix wall. For general risks related to construction pits and their environment, please refer to CUR 223 'Directive measuring and monitoring construction pits'. This directive has been created because increasingly often complex underground structures are being built in urban centres. The goal of this CUR directive "is improving the design of construction pits in relation to their environment and improving the use of monitoring and quality control in construction pits as a tool for quality and risk management".

Because of their vibration-free execution, soil mix walls are often used in urban areas, CUR 223 thus suits these conditions well. The handbook Diaphragm Walls (CUR 231) covers the design and construction of diaphragm walls. Given the similarities between diaphragm and soil mix walls in certain aspects, the Diaphragm Walls handbook may provide additional insight into the design, execution, environmental influence etc. of soil mix walls.

The risk table does not include the causes of the mentioned risks because multiple/various causes can result in the same risk. The quality of the soil mix material (compressive strength, stiffness and permeability) may be influenced by the following:

- the homogeneity of the soil
- the quality of the execution process
- the applied (and ratio of) aggregates
- etc.

The listed control measures are not exhaustive.

Risk	Consequence	How to monitor	Measures		Index
			Preventive	Corrective	
Bad execution of the wall	strength too low, stiffness, permeability properties of mix product; incorrect positioning of panels; incorrect positioning of the reinforcement; etc.	Conducting QC during execution by measuring and inspecting execution parameters and direct adjustment of the process.	Creating work plan, e.g. also coordinating grout flow rate and withdrawal speed of mixing shaft or mast; use lasers / guide beam	Adjust process on the basis of registration of execution parameters	Ch.7 / Ch. 8 in part 1 and §2.3.2 in part 2 of the handbook
Panels not at proper depth as a result of obstacles in subsoil	insufficient capacity of the wall	registration execution data	soil research, archive research	extend wall by means of jet grout columns, extra panel behind planned position	Ch. 7 in part 1 of the handbook
Inclusions of cohesive material in the wall / homogeneity of the wall	parts in wall of lower strength, potentially giving rise to holes, which will also affect the cut-off function	visual inspection during excavation / continuous sampling with 'liners'/ pumping tests / monitoring wells	execution process (type of wall / order of panels / mixing energy)	extra panel behind the wall or injection with suitable material (water glass or jet grout columns)	Ch. 8 in part 1 of the handbook
poor or no penetration speed as a result of soil composition	planning is not achieved, wall doesn't reach proper depth, quality hardened material reduced	registration installation process	soil and lab research, adjustment cutting head	extend wall by means of jet grout columns	Ch. 7 in part 1 of the handbook
presence of weak clay and peat layers	feasibility	prior soil investigation	attuning production method		Chapter 5.2 part 1 of the handbook
presence of stiff clay and loam layers	feasibility, homogeneity	prior soil investigation	attuning production method		Chapter 5.2 part 1 of the handbook

Risk	Consequence	How to monitor	Measures		Index
			Preventive	Corrective	
presence of dense sand layers	major use of grout	prior soil investigation	attuning production method		Chapter 5.2 part 1 of the handbook
presence of coarse sand and gravel layers	major use of grout	prior soil investigation	attune mixture		Chapter 5.2 part 1 of the handbook
presence of chemical substances	influences curing time and final strength	prior geochemical research	attune mixture		Chapter 5.2 part 1 of the handbook
too low compressive strength of hardened material	insufficient capacity of the wall as a whole, but also in more specific parts, e.g. at the location of the anchor seat	compressive strength tests	conservative assumption compressive strength	1) longer curing period 2) reduction of driving loads (e.g. relief trench, limiting top loads) 3) displacement and/or additional supports 4) drilling additional reinforcement 5) expanding anchor seat 6) extra wall behind the soil mix wall	Chapter 6 / §8.6 in part 1 of the handbook §2.3 in part 2 of the handbook
insufficient stiffness of hardened material	increase in deformation of the wall, the change to the risk profile with regards to the environment	compressive strength tests	conservative assumption compressive strength / stiffness	1) longer curing period 2) reduction of driving loads (e.g. relief trench, limiting top loads) 3) displacement and/or additional	Chapter 6 / §8.6 in part 1 of the handbook §2.3 in part 2 of the handbook
permeability hardened material too high	water hazard is too great	permeability tests, monitoring wells	aggregates (higher strength = lower permeability), use of geomembrane in wall	1) longer curing period 2) applying protection banier	§6.3 / §6.4.5 / §8.6 / §8.8 in part 1 of the handbook §2.3 in part 2 of the handbook

Risk	Consequence	How to monitor	Measures		Index
			Preventive	Corrective	
strength reduction as a result of geomembrane in the wall	large shear stresses in the geomembrane	-	only apply if water-retaining wall is embedded in the ground	-	§6.4 in part 1 of the handbook
Determining properties in lab with samples that are not sampled, transported, stored and prepared in accordance with the procedures	unreliable test of strength and stiffness properties	supervision on construction site, contra expertise	contractually impose procedure, follow alternative procedure and compare data sets	contra expertise	§2.3.10 and §2.7.6 in part 2 of the handbook
representativeness of the cored samples for determining properties of the entire wall	soil properties that are not representative of the wall		sufficiently sized BRN	follow sample selection procedure. In practice, local inclusions can occur.	§2.3.11 in part 2 of the handbook
overestimation of long-term strength	failure of permanent walls in the long term	-	in calculation, use stiffness and strength of steel beams, not the composite stiffness	-	1 - Table 3.1 / §6.14 in part 1 of the handbook
mutual overlap between panels/ columns	no continuous wall, risk of water and soil permeability of the wall	registration installation parameters	correct expansion of the wall, execution drawing, correct execution ('Pelgrimsgang'), use of lasers / guide frame during installation	extra panel behind wall or inject with suitable material (water glass or jet grout columns)	Chapter 7 in part 1 of the handbook
bad connections of walls in corners	leakage, failure	construction pit monitoring (CUR 223)	engineer corner solution in advance, taking into account formation of pressure arches between steel beams	soil mix panels behind the wall, apply injection techniques	§6.11.1 in part 1 of this handbook

Risk	Consequence	How to monitor	Measures		Index
			Preventive	Corrective	
cracking in external corners	crack formation due to displacements with a loss of strength and leakage as a result	construction pit monitoring (CUR 223, 2010)	engineer corner solution in advance, taking into account formation of pressure arches between steel beams, align deformations (EI)	soil mix panels behind the wall, apply injection techniques	§6.11.1 in part 1 of this handbook
Incorrect positioning of reinforcement in wall	insufficient capacity of the wall as a whole but also in specific parts e.g. at the location of the anchor seats	proper reporting during installation, if reinforcement protrudes above wall (local) visual inspection is possible, however there is a risk in regards to the placement at depth, cover measurements, CUR 223	high quality execution, work with guide beam when applying reinforcement, take into account placement tolerance in design phase, in accordance with EN 1538 and CUR 231, reinforcement through bottom of wall	1) longer curing period 2) reduce driving loads 3) displacement and/or additional supports	§6.7 / §6.11.2 / Ch. 7 in part 1 of the handbook
applicability of the wall in relationship to contaminations / chemical substances in the soil	1) No acquisition of permit due to building materials decree 2) influence on curing time / durability / quality of the wall		1) consultation with authorised personnel, create work plan, create product certificates 2) lab research / adjust aggregate material / w/c factor / mixing energy		Ch. 5 / Ch. 9 in part 1 of the handbook §2.3 and §2.4 in part 2 of the handbook

45

Risk	Consequence	How to monitor	Measures		Index
			Preventive	Corrective	
drying of the wall	corrosion due to local loss of protection of steel construction components	visual inspection	minimisation of shrinkage by adjusting wall composition, prevent exposure to sunlight and heat sources, prevent major wall deformations, apply cathodic protection	wall surface water-retaining or water-proof treatment, apply cathodic protection	Ch. 9 in part 1 of the handbook
damage to cables and/or pipes	damage, delay	attentiveness with respect to damage occurring during execution (CUR 223)	click report, application of trial trenches, archive research	restoration of damage, continuation of work activities at different location	Ch. 7 in part 1 of the handbook
influence on adjacencies	influence on adjacencies as a result of installation process	construction pit monitoring (CUR 223)	carrying out proper inventory of adjacencies (foundation research, damage category), risk analysis and damage prediction by means of finite element calculations	foundation recovery, compensation grouting etc.	§6.10 / §7.7 in part 1 of the handbook

Chapter 5 Preliminary research and preparation

5.1 Introduction

The preliminary research aims to collect information about the project location that is required for designing and executing a soil mix wall and to determine whether the local soil properties may influence this process. For example, in the presence of certain chemical substances and/or minerals, it may be necessary to adjust the composition of the injected mixture. To illustrate this, the example of the railway tunnel in Delft (see CUR 231 'Handbook Diaphragm Walls') is mentioned here again, in which the groundwater contained high concentrations of ammonium up to depths greater than 8m, which naturally is an 'aggressive environment' for concrete.

Preliminary research may partly make use of the soil investigation that must be carried out for the design of the project. In the Netherlands, this soil investigation must comply with the requirements of NEN 9997-1 and NEN-EN 1997-2. There are also additional documents available, such as CUR report 2003-7, CUR report 2008-2 and CUR publications 231 and 247. CUR publication 247 (Risk-driven soil investigation from planning phase to realisation) also includes excellent descriptions about what type of soil investigation may be conducted in what specific project phase, depending on the type of construction. In Belgium, the soil investigation must be conducted according to NBN EN 1997-1 and NBN EN 1997-2 and the national annexes. Additional considerations are mentioned in BGGG (2015a).

The preliminary research provides location-specific information that both positively influences the design and execution. Preliminary research may also be important in the event of legal disputes.

5.2 Preliminary research

5.2.1 Generalities

With regards to soil mix walls, special attention must be paid to:
- presence of peat and soft clay layers
  (with regards to feasibility/homogeneity of the wall)
- presence of stiff clay/loam layers
  (e.g. clay loam, pot clay, with regards to feasibility/homogeneity of the wall)
- presence of densely packed sand layers (with regards to grout usage)
- presence of coarse sand and gravel layers
  (with regards to mixture composition and/or grout usage)
- presence of chemical substances (with regards to curing, final strength).

The first 4 aspects are generally examined by means of penetration tests with measurement of the local frictional resistance, optionally combined with measurement of the water tension.

The coarse sand and/or gravel layers may, if necessary, be more closely examined by means of multiple mechanical bores (with (undisturbed) sampling) with regards to the grain size distribution and grain shape.

The preliminary research that may be conducted when one intends to install a soil mix wall consists of both desk research and field research. In certain circumstances it may also be appropriate to conduct a suitability study.

5.2.2   Desk Research

Desk research consists of:
- archaeological research
- historical research
- research related to the potential presence of explosives
- the assessment of obstacle risks
- research into the presence of utilities (KLIC or KLIP in case of public domains)
- research related to built-up and undeveloped environments.

5.2.3   Field Research

The field research consists of a soil investigation, groundwater survey and environmental soil survey, which are discussed in more detailed below:
- Soil investigation in accordance with the applicable standards and guidelines (see §5.1). For soil mix walls it is advisable to provide at least one penetration test per excavation pit side, preferably with measurements of the local frictional resistance (if desired, combined with measurement of the water tension) and up to 5 m below the intended point level of the wall. In certain cases it may be appropriate to bore up to the point level of the wall with disturbed samples for determining the grain size distribution and grain shape for each distinct layer of sand. The number of bores is dependent on the variability of the soil that is encountered during the penetration tests and which is to be expected on the basis of local knowledge.
- Groundwater survey:
  ○ determining the phreatic groundwater level and the piezometric hydraulic head in the sand layers;
  ○ analysis of the degree of acidity pH, sulfate SO_4^{2-}, ammonium NH_4^+, magnesium Mg_2^+, chloride Cl^- and carbon dioxide CO_2.

- the exploratory environmental soil survey in accordance with NEN 5740. (in the Netherlands). In Belgium, the individual regions are responsible for this matter and one may solicit the assistance of Standaardprocedures van OVAM (Flanders, http://www.ovam. be/standaard- procedures), le Code Wallon de Bonnes Pratiques (Wallonia, https:// dps.environnement.wallonie.be/home/sols/sols-pollues/code-wallon-de-bonnes-pratiques--cwbp-.html) of de Codes van Goede Praktijk (Brussels , http://www. leefmilieu.brussels/themas/bodem/bodemveront-reiniging-identificeren-en-behandelen /codes-van- goede-praktijk).

More information about the impact of certain chemical substances on the setting and hardening process is available in §5.3.

5.2.4   Suitability study

In certain cases, before commencing the execution process of the soil mix wall, it may be useful to gain insights into the influence of certain parameters (binder type, content, ...) on the characteristics of the soil mix material and/or to determine an optimum design mixture. These may be adjusted depending on the soil in which the work activities will take place. The design mix should thus have such a composition and consistency that the final product meets the requirements with regards to permeability, compressive strength and modulus of elasticity.

Suitability tests are especially interesting in cases where one aims for (a) specific (combination of) properties (e.g. extremely low permeability, minimum or maximum strength, ...), or if components are present in the soil that may potentially influence the curing time and final strength and stiffness. In addition, it is also useful to carry out suitability tests if one uses new/ adapted binder types/aggregates, or if one uses soil mix technology in soils of which little to no empirical data is available. To determine the design mix, one can conduct suitability study. This consists of:

- **Conducting and assessing the soil investigation:**
  the soil investigation includes penetration tests, among other things. If the various penetration tests show extremely variable soil conditions, additional penetration tests will be necessary. Because penetration tests can only show an indicative soil classification, it is recommended to perform complementary mechanical drilling, in which samples are taken from the characteristic soil layers or at least 1 per metre. The non-cohesive samples are then analysed, after which the grain size distribution can be further assessed through a sieve analysis. For cohesive soil samples one can determine volumetric weights and Atterberg limits. This process allows one to determine the exact soil classification. Moreover, for the purpose of the soil analysis, it is recommended to conduct 1 to 2 bores per project location. One should also keep the soil samples obtained from geotechnical survey so that these can be used in the creation of lab mixtures.

- **Assessment of chemical analysis:**
  This examines the contaminants and substances that affect the cement curing process. In addition, the chloride content is assessed for the amount of bentonite in the drilling mud.
- **Creating and testing lab mixtures:**
  From the bores, sample material is composed which is in proportion to the soil at the location of the future wall. This sample material is tested to determine the influence of certain parameters on the characteristics of the soil mix material and to create an optimum recipe for the required permeability and strength.

Bear in mind, the absolute values of the characteristics resulting from the suitability study on the basis of lab mixtures must always be observed with caution. The mixing method employed by the laboratory, the way in which the lab mixture is applied and compacted into the moulds, the dimensions of the test specimens, etc. may all have significant influence on the test results. In order to obtain mutually similar results, this process should be conducted in a standardised fashion. Unfortunately, the current performance standard EN 14679 "Execution of special geotechnical works - Deep Mixing" does not yet include such a standardised approach. The literature may provide certain guidelines (e.g. Terashi and Kitazume, 2011; Eurosoilstab, 2002; CUR-report 2001-10), although these can also vary significantly from each other.

A verification of the in situ realised characteristics is therefore always necessary to correlate the results of the suitability tests with the actual in situ performance of the soil mix material.

- **Realisation of test elements on the field:**
  Finally, in certain circumstances it may be preferable to create in situ test columns or panels in advance.

5.2.5   Limited versus extensive preliminary research

By conducting the above-mentioned preliminary research, one can obtain a complete overview of the capabilities and limitations of the soil mix wall.

If one decides to conduct limited preliminary research, one must at least conduct the soil investigation from §5.2.3 (penetration tests and, if desired, mechanical boring up to the point level of the future soil mix wall) as well as the desk research from §5.2.2. Because such research means information will be considerably more limited and uncertainties greater, one must take this into account during the design phase by starting from more conservative assumptions and characteristics of the wall. Through intensive monitoring during execution, the underlying assumptions can then be checked. At the same time, the monitoring activities are highly dependent on the project environment. Not only should the influence of the environment be monitored, but also the quality of the wall itself. The mixture (soil mix) must be analysed on the basis of sample checks. This also applies to the final product itself.

If extensive preliminary research is conducted, including the aforementioned suitability study (§5.2.4), then a less intensive monitoring programme is required during execution. In general, this will mean that the wall will be significantly less conservative than in case of limited preliminary research. The amount of monitoring tasks for determining the influence on immediate surroundings as well as those of the wall itself need to be carried out less intensively. The complete omission of all monitoring operations is not recommended precisely because the quality of the wall and the corresponding operations can be demonstrated through such monitoring.

5.3   Chemical substances

Research into chemical substances in the soil will most likely have to be carried out separately in the preparatory phase. Environmental research required to obtain construction permits – the so-called NEN packages in the NEN-5740 or the standard procedures or codes of good practice of the different regions in Belgium (see §5.2.3) -, has often not yet been conducted in this phase. Moreover, these NEN packages or procedures do not include all important substances. It is recommended to also analyse the soil and groundwater on acidity pH, sulfate SO_4^{2-}, ammonium NH_4^+, magnesium Mg_2^+, chloride Cl^- and carbon dioxide CO_2 (typical substances in aggressive groundwater).

CUR 199 (published in 2001) lists the following substances that may affect the formation of a soil-cement mixture:
- retardant: ZnO, PbO, $As_2 O_3$, CdO, CaO, $PbNO_3$, $Pb(OH)_2$, PbS, $Na_3 PO_4$, NaF, borates and phosphates
- accelerating: $CaCl_2$ and $CrCl_3$.

The aforementioned substances may end up in the soil due to human activity, wherein the oxides generally only appear in unsaturated zones. While conducting the historical research, one may find clues for the presence of these substances. Information about minimum concentrations that may disrupt the setting process is not available.

Depending on their concentration, zinc salts can have both a retarding or accelerating effect on the setting behaviour. Ettringite formation may occur in the event of an excess of sulfates and may result in crack formation.

Both in the literature and in practice, there are currently few examples of environmental analyses that have been conducted for the purpose of the realisation of a soil mix wall. In 2012, at the 'Haak om Leewarden Zuid' (connecting the A31 at Marsum and N31 at Hemriksein, near Leeuwarden, the Netherlands), the aggressiveness of the groundwater was analysed in relation to the realisation of soil mix walls, during which high amounts of carbon-dioxide were discovered. Generally speaking, it is believed that contaminations – especially heavy metals and oil-like substances – are enclosed by the cement and therefore immobilised.

5.4 Relationship with the environmental precautions

In the Netherlands, a soil mix wall falls under the 'Besluit bodemkwaliteit' (Soil Quality Decree, Bbk, formerly: Bouwstoffenbesluit). The Bbk places requirements on the compositional and emission values of stony construction materials (other than soil and dredged material). When these values are met, the construction materials may be applied in the soil.

Soil mix walls are unique construction materials, because the construction site soil is not moved or removed, but is mixed in place with a cement paste. The existing soil is applied to the same location as the raw material for creating the construction material, being a stone-like material. This in situ conversion – from soil to a sustainable, form-retaining material – is not explicity described in the Bbk.

The Decree is based on a risk-based approach, in which the following risks in (direct) relation between the (chemical) quality and the use of the soil are central:
- the risk of affecting people's health;
- the risk of affecting ecosystems, such as effects on plants and animals and disruption of natural process in the soil;
- the risk of spread of contamination through groundwater;
- the probability of influences on agricultural production, such as the yield, health of livestock and transgression of Warenwet (Commodities Act) or livestock standards.

The goal of this is to prevent unwanted spread of substances into the environment.

Because the composition of the soil (substances) is only modified by the addition of cement paste, the compositional and emission values of the formed construction substances can only be affected by this addition. On the basis of current information, the risk of contamination of the soil is present when the soil has been contaminated by metals. By adding cement paste (and thus increasing the pH value), some metals (such as copper, for example) are mobilised, so that these can egress. When metal contaminations are observed in the soil, further assessment and leaching tests are required in order to determine whether the maximum permissible emission values are met.

In the absence of metal pollution in the soil, one must ask the question whether the execution of soil mix walls falls under the exceptions of the required quality determination according to the Bbk. In principle, the quality of all construction materials must be determined and they must be assigned an environmental statement. Exceptions include situations in which the financial and administrative burdens of determining the quality of a construction material are disproportionately large relative to the risk of the construction material not meeting the quality standards.

The Bbk lists the following examples:

- the application of mortar or natural stone products, with the exception of rubble and crushed stone;
- the use of form-giving construction materials such as concrete, ceramics, natural stone and bricks under the same conditions and without processing.

Given the material scientific comparison between, on the one hand, soil mix material and mortar/concrete, and on the other hand the high degree of certainty that the addition of cement paste (in the absence of metal contaminants) involves no risk under the Bbk, the claim to an exception to the quality standard seems justified. Reports on this matter are not yet available, however.

For this reason, one should bear in mind that the competent authority may require an environmental hygiene statement. It therefore advisable to discuss these matters in advance with the competent authority. This should be done in order to avoid issues relating to the Bbk from occurring.

To prevent arguments with the competent authority per project, it is recommended to discuss these matters with the soil quality implementation team, with the goal of adjusting the soil quality regulations for soil mix walls, so that the soil is converted in situ into a construction material.

With regards to the situation in Belgium, the regulations differ per region. As a result, the soil mix wall is not subject to specific laws or regulations. This is partly due to the specific nature of the soil mix wall.

After all, the soil mix wall is processed on the spot at the construction site, without soil being moved, removed or dug up. Through additions and mixing with a cement paste, one obtains a construction element. However, this construction element is not an element that is available on the market. As such, the construction element is not subject to the federal law of 21 December 1998 relating to product standards promoting production and consumption patterns and protecting the environment, public health and employees. Incidentally, for the same reasons, soil mix is not subject to the construction product regulations. In the terms of these construction regulations, it is not a construction product, but instead an element that will be part of the eventual construction.

With regards to the construction or its elements, there are currently few to no legal requirements in Belgium. The requirements that are legally regulated are mainly related to the energy performance (e.g. the EPB regulation), fire safety and accessibility of the buildings. In addition, there are requirements for indoor environments, often through specific decrees or general housing codes. Not a single one of these legal provisions includes requirements that would be applicable to soil mix walls, and therefore neither in an environmental hygiene sense.

Obviously, Belgium (through the regional decrees and legislation) has regulations covering soil remediation and protection.

In fact, each region has its own regulations:
- In Flanders this is the Bodemdecreet (Soil Decree) of 27 October 2006 and the related VLAREBO (the so-called Flemish regulation with regards to soil remediation and protection).
- In the Walloon region this is the Bodemdecreet (Soil Decree) of 5 December 2008 and the related Bodembesluit (Soil Order) of 27 May 2009.
- In Brussels, this aspect is regulated by the Ordinance of 5 March 2009 on the management and remediation of contaminated soils and a series of associated executive orders.

In the context of this subject, it is relevant that this regulation covers what is meant by soil contamination. It also stipulates which procedures must be followed in the event of contamination of the soil. The focus of these regulations is therefore on the quality of the soil, and not the constructions that are present in it.

An important issue included in this regulation is the concept of 'excavated soil'. Given the description of the concept, the soil mix wall cannot be categorised by this terminology. After all, the soil used for creating a soil mix wall is not excavated or moved, but is instead mixed in its initial location. The requirements regarding the handling of excavated soil are thus not applicable to soil mix material.

To summarise, one can conclude that the legislation with regards to soil does not necessarily include any requirements with regards to construction materials or elements used therein. The same thus applies to soil mix walls, which once processed are no longer part of the soil, but instead the structure itself.

However, there are requirements that are indirectly placed that ensure sufficient soil quality. Should the creation of the soil mix wall thus contribute to (or amplify the already present) contamination, then, in the context of the liability regulation, there may be certain responsibilities and research/remediation costs at play. However, it should be understood that this only goes for exceptional (and more than likely hard to demonstrate) cases (and, referring to the Dutch part, only in cases where the soil is already contaminated with metals).

One final element that should be addressed is the raw material problem. Strictly speaking, the application of the on-site soil is not covered by waste legislation. Referring to the Waste Framework Directive, one never disposes of/moves the substance (in this case the soil). Instead, the soil is processed and mixed in situ. For this particular case, the Flemish materialendecreet (Material Decree) clarifies this even further: Artikel 3 clearly states that 'unexcavated soil, including structures permanently connected to the soil' are not considered waste substances.

Naturally, this changes once the soil mix wall ends up in the waste substances cycle at the end of its life (e.g. after demolition). At that stage, the degraded soil mix wall originating from contaminated soil may have to be processed differently than when originating from non-contaminated soils. From a formal and technical legislation perspective, in both cases they will perhaps no longer be applicable as soil (after all, as soil mix they are no longer soil but part of the structure). Of course, as long as they meet the applicable standards of shaped or non-shaped construction materials, they may find further use as a recycled product originating from construction and demolition waste. It is evident that the use of raw materials originating from soil mix walls implemented from non-contaminated zones will be able to find a broader market than those from contaminated zones.

Chapter 6 Design

6.1   Notations

For complete understanding of Chapter 6, please refer to the symbols and definitions presented in the Notations chapter.

6.2   Safety philosophy and approach to design

6.2.1.   *The situation in Belgium*

Standards and guidelines
The geotechnical design in Belgium is described in part 1 of Eurocode 7, – NBN EN 1997 – Part 1 Algemene regels (General rules) (+AC: 2009) -, and in the Belgian national Annex NBN EN 1997-1 ANB:2014.

For the geotechnical calculation method of retaining structures and for the computation of the axial vertical bearing capacity, the Belgian national annex (ANB) respectively refers to:
- "De richtlijnen voor de toepassing van de Eurocode 7 in België; het grondmechanische ontwerp van kerende constructies". A preliminary version of this document was issued by (BGGG, 2015b). Over the course of 2016, this document will be officially published as a BBRI publication.
- "De richtlijnen voor de toepassing van de Eurocode 7 in België volgens de NBN EN 1997-1 ANB. Deel 1: het grondmechanische ontwerp in uiterste grenstoestand (UGT) van axiaal belaste funderingspalen op basis van statische sonderingen (CPT's)". BBRI report no. 19 (BBRI, 2016).

For the general principles and the determination of loads, both Eurocode 0 (EN-1990 and national annexes) and Eurocode 1 (EN 1991-1-1 to EN 1991-2 including national annexes) are applicable.

With regards to the structural calculation, the structural Eurocodes and their national annexes are applicable, particularly EN 1992-1-1, EN 1993-5 and EN 1994-1-1.

General guidelines with regards to construction pit including issues in execution and design are also provided in the 'Handboek beschoeiingen (Construction pit Handbook), draft version 13 June 2013' (BGGG, 2013). General guidelines related to ground water lowering are included in the 'Richtlijn Bemalingen' ('Dewatering Guideline') (Van Calster and others, 2009).

Testing and approach to design

The geotechnical design of a retaining structure consists of verifying the stability and deformations. Limit states that may occur and should therefore be assessed are:

- the collapse of the retaining wall as a result of exceeding the ground resistance, the structural failure of the wall or the failure of one or multiple anchor(s) or strut(s);
- the failure of the anchoring due to collapse of the soil next to an upper straight slip plane (Kranz stability);
- the occurrence of soil rupture (in weak soils);
- the loss of overall stability;
- collapse as a result of hydraulic gradients, e.g. internal erosion (piping), loss of effective stress, the rise of an "impervious" layer;
- deformations of the retaining wall and the back soil massifs.

(BGGG, 2015b) provides the calculation method for the failure mechanisms in the classes GEO and STR. Every failure mechanism must be calculated on the basis of its ultimate limit state (ULS = UGT in Belgium), where $E_d < R_d$, and its serviceability limit state (SLS = GGT in Belgium, BGT in the Netherlands), where $E_d < C_d$ with,

- E_d the design value of the (consequences) of the actions;
- R_d the design value of the resistance;
- C_d the limit value for the design value of the load effect.

In Belgium, the calculation of the ULS is carried out in accordance with design approach 1 (DA 1). (BGGG, 2015b) includes guidelines for determining:

- the design value of the loads $F_d = \gamma_F.F_{rep}$,
- the design value of the soil parameters $X_d = X_k / \gamma_M$,
- the design value of the constructive parameters;
- the design value of the excavation level;
- the design value of the water level;
- the characteristic values of the geotechnical parameters: the cohesion c', the friction angle φ', the undrained cohesion c_u, the wall friction angle δ, the horizontal modulus of sub-grade reaction k_h, the modulus of elasticity E, and the Poisson coefficient.

I - Table 6.1. Partial safety factors in the calculation ULS (DA 1/1 and DA 1/2) and SLS for calculations of construction pit in geotechnical category 2 – RK2.

	Load γ_F				Soil parameters γ_M			
	γ_G (permanent load)		γ_Q (variable load)		γ_γ	γ_φ	γ_c	γ_{cu}
	unfavourable	favourable	unfavourable	favourable				
ULS								
DA 1/1	1,35	1,00	1,50	0,00	1,00	1,00	1,00	1,00
DA 1/2	1,00	1,00	1,10	0,00	1,00	1,25	1,25	1,40
SLS	1,00	1,00	1,00	0,00	1,00	1,00	1,00	1,00

The partial safety factors γ_F and γ_M are summarised in I - Table 6.1.

In certain circumstances and if justified, according to EN 1990, it is permitted to calculate using partial safety factors as long as one expects/permits a higher or lower reliability of the construction pit. In Belgium there are currently three risk categories: RK1, RK2 and RK3. For determining the risk category of a construction pit, one may take the proposal of an assessment methodology in informative annex 1 of the BBRI (2016) as an example.

In principle, the specifications should indicate under which risk categorty a specific construction pit may be classified. If the specifications do not include such information, one should assume that the construction pit falls under risk class 2 (partial safety coefficients from I - Table 6.1).

Spring models, finite elements or finite difference methods may be involved in carrying out the wall calculations. In the spring model, the algorithm is based on design approach 1 in Eurocode 7, part 1. The calculation is made as follows (shown schematically in Figure 6.2):

1. A first calculation of the wall in UGT-DA1/2 for the most determining stage, for determining the required wall length (embedment). Here, the criterion that the passive resistance is mobilised up to 100% is applied.
2. For all phases: a simplified calculation for the design of the wall and the horizontal support, with the factors from the SLS and with a factor α_{ver} to 1.1[1*] that is applied to the variable loads. From these calculations, the following values are derived for the different phases:
   - by multiplying by a factor of 1.35[1*]: the calculation values in ULS DA1/1 of the anchor and strut forces ($F_{a,1}$), further used for the structural and geotechnical design;
   - by multiplying by a factor of 1.35[1*]: the calculation values in ULS DA1/1 of the internal forces, to be used for the structural design of the wall;
3. From the SLS calculations, the following values are derived for the different phases:
   - the wall deformations, displacements, anchor and strut forces ($F_{a,2}$).

[1*] Values for RK 2; in case of RK 1, other values may be applied, in case of RK 3 other values must be applied

When applying **finite elements or finite difference models for the design of the wall and horizontal support, all phases are calculated with** the factors of the SLS and with the factor α_{ver} to 1.1 by which the variable loads must be multiplied. In the same way as with the spring model, the anchor or strut forces and internal forces (M, N, Q) are verified for each phase. A φ-c reduction is applied to the governing/critical phase, where a safety of at least 1.25* should be achieved.

Subsequently, an SLS calculation is carried out for all phases, which is done in order to determine the permissible deformations and displacements.

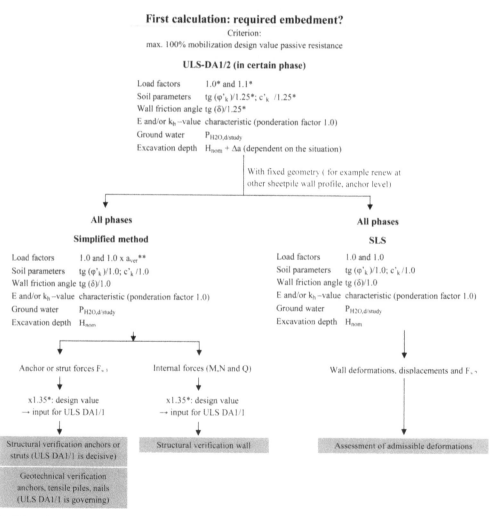

*: Values for RK 2; in case of RK 1 other values may be applied, in case of RK 3 other values must be applied

I - Fig. 6.2. Simplified flow chart with a spring model for calculating the different phases (based on design approach 1 of the EC7).

Design approach

In Belgium, the following design plan may be applied for the design of construction pit with a spring model:

Step 1	Determining the relevant principles. Cross-section, level horizontal support(s), side-loads, water pressures, risk category, phasing.
Step 2	Determining the characteristic values of the following parameters. Geotechnical parameters, stiffness parameters of the wall and support point(s), loads, geometrical parameters, excavation level and water levels.
Step 3	Determining the design values of the parameters for the most critical phase (usually deepest excavation). To this end, the load and material factors in ULS DA1/2 are applied (see previous text).
Step 4	First calculation for determining the minimum embedding depth on the basis of the design values of the parameters from step 3. This involves an assessment of the geotechnical (GEO) limit state.
Step 5	The aforementioned steps may be repeated with, for example, other geometries, etc., to determine whether a more economical design may be realised. Determining the final geometry and embedding depth.
Step 6	Subsequently, every phase is calculated separately. To this end, a simplified method for determining the design values of the parameters is applied, in which the load and material factors are applied in SLS and a factor α_{ver} on the variable loads. Following these calculations, one will end up with the design values of the anchor or strut forces ($F_{a,1}$) and the internal forces in the wall (moments, shear forces, normal forces) for each phase. Determining the critical anchor or strut forces and the critical internal forces.
	Subsequently, a calculation in the SLS is carried out for each phase, from which the critical deformations and displacements of the wall can be derived. From this SLS calculation one will also derive anchor or strut forces ($F_{a,2}$). $F_{a,2}$ can be introduced in the design of the ground anchors on the basis of tests. If no calculation in the SLS is carried out, e.g. because the displacements are not critical/relevant, then one must assume that $F_{a,1} = F_{a,2}$.
Step 7	Verification of the internal forces (STR). The critical internal forces from step 6 are multiplied by a factor of 1.35. In this way, one can obtain the design values of the internal forces in the wall (moment, shear force, normal force) in ULS, which are further used for verification in accordance with the relevant structural Eurocodes.

Step 8	Verification of the anchor or strut force (STR). The critical anchor or strut forces from step 6 are multiplied by a factor of 1.35. In this way, one can obtain the design values of the normal forces in the anchor or strut in ULS, which are further used for the structural verification in accordance with NBN EN 1993-5 and the ANB. In most cases, NBN EN 1993-5 and the ANB are also applicable for the structural verification of tensile micropiles or nails. Anchor plates, purlins and connecting elements are also verified in accordance with the method in NBN EN 1993-5 and the ANB.
Step 9	Verification of the anchor force (GEO). The critical anchor or strut forces from step 6 are multiplied by a factor of 1.35. This allows one to obtain the design value of the anchor force for verification of the geotechnical bearing capacity of the grout body. The geotechnical design is executed in accordance with the methods in the NBN EN 1997-1 and the ANB; for ground anchors and tensile (micro)piles a Belgian application document is currently still pending.
Step 10	Verification of deformations. The critical deformations and displacement of the wall in the SLS that follow from step 6 are verified on the basis of the applicable limits: $E_d \leq C_d$
Step 11	Verification of other mechanisms. • Kranz stability • Soil rupture • Stability deep glide plane • Piping, hydraulic soil rupture (HYD), bursting/uplift of the "impervious" layer (UPL) • Vertical bearing capacity
Step 12	Execution aspects Redaction of the instructions for the execution and description of QA/QC – process (execution monitoring, control tests, required monitoring, …).
Step 13	Verification of decisions made. Verification of potentially adjusted critical principles during the design process, parameter determination, algorithms and embedding depth compared to the original principles and assumptions.

With the use of **finite elements or finite difference models**, for designing the wall and horizontal support, all phases are calculated with the factors of the SLS and the factor α_{ver} equal to 1.1[*] by which the variable loads should be multiplied. In the same way as with the spring model, the anchor or strut forces and internal forces (M, N, Q) are verified for each phase. A φ-c reduction is applied to the critical phase, where a safety of at least 1.25[*] should be achieved. Subsequently, an SLS calculation is carried out for all phases to verify if the deformations and displacements are admissible.

[*] Values for RK 2; in case of RK 1 other values may be applied, in case of RK 3 other values must be applied

62

6.2.2 Situation in the Netherlands

Standards and guidelines

In the Netherlands, the Building Act 2012 stipulates that Eurocode 7 is applicable for geotechnical calculations. The standard NEN 9997-1 'Geotechnisch ontwerp van constructies - Deel 1: Algemene regels' ('Geotechnical design - Part 1: General rules') includes the applicable regulations of NEN EN 1997-1, Nationale Bijlage (National Annex) and normalised additional provisions from NEN- and CUR publications.

One of the CUR publications, certain elements of which are copied over to NEN 9997-1, is CUR 166 'Damwandconstructies' ('Sheetpile Walls'). In the Netherlands, the safety philosophy and calculation methods in accordance with CUR 166 are applied for retaining walls.

Also applicable for the design and constructive verifications are:

- Eurocode 0: NEN-EN 1990 incl. NA, change and correction sheet for the foundations of the design.
- Eurocode 1: NEN-EN 1991-1-1 to 1991-2 incl. NA and correction sheets for determining loads.
- Eurocode 2: NEN-EN 1992-1-1 incl. NA and correction sheet for concrete calculations.
- Eurocode 3: NEN-EN 1993-1-1 and NEN-EN 1993-5 incl. NA and correction sheets for steel calculations.
- Eurocode 4: NEN-EN 1994-1-1 incl. NA and correction sheet for calculations of steel-concrete constructions.

Assessment, design approach and risk class

One must check whether the assessment criteria for the ultimate limit state ULS and the serviceability limit state SLS are met.

In the Netherlands, assessment of structural (STR) and geotechnical (GEO) (ultimate) limit states should be conducted on the basis of design approach 3 (OB3). The following should apply: $E_d \leq R_d$, with:
- E_d the design value of the loads or load effects.
- R_d the design value of the resistance.

Factors are applied to the loads or load effects from the structure and on the material parameters. One must make sure a limit state does not occur as a result of failure or excessive deformation with the following combination of partial factors: A2 + M2 + R3 (for a retaining wall with geotechnical load). With :
- A2 being the collection of partial factors on the loads or load effects.
- M2 being the collection of partial factors on the soil parameters.
- R3 being the collection of partial factors on the resistance.

These factors are included in NEN 9997-1 (I - Table A.3, A.4b, A.4c and A.13) and CUR 166 for the various safety classes.

For the serviceability limit state SLS the occurring deformations must be assessed on the basis of: $E_d \leq C_d$, with:
- E_d the design value of the deformation in the serviceability limit state
- C_d the limit value for the design value of the deformation

NEN EN 1990 defines the 3 reliability classes (formerly also called "safety classes") RC1 to RC3, under which retaining structures may be classified. Additionally, CUR 166 indicates CUR class I. These classes are related to a value for the reliability index β (see I - Table 6.2 and NEN EN 1990 annexes B and C).

I - Table 6.2. Reliability classes according to NEN-EN 1990.

CUR class I: ($\beta \approx 2.5$)	Relatively simple structures, no personal safety risks in the event of failure, relatively limited damage in the event of failure, e.g. a constrution pit.
RC1: ($\beta \approx 3.3$)	Minor consequences regarding the loss of life and/or small or negligible economic or social consequences or effects on the environment, e.g. a shallow construction pit (1 underground level) and a quay wall with limited retaining height (max. 5 m).
RC2: ($\beta \approx 3.8$)	Moderate consequences regarding the loss of life and/or substantial economic or social consequences or effects on the environment; e.g. a deep construction pit (2 or more undergrond levels) in urban areas, a retaining wall along inland water and a quay wall of a sea port with a great retaining height (> 5 m).
RC3: ($\beta \approx 4.3$)	Major consequences regarding the loss of life and/or significant economic or social consequences or effects on the environment, e.g. with unique structures and (retaining walls in) a primary water barrier.

A reliability class should be employed that is consistent with the applicable situation, or one prescribed by a client (determined on the basis of risk assessment in respect of life and economic consequences) or additional guidelines.

Load and material factors
Table 6.3 summarises the partial load and material factors for a retaining wall
- For the subsoil constants a distinction is made between a low average of the characteristic value ($k_{h;low}$) and a high average of the characteristic values ($k_{h;high}$). In between, a factor of 2.25 is applied.
- The factor γ_R for the resistance of retaining structures is 1.0.
- In addition to the aforementioned parameters, CUR 166 also defines a number of important geometrical supplements Δa, which relate to the retaining height and the to be calculated (ground)water level difference.
- In addition to the partial factors for critical moment, shear force and maximum anchor/ strut forces resulting from the partial factors, for determining the design value of the internal forces in question, a minimum load factor of $\gamma = 1.2$ towards the representative value $M_{s,rep}$, $D_{s,rep}$ and P_{rep} must be applied.
- In addition, an extra load factor is applied to the anchor/strut force for verifying the purlins ($\gamma = 1.1$), geotechnical bearing capacity of grout bodies of anchors ($\gamma = 1.1$), vertical equilibrium ($\gamma = 1.1$), steel capacity anchors ($\gamma = 1.25$) and steel capacity struts ($\gamma = 1.25$).

1 - Table 6.3. Partial load and material factors according to NEN EN 1990.

	Load γ_F				Soil parameters γ_M				
	γ_G (permanent load)		γ_Q (variable load)		γ_γ	γ_φ	$\gamma_{c'}$	$k_{h,low}$	$k_{h,high}$
	unfavourable	favourable	unfavourable	favourable					
Class 1	1.00	1.00	1.00	0.00	1.00	1.05	1.00	1.30	1.00
RC1	1.00	1.00	1.00	0.00	1.00	1.15	1.15	1.30	1.00
RC2	1.00	1.00	1.10	0.00	1.00	1.175	1.25	1.30	1.00
RC3	1.00	1.00	1.25	0.00	1.00	1.20	1.40	1.30	1.00
SLS	1.00	1.00	1.00	0.00	1.00	1.00	1.00	1.00	N/A

Design approach
In the Netherlands, design calculations of soil retaining walls are carried out on the basis of the design plan as described in CUR 166.

A short summary of this plan:
The following page shows this plan in the form of a flow chart.

Step 1	Determining the relevant principles Cross-section, side-loads, water pressures, risk class and phasing.
Step 2	Determining the characteristic values of the parameters. Soil parameters, stiffness parameters of the wall and support point(s), loads, geometrical parameters, water levels.
Step 3	Determining the design values of the parameters. Application of the load and material factors, as discussed in par. 6.1.2 .
Step 4	Choose between algorithm A or B (A = design values in all phases; B = only design values for the to be assessed phase; other phases with characteristic values)
Step 5	Calculation of the minimal embedding depth on the basis of the design values of the parameters. This concerns assessment of the geotechnical (GEO) limit state.
Step 6	Varying the embedding depth, to determine whether a more economical design may be realised. Determining the embedding depth. Design calculations on the basis of the selected embedding depth, in which the critical internal forces and deformations are determined. This involves 4 calculations for the ULS (Step 6.1 to 6.4) with variation in $k_{h;low}$ resp. $k_{h;high}$, retaining height Δa and groundwater levels Δa. In addition, a calculation is made for the SLS (Step 6.5) with $k_{h;low}$ and geometric supplements Δa equal to 0.
Step 7	Verification of moment in the wall (STR): $\max (M_{E;d}; 1.2 M_{E;rep}) \leq M_{R;d}$
Step 8	Verification of shear forces in the wall (STR): $\max (D_{E;d}; 1.2 D_{E;rep}) \leq D_{R;d}$ Verification of normal forces in the wall: $N_{E;d} \leq N_{R;d}$
Step 9	Verification of anchor and strut forces. This involves assessment of the anchor bar, bearing capacity grout body, anchor plate, purlin and strut. Additional load factors 1.1 resp. 1.25 are applicable, such as described in par. 6.1.2 .
Step 10	Verification of deformations. The calculated wall displacement in the SLS: u_{max} (or E_d) must be assessed on the basis of the applicable u_{limit} (or C_d): $u_{max} \leq u_{limit}$
Step 11	Verification of other mechanisms: • Kranz stability • Soil rupture • Stability deep glide plane • Piping, hydraulic soil rupture (HYD), bursting/uplift of an "impervious" layer (UPL) • Vertical geotechnical bearing strengt
Step 12	Execution aspects Redaction of the instructions for the execution and description of the required monitoring
Step 13	Verification of decisions made. Verification of potentially adjusted critical principles during the design process, parameter determination, algorithms and embedding depth compared to the original principles and assumptions.

For each step, the applicable paragraphs and reference documents of this handbook are indicated.

Flow chart
Design according SBRCURnet/BBRI Handbook of Soil mix walls – design and execution

Flow chart 'Design plan Belgium'

Step 1 Determine the critical/relevant principles

    Safety class →   NBN-EN-1997, NBN-EN-1997-1-ANB, CUR166 & CUR/BBRI §6.2.1

    Reference period →   CUR/ BBRI §6.2.3 & §6.13

    Partial factors →   NBN-EN-1997, NBN-EN-1997-1-ANB, CUR166 & CUR/ BBRI §6.2.1

    Temporary construction vs. permanent construction →   CUR/ BBRI

Step 2 Determine the characteristic value of the parameters

    Soil mix material

        Wall geometry →   CUR/ BBRI §6.7

        General factors →   NBN-EN-1997, NBN-EN-1997-1-ANB, CUR166 & CUR/ BBRI §6.3

        Compressive strength →   NBN-EN-1997, NBN-EN-1997-1-ANB, CUR166 & CUR/ BBRI §6.3

        Tensile strength →   NBN-EN-1997, NBN-EN-1997-1-ANB, CUR166 & CUR/ BBRI §6.3

        Shear strength →   NBN-EN-1997, NBN-EN-1997-1-ANB, CUR166 & CUR/ BBRI §6.3

        Adhesion soil mix material - steel →   NBN-EN-1997, NBN-EN-1997-1-ANB, CUR166 & CUR/ BBRI §6.3

        Stiffness →   CUR/ BBRI §6.3 & §6.7

    Steel beam

        Beam →   NEN-EN-1993-5-ANB

        General factors →   NEN-EN-1994-1-1-ANB

        Strength →   NEN-EN-1994-1-1-ANB

        Stiffness →   CUR/ BBRI §6.7

        Cover →   NEN-EN-1994-1-1-ANB

    Top loads

    Soil mix wall

        Determine the effective width →   CUR/ BBRI §6.7.2

        Determine the moment capacity of the steel beams only →   CUR/ BBRI §6.7.3

        Determine the moment capacity of the composite section →   CUR/ BBRI §6.7.4

        Pressure & bending of the composite section →   CUR/ BBRI §6.7.5

        Determine the shear force capacity →   CUR/ BBRI §6.7.7

        Determine the stiffness of the composite section →   CUR/ BBRI §6.6

Step 3 Determine the design values of the parameters

    No different to any other retaining wall →   NBN-EN-1997, NBN-EN-1997-1-ANB, CUR166

Step 4 Calculate the minimal embedding depth GEO

        No different to any other retaining wall   ⟶   NBN-EN-1997, NBN-EN-1997-1-ANB & CUR166

Step 5 Repeat step 1 to 4 until an economic design is realised

        No different to any other retaining wall   ⟶   NBN-EN-1997, NBN-EN-1997-1-ANB & CUR166

Step 6 Design calculations, separate for each phase

        No different to any other retaining wall   ⟶   NBN-EN-1997, NBN-EN-1997-1-ANB & CUR166

Step 7 Verification of internal forces

        No different to any other retaining wall   ⟶   NBN-EN-1997, NBN-EN-1997-1-ANB & CUR166

Step 8 Verification of anchor & strut forces STR

        No different to any other retaining wall   ⟶   NBN-EN-1997, NBN-EN-1995-5-ANB & CUR166

Step 9 Verification of anchor & strut forces GEO

        No different to any other retaining wall   ⟶   NBN-EN-1997, NBN-EN-1997-1-ANB & CUR166

Step 10 Verification of deformations

        No different to any other retaining wall   ⟶   NBN-EN-1997, NBN-EN-1997-1-ANB & CUR166

Step 11 Verify other mechanisms

        No different to any other retaining wall   ⟶   NBN-EN-1997, NBN-EN-1997-1-ANB & CUR166

        Assessment anchor seat construction   ⟶   CUR/BBRI §6.8.10

        Vertical bearing capacity   ⟶   CUR/BBRI §6.9

        Verification pressure arch   ⟶   CUR/BBRI §6.4

        End beam underside   ⟶   CUR/BBRI §6.11.2

Step 12 Verify the execution aspects

        Influence on the adjacent foundations   ⟶   CUR/BBRI §6.10

        Execution   ⟶   CUR/BBRI Chapter 7

Step 13 Verify decisions made

        No different to any other retaining wall   ⟶   NBN-EN-1997, NBN-EN-1997-1-ANB & CUR166

Flow chart
Design according SBRCURnet/BBRI Handbook of Soil mix walls – Design and execution

Flow chart 'Design plan The Netherlands'

Step 1 Determine the relevant principles

Safety class	NEN9997-1,CUR166 & CUR/BBRI §6.2.2
Reference period	CUR/ BBRI §6.2.3 & §6.13
Partial factors	NEN9997-1,CUR166 & CUR/ BBRI §6.2.2
Temporary construction vs. permanent construction	CUR/ BBRI

Step 2 Determine the characteristic value of the parameters

Soil mix material

Wall geometry	CUR/ BBRI §6.7
General factors	NEN9997-1,CUR166 & CUR/ BBRI §6.3
Compressive strength	NEN9997-1,CUR166 & CUR/ BBRI §6.3
Tensile strength	NEN9997-1,CUR166 & CUR/ BBRI §6.3
Shear strength	NEN9997-1,CUR166 & CUR/ BBRI §6.3
Adhesion soil mix material - steel	NEN9997-1,CUR166 & CUR/ BBRI §6.3
Stiffness	CUR/ BBRI §6.3 & §6.7

Steel beam

Beam	NEN-EN-1993-1-1-NB & NEN-EN-1993-5-NB, CUR166
General factors	NEN-EN-1994-1-1-NB
Strength	NEN-EN-1994-1-1-NB
Stiffness	CUR/ BBRI §6.7
Cover	NEN-EN-1994-1-1-NB

Top loads

Soil mix wall

Determine the effective width	CUR/BBRI §6.7.2
Determine the moment capacity of the steel beams only	CUR/ BBRI §6.7.3
Determine the moment capacity of the composite section	CUR/ BBRI §6.7.4
Pressure & bending of the composite section	CUR/ BBRI §6.7.5
Determine the shear force capacity	CUR/ BBRI §6.7.7
Determine the stiffness of the composite section	CUR/ BBRI §6.6

Step 3 Determine the design values of the parameters

No different to any other retaining wall	NEN9997-1, CUR166

69

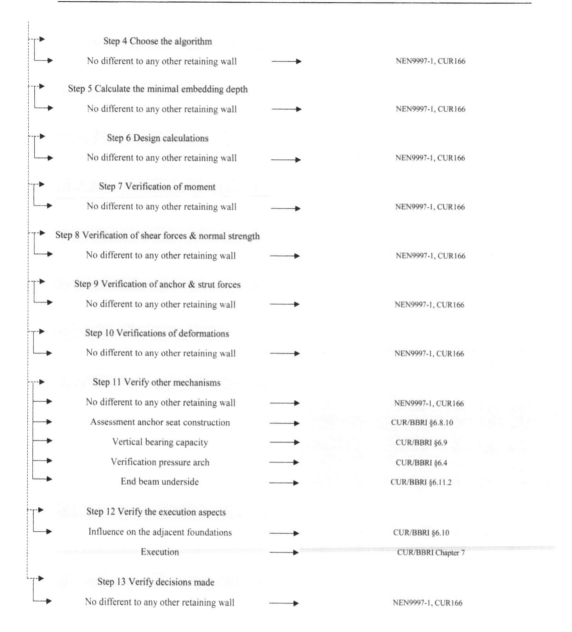

Step 4 Choose the algorithm

No different to any other retaining wall ⟶ NEN9997-1, CUR166

Step 5 Calculate the minimal embedding depth

No different to any other retaining wall ⟶ NEN9997-1, CUR166

Step 6 Design calculations

No different to any other retaining wall ⟶ NEN9997-1, CUR166

Step 7 Verification of moment

No different to any other retaining wall ⟶ NEN9997-1, CUR166

Step 8 Verification of shear forces & normal strength

No different to any other retaining wall ⟶ NEN9997-1, CUR166

Step 9 Verification of anchor & strut forces

No different to any other retaining wall ⟶ NEN9997-1, CUR166

Step 10 Verifications of deformations

No different to any other retaining wall ⟶ NEN9997-1, CUR166

Step 11 Verify other mechanisms

No different to any other retaining wall ⟶ NEN9997-1, CUR166

Assessment anchor seat construction ⟶ CUR/BBRI §6.8.10

Vertical bearing capacity ⟶ CUR/BBRI §6.9

Verification pressure arch ⟶ CUR/BBRI §6.4

End beam underside ⟶ CUR/BBRI §6.11.2

Step 12 Verify the execution aspects

Influence on the adjacent foundations ⟶ CUR/BBRI §6.10

Execution ⟶ CUR/BBRI Chapter 7

Step 13 Verify decisions made

No different to any other retaining wall ⟶ NEN9997-1, CUR166

6.2.3 Points of attention related to permanent applications (= use phase - NL)

In cases where the soil mix wall must perform a specific function (= long-term function, in the Netherlands this is referred to as the 'use phase'), a number of important aspects should be taken into account. For example, for the wall calculation in the final phase one should determine whether the design should be performed with adjusted horizontal soil pressures (neutral).

Other factors to consider include:
- the distribution of pressures and transfer on the covering wall;
- potential remaining moments from the short-term phase;
- the long-term water levels;
- any expected additional long-term loads;
- soil mix material creep;
- soil mix material durability;
- the corrosion of the steel reinforcement.

With regards to the long-term effects on the characteristics of the soil mix material, the interaction of steel and soil mix, etc., please refer to §6.14.

6.2.4   Flow charts, design strategy and design approach soil mix walls

The first flow chart below provides an overview of the design strategy to be followed. The various focus points, aspects and risks that must be taken into account for the design of soil mix walls in their interdependence are included in this chart.

The second chart gives a general overview of the design of soil mix walls, depending on the function(s) the wall is intended for (see §3). For the various design aspects and aspects related to quality control, this chart refers to the relevant paragraphs in this handbook. Finally, a separate, third, chart has been created for the horizontal support.

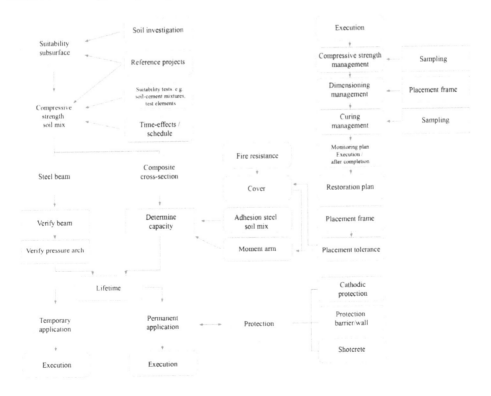

Chart flow
Design according to
SBRCURnet/BBRI Handbook Soil Mix Walls

Permanent vertical bearing wall

Temporary vertical bearing wall

Permanent cut-off wall

Temporary cut-off wall

Permanent water-retaining vertical bearing wall

Temporary water-retaining vertical bearing wall

Permanent earth retaining wall

Temporary earth retaining wall

Permanent earth retaining vertical bearing wall

Temporary earth retaining vertical bearing wall

Permanent earth retaining water-retaining wall

Temporary earth retaining water-retaining wall

Permanent earth retaining water-retaining vertical bearing wall

Temporary earth retaining water-retaining vertical bearing wall

Points of attention & minimal characteristics of the soil mix material

Suitability tests

Durability

Water tightness

Corner solutions

Beam end

Finishing aspects permanent wall

Horizontally supported

Quality control soil mix material

Verify wall deformations

Determine water permeability wall

Leak detection wall

Calculation

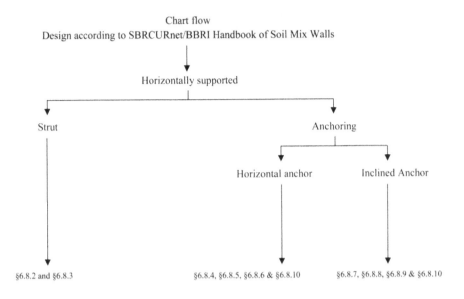

6.3 Material properties

6.3.1 Soil mix parameters

The basic parameters that are important during the geotechnical and structural verification of the horizontal and/or vertical stability of the soil mix wall are:

- the unconfined compressive strength (UCS) of the soil mix material over the height of the wall, either the average values $f_{sm,m}$, or characteristic values $f_{sm,k}$
- the tensile strength $f_{sm,t}$ of the soil mix material
- the shear resistance τ_{Rsm} of the soil mix material
- the modulus of elasticity E_{sm} of the soil mix material
- the adhesion f_{bd} between the soil mix material and the reinforcement

Note
The design below applies the cylinder compressive strength on cored samples on samples with H/D=1.0, which in accordance with EN 206 is similar to the cube compressive strength. For deviating H/D ratios, no systematic deviations were observed from the available test databank compared to the cylinder compressive strength at H/D=1.0.

The following parameters are required for determining the calculation magnitudes of the wall that must be taken into account in the stability calculations:

- the moment of resistance W, which is decisive for the bending moments (optionally in combination with shear forces and normal forces) that can be taken by the wall in the short and long term;
- the bending stiffness EI, which is decisive for the deformations of the wall and – to a lesser degree – the internal forces in the various construction phases.

With regards to the functionality of the wall, there may be other important factors, such as:

- the permeability coefficient or the hydraulic resistance of the soil mix wall;
- the resistance of the wall against climatic influences, contaminants, ...

In the above, a distinction can be made between:

- the short term behaviour versus the long term behaviour;
- the pre-design phase (PDP) versus the final design phase (FDP) and the execution phase (EP).

6.3.2 Average and characteristic compressive strength of the soil mix material

This section provides recommendations on the estimation/measurement of the compressive strength resistance of the soil mix material that can be implemented in the design. For more background information in connection with the method for determining the characteristic value of the compressive strength from a dataset of test data, please refer to Appendix 1.

Influencing parameters

For the design of a earth retaining soil mix wall, a realistic assessment of the to be expected compressive strengths of the soil mix material must be made. The designer must be aware that these compressive strengths and the uniformity/heterogeneity thereof are influenced by many parameters, such as (also see §2.3.1 in part 2 of the handbook):

1. the soil characteristics: type of soil, pH, water content
2. the binding agent: type, additives, dosage and water-cement factor
3. the mixing method: equipment, mixing energy, degree of homogenisation
4. the curing conditions: temperature, age, moisture
5. any possible presence of contaminants or brackish water.

Design levels

Two design levels are defined for the prior estimates of the average or characteristic compressive strengths of the soil mix material in the various soil layers:

- **Level 1** this takes into account general empirical data, based on a large database in which a distinction is only made between major soil type classes (peat, clay, loam, sand)

- **Level 2** in this level the design parameters of the compressive strength are determined on the basis of:
  - either specific empirical data, derived from a collection of test data in comparable conditions: soil layers of equal geologic origin or comparable granular distribution treated with a similar execution process;
  - the results of prior suitability tests, gained from in situ mixtures or lab test mixtures.

Applicability
The required design level and the scope of the appended studies are dependent on the reliability classes, in this case risk class (RC in NL, RK in B).

The recommended design level depending on the risk class is presented in I - Table 6.4. For more information, please refer to Chapter 8 Quality Control.

I - Table 6.4. Required design level depending on the project phase purpose and risk class.

Project phase	Required design level depending on the risk class			
	CUR class I	RC1/RK1	RC2/RK2	RC3/RK3
Pre-design PD	1	1	1 or 2	1 or 2
Detail design FDP	1	1 or 2	2	2
Execution Design EP	1 or 2	2	2	2

Method of determination
Details on the methodology for determining and deriving the average and characteristic compressive strength on the basis of a collection of test data are provided in Appendix 1. On the basis of this, below are the various approaches for the different design levels:

1. Level 1
Target values of the compressive strengths may, for example, be derived from statistical processing of a great amount of test data in a variety of soils. Examples of this are provided in the histograms in figures 2.22 to 2.27 in §2.3.2 in part 2 of the handbook that are based on the research programme of the BBRI, or in the summarising diagrams in figure T1.1 and T1-2 of Appendix 1. The wide and narrow ranges of the compressive strengths that can be derived on the basis of this BBRI test data are summarised in I - Table 6.5. These ranges are merely listed indicatively and informatively; it is up to the designer to estimate the compressive strengths he applies in design level 1.

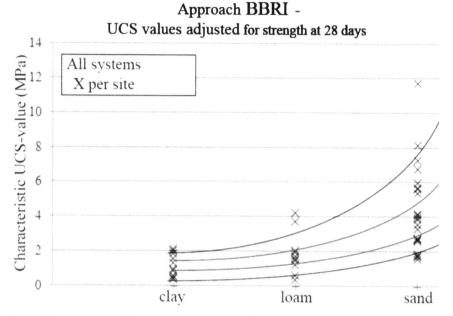

I - Fig. 6.3. Characteristic UCS value per soil type adjusted for strength at 28 days with the 15% D6 rule with regards to the soil inclusions (BBRI design approach, see Appendix 1).

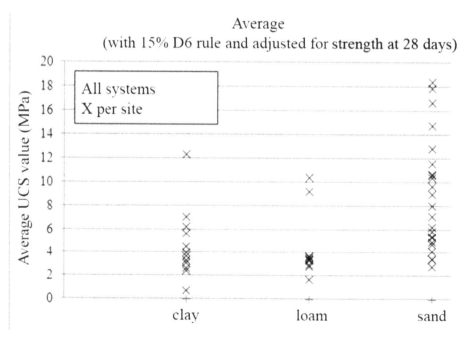

I - Fig. 6.4. Average UCS value per soil type adjusted for strength at 28 days with the 15% D6 rule with regards to the soil inclusions (BBRI design approach, see Appendix 1).

1 - Table 6.5. Average and characteristic compressive strengths per soil type with strength at 28 days and with the D6 rule.

Soil type	Average compressive strength $f_{sm,m}$		Characteristic compressive strength $f_{sm,k}$	
	Wide range	Narrow range	Wide range	Narrow range
Clay	1.5 - 4.0 MPa	2.0 - 3.0 MPa	0.5 - 2.0 MPa	1.0 - 1.5 MPa
Loam	2.0 - 5.0 MPa	2.5 - 4.0 MPa	0.75 - 3.0 MPa	1.25 - 2.0 MPa
Sand	4.0 - 16.0 MPa	6.0 - 10.0 MPa	2.0 - 8.0 MPa	3.0 - 5.0 MPa

In peaty soils the compressive strength of the soil mix material is very low. An adjusted working method, which also ensures proper homogenisation of the underlying and overlying layers, is required to obtain at least the compressive strength which is, for example, required for the arching effect.

2. Level 2

Other target values may be used by the designer if he possesses more specific test data for the specific soils, the applied mixing technique and the expected cement dosage. The statistical processing for determining the average and characteristic compressive strengths in various soil layers occurs as follows:

- the acquired test results are – if necessary – converted into a 28-day compressive strength by applying a correlation factor β
- the average compressive strength $f_{sm,m}$ is determined as the arithmetical average of the acquired UCS values
- the characteristic compressive strength $f_{sm,k}$ is determined by one of the methodologies in Appendix 1:
  - either as the 5% fractile of the cumulative curve of the 28-day compressive strengths
  - or = $\min(f_{sm,min} ; 0.7xf_{sm,m} ; 12\ \text{MPa})$.

It is important to consider, in accordance with the proposed procedure, the minimum value of the results of the pressure tests in the design procedure. This way the distribution of the results of the test samples is taken into account. An approach that is only based on the average value of the obtained compressive strength is insufficient for taking into account the distribution (variation) in determining a characteristic value of the compressive strength.

Verification after realisation

Depending on the risk classes and the function of the wall, compressive tests conducted on samples in the realised walls must confirm that the accepted design-compressive strengths are complied with. The verification tests required for this are described in Chapter 8. Quality control (see Table 8.1). The processing of the test results is carried out as described above for Level 2.

Notes – time dependency of the compressive strength

The procedure for determining the compressive strength of the soil mix material is based on results of compressive tests carried out after a reference period of 28 days of curing. In addition, the following remarks can be formulated:

- For applications and functions in which the curing period is shorter than 28 days one should apply reduction factors on the value of the compressive strengths that take into account the development of the compressive strength in the context of time (see §2.3.2 and I - Table 2.12 in part 2 of the handbook).
- Despite the fact that the compressive strength of the soil mix material may still increase significantly after a period of 28 days of curing, for safety reasons, no multiplication factors are permitted with longer ages.
- For long term applications, a reduction factor α_{sm} – in accordance with §3.1.6 of EN 1992-1-2 - is applied to the compressive strength.

The factors β and α_{sm} that are related to the time dependency of the compressive strength compared to the 28-day strength, are applied in the following paragraph when determining the design value of the compressive strength.

6.3.3   Design value of the compressive strength of the soil mix material

Once the characteristic value of the 28-day compressive strength is determined, the design value of the compressive strength $f_{sm,d}$ in the design phase in accordance with the methodology of EN 1992-1-1 is determined as follows:

$$f_{sm,d} = \alpha_{sm} \frac{f_{sm,k}}{\gamma_{SM} k_f} \beta \tag{6.1}$$

in which:

α_{sm} : the coefficient for long term effects (> 24 months); for temporary states α_{sm} is equal to 1.0, for long term situations equal to 0.85.

γ_{SM} : the partial factor for a material property; in the European standard EN 1992-1-1 defined as γ_C for concrete; for soil mix material we define the factor as γ_{SM}.

k_f : a pile factor defined in §2.4.2.5 (2) of the European standard EN 1992-1-1; the factor k_f is applied in the concrete procedure to make a difference between the conditions for curing in a lab and in situ; if the compressive strength is determined on the basis of samples from the in situ realised soil mix walls, this pile class factor is not applicable and k_f is equal to 1.0; in all other cases, e.g. starting from 'estimated' values, this factor is equal to 1.1.

β : the correction factor for a young age of the soil mix material; depending on various factors (see §2.3.2 in part 2 of the handbook) $\beta = 0.3$ to 0.7 after 7 days of hardening, $\beta = 0.6$ to 0.8 after 14 days of hardening.

It is expressly noted that the time factors and curing conditions are explicitly calculated using the above formula when determining the design value of the compressive strength. The design value of the compressive strength can therefore vary depending on the age of the wall.

The factors to be used must be consistent with the national regulations for the application of the Eurocodes. The factors to be used (which are the same in Belgium and the Netherlands) are summarised in I - Table 6.6.

I - Table 6.6. Factors for determining the design value of the compressive strength $f_{sm,d}$ of soil mix.

Applicable limit state	Factors to be used fo determining $f_{sm,d}$ (Belgium - The Netherlands)			
	γ_{SM}	α_{sm}	k_f	β
ULS DA1-1 (B) ULS-DA3 (NL)	1.50/1.20	1.00/0.85	1.10/1.00	Depending on age – see §2.3.2 in part 2 of the handbook

6.3.4 Design value of the tensile strength of the soil mix material

No tensile strength is considered for the soil mix material during design, except for determining the moment for which cracks arise in the soil mix and the adhesion with the reinforcement. The design value of the tensile strength $f_{sm,td}$ may be estimated at 10 to 15% of the design value of the compressive strength. It is better to use the correlation of I - Table 3.1 of EN 1992-1-1 for the 5% fractile characteristic value of the tensile strength:

$$f_{sm,tk}=0.7 \times f_{sm,tm} = 0.7 \times 0.30 \times f_{sm,k}^{2/3}=0.21 \times f_{sm,k}^{2/3} \qquad (6.2)$$

and therefore:

$$f_{sm,td}=\alpha_{sm}\frac{f_{sm,tk}}{\gamma_{SM} \cdot k_f} \beta \qquad (6.3)$$

The design values of the tensile strength calculated in this way, including α_{sm} = 1.0, γ_{SM} = 1.50 and k_f = 1.0, are indicated in I - Table 6.7.

Characteristic 28-day compressive strength	28-day tensile strength of soil mix	
$f_{sm.k}$	Characteristic value $f_{sm.tk}$	28 day design value $f_{sm.td}$
$= f_{sm.m}*0.7$	$= 0.21 f_{sm.k} (2/3)$	$= f_{sm.tk}/1.5$
1	0.21	0.14
2	0.33	0.22
3	0.44	0.29
4	0.53	0.35
5	0.61	0.41
6	0.69	0.46
7	0.77	0.51
8	0.84	0.56
9	0.91	0.61
10	0.98	0.65

6.3.5 Design value of the shear strength of the soil mix material

§2.3.5 in part 2 of the handbook provides information about the shear strength τ_{Rsm} of soil mix material. Data shows that soil mix material with compressive strengths up to 6 MPa have the shear strength $\tau_{Rsm,d}$ greater than 20% of the compressive strength.

Because little test data is available in Belgium and the Netherlands with regards to this characteristic, it is (quite conservatively) assumed that the design value of the shear strength of the soil mix material is equal to the design value of the tensile strength:

$$\tau_{Rsm,d} = f_{sm,td} \tag{6.4}$$

The design value of the shear strength determined in this way can then be employed in the verification of the arching effect.

6.3.6 Design value of the adhesion f_{bd} between steel and soil mix

The adhesion between the steel beam and soil mix material is, in accordance with EN 1994-1-1 partly dependent on:
- the tensile strength of the soil mix material
- the degree of covering of the beam, or – in other words – the thickness of the covering
- the nature of the steel beam's surface: whether coated or not, free of oil, grease or loose mill scale or rust
- the possible presence of elements increasing the adhesion.
- the normal stresses perpendicular to the shear area.

The following criteria are applicable for determining the design value of the adhesion f_{bd} (including a material factor of 1.5) between the steel beam and the soil mix material:

1. Criterion 1: in accordance with the relationship that was included in ENV 1992-1-1 (§5.2.2.2) at the time for smooth steel, a design value is based on the characteristic compressive strength of the soil mix material and the (equivalent) diameter of the steel beam via the relationship below:

$$f_{bd}=\eta_1\,\eta_2\,0.240 f_{sm,k}^{1/2} \tag{6.5}$$

(with f_{bd} and $f_{sm,k}$ in MPa) in which:
- η_1; a factor that takes into account the setting conditions; for concrete, this factor is dependent on the dumping direction of the concrete; the soil mix is, however, not deposited on the beam; one can thus assume this factor is equal to 1.0.
- η_2; a factor that determines the influence of the bar diameter.

$$\eta_2=\frac{132-\phi}{100} \quad \text{but } \eta_2\leq 1.0 \tag{6.6}$$

An equivalent diameter must be determined for diameter Ø; for a beam with a flange width b_a and flange thickness t_f the equivalent diameter can be estimated as follows:

$$\phi=\sqrt{\frac{4\,b_a\,t_f}{\pi}} \tag{6.7}$$

2. Criterion 2: up to 10% of the compressive strength of the soil mix material, or:

$$f_{bd} \leq 0.1 f_{sm,d} \tag{6.8}$$

3. Criterion 3: in accordance with EN 1994-1-1 (§6.7.4.3), the design value f_{bd} of the adhesion is also adjusted on the basis of the following values in the design calculations:

- For the calculation of the moment capacity: on 0.3 MPa for temporary situations and 0 MPa(*) for long term situations

(*) in 'suitable' circumstances, the client may authorise the use of an adhesion of up to 0.3 MPa for the moment capacity in the long term; a 'suitable' circumstance must at least comply with the following conditions:

- *construction in risk class 1 or 2*
- *protection of the wall against climatic influences (frost, thaw, leaching, drying, ...)*
- *absence of contaminants that may affect the integrity of the soil mix material over the course of time*
- *safeguards to prevent corrosion of the steel reinforcement*

- For the vertical load transfer: 0.3 MPa for temporary situations and long term situations(**)

(**) the adhesion between the reinforcing steel and the soil mix material (e.g. for the verification of the safety of punching of the reinforcement beam under the effective vertical load) may be taken into account in long-term situations, as long as the following 'suitable' conditions are met:

- *construction in risk class 1 or 2*
- *load transfer by adhesion between the reinforcement and soil mix material only to be taken into account from a characteristic compressive strength of the soil mix material of at least 3 MPa*
- *protection of the wall against climatic influences (frost, thaw, leaching, drying, ...)*
- *absence of contaminants that may affect the integrity of the soil mix material over time*
- *either by applying protection measures to prevent corrosion of the steel reinforcement, or – but only applicable below the excavation level – by calculating with a corrosion reduction, in accordance with EN 1993 (see §6.14.2).*

One may only deviate from the above 3 criteria in case of availability of extensive and target-oriented research and measurements or when using special steel elements enhancing the adhesion.

For additional information, I – Table 6.8 provides a comparative overview of the shear adhesion calculated according to the 3 above criteria, including the adjustment at 0.3 MPa.

The colour notations of the cells are also applied in the calculation I – Tables which are informatively used in §6.7.6 are applied.

For comparison, column 2 includes the adhesion which is calculated for ribbed steel in accordance with §8.4.2 of EN 1992-1-1; it is expressly noted that these values may not be applied for soil mix and require specific research with regards to the applicability of low compressive strengths.

1 - Table 6.8. Design value of the adhesion soil mix - steel beam according to 3 criteria.

Characteristic compressive strength of the soil mix	Adhesion soil mix - ribbed bars	Design value of the adhesion soil mix - smooth reinforcement beam at 28 days			
		Criterion 1		Criterion 2	Criterion 2 + 3
$f_{sm,k}$	2.25 f_{ctd}	$\eta_1\,\eta_2\,0.24\,f_{ck}^{(1/2)}$		10% $f_{sm,d}$	0.30 MPa
	$(x\eta_1\eta_2)$	$(\eta_1\eta_2) = 1.0$ (small beam)	$(\eta_1\eta_2) = 0.5$ (large beam)		
1	0.32	0.24	0.12	0.07	0.07
2	0.50	0.34	0.17	0.13	0.13
3	0.66	0.42	0.21	0.20	0.20
4	0.79	0.48	0.24	0.27	0.27
5	0.92	0.54	0.27	0.33	0.30
6	1.04	0.59	0.29	0.40	0.30
7	1.15	0.63	0.32	0.47	0.30
8	1.26	0.68	0.34	0.53	0.30
9	1.36	0.72	0.36	0.60	0.30
10	1.46	0.76	0.38	0.67	0.30
11	1.56	0.80	0.40	0.73	0.30
12	1.65	0.83	0.42	0.80	0.30

6.3.7 Modulus of elasticity of the soil mix material

The modulus of elasticity of the soil mix material can be determined or estimated via two approaches (see §2.3.3 in part 2 of the handbook):
- either by determining the E-modulus in compression tests on samples, with application of a specific test procedure; the number of tests for determining the E-modulus is specified in Chapter 8 Quality Control and is dependent on the function of the soil mix wall;
- or by derivation from the (estimated or measured) compressive strengths, with application of accepted correlations.

The following aspects must always be taken into account when assessing and determining the E-moduli that are used for the design:
- the bending tests conducted by the BBRI (see §2.3.13 in part 2 of the handbook), confirming step by step that the bending stiffness of a reinforced soil mix wall is greater than the bending stiffness of the reinforcement itself on the one hand, but smaller than the uncracked soil mix wall on the other hand;
- there should be consistency between the applied compressive strengths and the soil mix material in the various soil layers and the applied E-moduli;
- to a certain extent, the E-modulus of the soil mix material influences the global bending stiffness EI of the composite section (soil mix + reinforcement), and therefore also the wall calculations:
  ○ as a lower EI value is maintained, larger theoretically calculated deformations are calculated;
  ○ on the other hand, a higher EI value results in greater internal forces
- as with the compressive strength, the E-modulus should also take the time effect into account, both in the short term (<< 28 days) and long term (creep effects).

The issue of the bending stiffness that may occur in the wall calculations is critically assessed in Appendix 2. The calculation methodology that is used on the basis of this is described in detail in §6.6. One can apply the following rules of thumb for the design values of the E-moduli:
- the characteristic E values are taken as equal to the average E values;
- these average values are obtained from either specific compression tests or in correlation with the compressive strengths. Here it is necessary to take into account that the stress range in which and the way in which the elasticity modulus is determined (tangent or secant modulus) from the stress-strain relationship of the soil mix may have a crucial influence on the value of the E-modulus. The following correlation can be used for the tangent elasticity modulus in the stress range of 10 to 30% of the compressive strength of the soil mix material (see §2.3.3 in part 2 of the handbook):

$$E_{sm} = 1482 f_{sm,m}^{0.8}$$

(6.9)

$(E_{sm}$ and $f_{sm,m}$ in MPa)

- The literature also refers to proportions of 350 to 1000 between the secant elasticity modulus at a stress of 50% of the compressive strength (commonly referred to as E_{50}) and the compressive strength of the soil mix material. The correlations referred to above are derived from experiments during which the deformations were measured on the test samples themselves, and not between the pressure plates of the press, which often leads to much smaller ratios.
- for a younger wall ($\ll$ 28 days), a reduced E_{sm} modulus must be applied for the wall calculations in the serviceability limit state (BGT);
- in the long term, the average E-modulus of the soil mix material is reduced by a factor $(1+\Psi)$ of $\Psi=1,0$ (due to creeping behaviour), at least under the condition that the soil mix material meets the standards for "suitable" conditions as described in §6.3.6; if not, only the bending stiffness of the steel beams will be taken into account in the long term;
- if, for example, due to the ground layers or mixing procedure, large 'layerwise' differences in the E-modulus are expected, it is recommended to also take this variation into account in the wall calculations.

6.3.8 Geometrical data

The dimensions of the soil mix wall result from the dimensions of the mixing tools used, the overlap of elements and execution depth. The following rules apply:
- no over-width is applied to the dimensions of the mixing tools
- the base depth of the wall is the depth at which the wall has its full section
- the design value of the wall geometry is taken as equal to the representative value.

Potential applicable reductions of the wall thickness, e.g. by milling during the construction phase, are explicitly included in the design.

contact surface width (overlap): $d_3 = \sqrt{d_1^2 - d_2^2}$ (6.10)

half thickness overlap: $h = (d_1 - d_2)/2$ (6.11)

gross cross-section column: $A_0 = \dfrac{\pi d_1^2}{4}$ (6.12)

net cross-section secant column: $A_1 = A_0 - 2\left[\dfrac{h}{6d_3}\,(3h^2 + 4d_3^2)\right]$ (6.13) equivalent height wall: $h_{sm,eq} = A_1 / d_2$ (6.14)

moment of inertia wall/m: $I = \dfrac{\left[\dfrac{\pi d_1^4}{64} \dfrac{A_1}{A_0}\right]}{d_2}$ or also $I = \dfrac{h_{sm,eq}^3}{12}$ (6.15)

For a column wall (diameter d_1 /axis on axis distance d_2) the following values can be obtained (Figure 6.5):

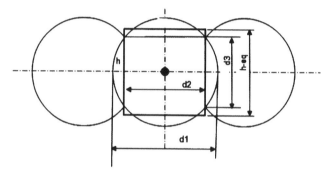

I - Fig. 6.5. Geometrical data of a column wall.

The equivalent height $h_{sm,eq}$ can be applied in the calculation of the bending stiffness of the soil mix wall, in determining the moment capacity of the composite section, wherein only the reinforced column is included, and in the calculation of the structural and geotechnical vertical bearing capacity.

The dimensions and characteristics of the steel beams that are applied in the design are in accordance with the specifications of the manufacturer or supplier. The design should take into account some tolerances in the placement of the steel beams:
- the deviation perpendicular to the wall (eccentricity of the beams); for the influence on the moment capacity we refer to the sensitivity analysis in §6.5.2 and §7.4 with regards to the execution tolerances; in calculating the bending stiffness and arching effect no eccentricity should be calculated (with the exception of important deviations).
- the rotation of the steel beams; from the executed sensitivity analysis (see §6.5.3) shows that excessive/exceptional rotations are required to result in an appreciable reduction of the moment of resistance W and consequently, also of the moment capacity;
- the deviation in parallel with the wall; the adverse influence on the moment capacity of a continue wall will usually not be decisive; an inspection of, for example, the arching effect and transfer of the anchor force is however required.

6.3.9   Hydraulic resistance of the soil mix wall

The water permeability of the soil mix material, variations over the wall and the leakage flow rate of the entire system is heavily dependent on various factors. Important factors include:
*   Soil stratification; presence of a homogeneous or highly heterogeneous layers of sand, clay, loam or local peat.
*   Mixing method and procedure. Here, it is especially important that the mixing is carried out per depth/soil layer (panel walls), or that mixing takes place across the entire element (column walls).
*   Type of additives and mixing ratio (mixture composition).
*   Curing period; the water permeability will decrease over time as a result of the hardening of the samples and/or wall.
*   Detailing of wall, presence of leaks at the location of the connection of different columns/panels, right-angle wall connectors, anchoring structures, or connecting floors.

Depending on the above factors, the water permeability of soil mix material lies in a range of 10^{-7} to 10^{-11} m/s. Table 2.13 of part 2 of the handbook includes typical in situ permeability coefficients that depend on the soil type and cement content (according to Topolnicki, 2004). The above range of permeability is large. In projects where the hydraulic resistance of the walls is considered to be important, it is recommended to pay ample attention to the aforementioned factors.

6.4   Analysis of the arching effect

6.4.1   Pressure arch system

Within the calculation model of the soil mix wall, the to be incorporated soil and water pressure on the soil mix material should be transferable to the steel H-beams. This verification must take place for the most detrimental combinations of acting pressures (both along the active and passive sides), intermediate distance from the steel beams and compressive strength of the soil mix material. The verification in vertical direction may be averaged over a height equal to the thickness of the wall.

In order to prevent an undesired behaviour (type beam effect) of the material between the steel H-beams, with corresponding vertical cracks as a result, the axis-to-axis distance between the steel H-beams must be limited to such a degree that the load can be transferred to the steel H-beams via an arching effect. Figure 6.6 shows the corresponding principle.

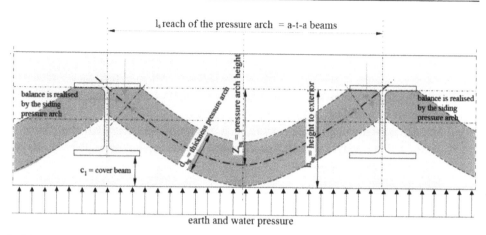

I - Fig. 6.6. Arching effect in a soil mix wall with steel H-beams.

6.4.2 When can one expect a pressure arch?

A pressure arch may occur if the construction is compact. If this is not the case, there is a beam effect with corresponding force effect and crack formation in the transverse direction of the wall.

EN 1992-1-1 indicates that one can assume a construction is compact if:

$$a < 3h_{bg} \tag{6.16}$$

with:
- a: reach of the pressure arch = axis-to-axis distance between the beams [mm].
- h_{bg}: maximum height available for the pressure arch = height to exterior of wall [mm].

Only with sufficient substantiation (e.g. on the basis of theoretical FEM detail modelling or tests) can a designer deviate from these pressure arch boundaries and implement a 'leaner' pressure arch.

6.4.3   Geometry of the pressure arch

Height of pressure arch
The height is expressed as level of the pressure arch (z_{bg} [mm]). This is determined in accordance with the Dutch national annexes of EN 1992-1-1 §6.1 (10):

$$z_{bg} = 0.2\,a + 0.4\,h_{bg} \leq 0.6\,a \tag{6.17}$$

In addition, for geometrical reasons:

$$z_{bg} \leq h_{bg} - d_{bg;mid}/3 \tag{6.18}$$

with $d_{bg;mid}$ = thickness pressure arch in the middle cross-section [mm].

The factor 3 results from the apparent quasi-linear course of the compressive stress in the middle cross-section.

For $d_{bg,mid}$ one should initially make an estimate on the basis of a further estimated angle α. (see below). Taking into account: $d_{bg;mid} = b_f \tan α$ with:

b_f: 	the flange width of the steel beam
α: 	the angle of the arch as defined in more detail below

The geometry of the pressure arch must thus be determined on the basis of an iterative calculation.

Angle at the location of base of arch
The angle α [°] at the location of the base of the arch is determined as follows.

The angle is considered parabolic, with a dip at the location of the deflection (height) of the pressure arch and cutting through the heart of the engagement point of the pressure arch on the flange of the steel beams.

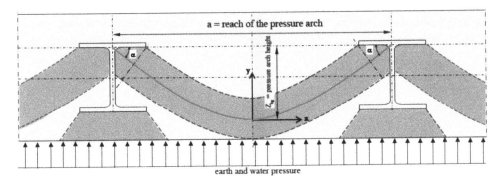

I - Fig. 6.7. Applied parabolic profile of the pressure arch system.

The equation for the correlated function of the pressure arch (red line in figure 6.7) can be represented as follows:

$$y = (4z_{bg} / a^2) x^2 \qquad (6.19)$$

The slope of the arch at the point of engagement of the flange of the steel beam is determined on the basis of the derivative of funtion:

$$y' = (8z_{bg} / a^2) x \qquad (6.20)$$

For x = 0.5 a, the following applies:

$$y' = \tan a = 4z_{bg} / a \qquad (6.21)$$

$$a = \arctan (4z_{bg} / a) \qquad (6.22)$$

Thickness of pressure arch at the beam
The thickness of the pressure arch is determined on the basis of angle a of the pressure arch with the flange of the beam and considering the width of the flange in question (applied as 0,5 b_f).

The following applies:

$$0.5 \, d_{bg} = 0.5 \, b_f \sin a \qquad (6.23)$$

$$d_{bg} = b_f \sin a \qquad (6.24)$$

Thickness of pressure arch in the middle cross-section
By means of calculations with applications such as Plaxis (see Appendix 3) a larger thickness of the pressure arch was found at the location of the middle cross-section. A good approximation of the occurring stresses is found if a thickness is applied equal to:

$$d_{bg;mid} = b_f \tan a \qquad (6.25)$$

6.4.4 Determining governing force effect in the pressure arch

A pressure arch has a non-linear strain distribution. In accordance with EN 1992-1-1, for determining the force distribution, one may use a strut and tie model.

This strut and tie model consists of:

- 3 pressure rods that represent the compressive stress in the pressure arch.
- A pressure force, which represents the force perpendicular to the web of the beam.
- A pressure force, which represents the force perpendicular to the flange of the beam.

The elements of the strut and tie model are designed in accordance with the calculation rules, listed in §6.5 of EN 1992-1-1. The following model has been applied, supported by finite element calculations:

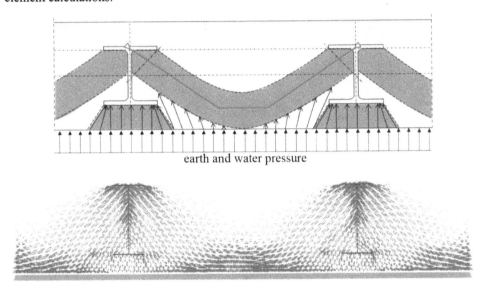

I - Fig. 6.8. Strut and tie model for the pressure arch of a soil mix wall and FEM of the arching effect.

Determining the maximum compressive stress in the pressure arch at the connection with the beam

The executed finite element calculations show that the stress distribution at the location of the base of the beam is equally divided. Therefore, the maximum compressive stress can be determined on the basis of a calculated normal force ($N'_{pressure\ arch}$) and the established thickness of the arch (d_{bg}).

Part of the earth and water pressure applying on the wall is directly absorbed by the outer flange of the steel beam. The remaining part between the beams is transferred via the pressure arch. If one assumes that loads below 45° can be transferred to the outer flange, the design value (in ULS) of the load to be borne by the pressure arch is:

$$F_{hor;pressure\ arch} = F_{hor;tot} - F_{hor;flange\ outer} = \sigma_{hor}\ (a - b_f - 2c_1) \qquad \text{[N/mm]} \qquad (6.26)$$

with:

σ_{hor} : total design value of the earth and water pressure [N/ mm²];
this is equal to 1.35 $\sigma_{hor;rep}$.
c_1 : cover on the beam with respect to the exterior wall [mm]
(depending on the direction of σ_{hor} this may also be c_2).

On the basis of an angle α between the pressure arch and a flange beam, the occurring normal force at the base of the flange is:

$$N'_{base} = 0.5\ F_{hor;pressure\ arch}\ /\ \sin a = 0.5\ \sigma_{hor}\ (a - b_f - 2c_1)\ /\ \sin a \qquad \text{[N/mm]} \qquad (6.27)$$

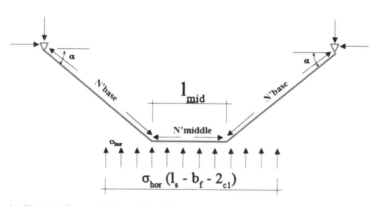

1 - Fig. 6.9. Strut and tie model of the pressure arch.

The occurring compressive stress in the arch is:

$$\sigma'_{arch;base} = N'_{base}\ /\ d_{bg} = 0.5\ \sigma_{hor}\ (a - b_f - 2c_1)\ /\ \sin \alpha/b_f \sin a$$
$$= 0.5\ \sigma_{hor}\ (a - b_f - 2c_1)\ /\ (b_f \sin^2\alpha) \qquad \text{[N/mm}^2\text{]} \qquad (6.28)$$

Determining the maximum compressive stress in the pressure arch in the centre of the reach of the arch

In contrast to the base of the arch, the executed finite element calculations show that a substantially linear stress distribution is formed in the centre of the pressure arch.

I - Fig. 6.10. Stress distribution in the centre of the pressure arch.

This is calculated on the assumption that the soil mix material cannot absorb any tension. The width of the horizontal beam of the strut and tie model (l_{mid}) is:

$$l_{mid} = a - 2z_{bg} / tan(\alpha) = a - 2z_{bg} / (4z_{bg}/a) = 1/2\ a \qquad [mm] \qquad (6.29)$$

The shear force at the location of the bend in the strut and tie model is then:

$$V_{mid} = 0.5\ l_{mid}\ \sigma_{hor} = 0.25\ a\ \sigma_{hor} \qquad [N/mm] \qquad (6.30)$$

The resultant shear force and normal force in the horizontal beam is then equal to the normal force in the diagonal beam. This is equal to the previously determined N'_{base}. The following thus applies:

$$N'_{middle} = \sqrt{(N'_{base}{}^2 - V_{mid}{}^2)} \qquad [N/mm] \qquad (6.31)$$

In the centre portion, a thickness of the pressure arch is applied:

$$d_{bg;mid} = b_f\ tan\ \alpha \qquad [mm] \qquad (6.32)$$

As such, the average stress in the middle cross-section of the pressure arch can be determined:

$$\sigma'_{arch;mid;avg} = N'_{middle} / d_{bg;mid} \qquad [N/mm^2] \qquad (6.33)$$

However, the stress in the middle cross-section is not uniform. One may assume that, as a result of the limited moment activity, the maximum absorbable tensile stress is reached, which is set equal to 0. This means that, on the tensile side, the normal compressive stress as a result of N'_{middle} is completely counterbalanced by the moment. At the compressive side, the compressive stress is then doubled.

$$\sigma'_{arch;mid;max} = 2\ \sigma'_{arch;mid;avg} \qquad [N/mm^2] \qquad (6.34)$$

Determination of the maximum shear stress in the pressure arch cross-section

On the basis of the strut and tie model, the maximum shear force in the pressure arch cross-section is found at the bend in the model. This is:

$$V_{mid} = 0.5\, l_{mid}\, \sigma_{hor} \tag{6.35}$$

The design value of the maximum shear stress is:

$$\tau_{Ed} = 1.5\, V_{mid}\, / d_{bg,mid} = 0.375\, a\, \sigma_{hor}\, cos\, \alpha\, / d_{bg} \qquad \text{[N/mm}^2\text{]} \tag{6.36}$$

This value should be compared with the design value of the shear resistance, as described in §6.3.5.

6.4.5 Admissible stresses in the pressure arch

Distinction is made between the admissible compressive stress in the middle cross-section (considered as compressive rod) and the admissible compressive stress at the location of the base (considered as node).

Admissible compressive stress in the middle cross-section

The pressure arch in the soil mix wall is characterised by a compressive stress in the horizontal plane and a compressive stress or tensile stress in the vertical direction (because of the vertical moment in the wall).

In accordance with §6.5 of EN 1992-1-1, based on the condition that there is a compressive rod with tensile stress in the opposite direction, the design value of the admissible compressive stress in the centre of the pressure arch is determined by:

$$\sigma_{RD,max} = 0.6\, v'\! f_{sm,d} \qquad \text{[N/mm}^2\text{]} \tag{6.37}$$

with $v' = 1 - f_{sm,k} / 250$ en $f_{sm,d}$ as determined in §6.3.3.

Admissible compressive stress at the base

For the insertion point/base of the beam, tensile stresses may occur in the vertical direction if the flange of the beam lies within the tensile zone.

In accordance with §6.5 of EN 1992-1-1, the following applies for the design value of the admissible compressive stresses:

$$\sigma_{RD,max} = k_2\, v'\! f_{sm,d} \qquad \text{[N/mm}^2\text{]} \tag{6.38}$$

with $k_2 = 0.85$ (in accordance with §6.5.4 of EN 1992 -1-1) and $f_{sm,d}$ as determined in §6.3.3.

Admissible shear stress

The admissible shear stress in the pressure arch is equal to the design value of the shear strength $\tau_{Rsm,d}$, as determined in §6.3.5.

In accordance with EN 1992-1-1 this may be increased by the proportion of the compressive stress in the arch, equal to $0.15 \, \sigma'_{arch;mid;avg}$.

6.4.6 *Example calculations*

Control calculations of the arching effect are included in Appendix 4, in which all calculation aspects for a soil mix wall are described by means of 2 case studies.

In addition, a specific case study of the arching effect on the basis of a finite element calculation in Plaxis can be found in Appendix 3.

6.5 Specific parameters for the design calculations

For the sake of completeness, below are a number of parameters that may be important in the design calculations.

6.5.1 *Wall friction between the soil mix wall and the soil*

The following rules apply for the characteristic values of the wall friction angles δ_a and δ_p between soil mix walls and the soil:

for straight sliding planes: $\quad |\delta| \leq 2/3 \; \varphi' \; with \; \varphi' \leq 35°$ (6.39)

for curved sliding planes: $\quad |\delta| \leq \varphi'$ (6.40)

For the applicability of straight, or curved sliding planes, please refer to NEN 9997-1 and CUR 166.

6.5.2 *Influence of the eccentricity of the beam due to placement*

The structural verification of the reinforced soil mix wall when bending is discussed in more detail in §6.7. With application of the described calculation methodology, both the steel beam and compressed soil mix zone are taken into account for the total moment capacity of the wall, at least temporarily. Consequently:

- eccentric placement, during which the beam shifts towards the compressed zone of the wall, results in a reduction of the compressed soil mix zone, and consequently also to a reduction in moment capacity;
- alternatively, a shift of the beam towards the tensile zone of the wall results in an increase of the moment capacity.

The influence of the eccentric placement on the moment capacity can be determined fairly easily by theoretically reducing the height h_{sm} of the soil mix wall or increasing it by 2 x the eccentricity, so that, theoretically, a 'centric' positioning of the bent beam can be achieved. For illustration of the influence of an 'unfavourable' eccentricity, the reduction of the design value of the moment capacity (per axis-to-axis distance a) has been research for the following configuration and assumptions (see I - Table 6.9):
- a soil mix wall with a thickness of 550 mm and a characteristic compressive strength of 4 MPa;
- a steel beam IPE300 or HEA 300, steel quality S235;
- an eccentricity varying from 0 mm to a maximum of 125 mm in the direction of the compression zone;
- In accordance with §6.7, the moment capacity M(Rd,4) has been determined for the composite section, taking into consideration both the plastic resistance moment of the reinforcement and the moment of resistance of the soil mix material;
- the effective width of the soil mix is determined by means of the calculation methodology described in §6.7.2; here a length L_e between the zero moment points of 5 m is assumed;
- for the reinforcement and for the soil mix material, material factors of 1.0 or 1.5 resp. are applied; the time-related factors α_{sm} and β are =1.0.

These comparative calculations show that the adverse effect of an eccentrically placed beam is fairly limited, with a reduction of the composite moment capacity of no more than approx. 4 to 5% at an eccentricity of 50 mm and 8 to 10% at a (quite considerable) eccentricity of 125 mm; the latter value means that the compressed beam flange is located against the outside of the soil mix wall and the favourable interaction is therefore almost completely lost.

I - Table 6.9. Influence of eccentricity on the moment capacity of the soil mix wall.

	Only the reinforcement		Composite section: reinforcement + soil mix					
	M(Rd,a,el)	M(Rd,a,pl)	M(Rd,4)					
h_{sm} (mm)			550	500	450	400	350	300
exc. (mm)			0	25	50	75	100	125
IPE300	131 kNm	148 kNm	166 kNm	162 kNm	158 kNm	154 kNm	151 kNm	149 kNm
	79%	89%	100%	97%	95%	93%	91%	90%
HEA300	296 kNm	325 kNm	349 kNm	342 kNm	336 kNm	330 kNm	314 kNm	320 kNm
	85%	93%	100%	98%	96%	95%	90%	92%

For light design tolerances, with eccentricities of up to approx. 50 mm, one may assume that the limited loss of moment capacity is sufficiently covered by the design by means of various safety factors and safe parameter assumptions. Only with high eccentricities and/ or in combination with high vertical wall loads (which, furthermore, lead to an important eccentric moment on the wall), it is recommended to take into account a control calculation in the design.

Finally: the beneficial effect of a deliberately chosen eccentric placement of the reinforcement towards the tensile zone may also be taken into account for the design, provided that the desired eccentric positioning can also be demonstrated and guaranteed.

6.5.3   Influence of rotation of the beam due to placement

Rotation of the beam during placement has a negative effect on the moment capacity. Some illustrative calculations related to this potential influence are provided below.

For simplification, the calculations assume that no bending may occur around the weak axis of the steel beam, as this is prevented by the surrounding soil mix. Because the steel beam is twisted, the lever arms of the flanges are reduced in size. If one only takes into account the reduction of the level arms and there are no bending moments in the weak direction, then one ends up with the following adjusted I - Table 6.10 (the beam here has been simplified as 3 rectangular plates).

I - Table 6.10. Moment capacity M(Rd,a,pl) due to rotation of the beam.

Rotation	IPE200		IPE300		HEA200		HEA300	
	kNm	-	kNm	-	kNm	-	kNm	-
0°	49	100%	142	100%	96	100%	307	100%
5°	49	100%	141	99%	95	99%	306	100%
10°	49	100%	139	98%	94	98%	302	98%
20°	46	94%	133	94%	90	94%	288	94%
30°	43	88%	122	86%	83	86%	266	87%

On the basis of this approach, a rotation of the beam – even up to approx. 20° – only unfavourably influences the moment capacity to a limited extent. For the design it is therefore wise to NOT take into account the potential limited rotational deviations of the steel beams in advance. If, however, important rotations are observed during execution, it may be necessary to carry out a control calculation of the moment capacity.

6.6 Determination of the horizontal bending stiffness EI

6.6.1 Applied notations

$E_a I_a$	bending stiffness of the reinforcement, per axis-to-axis beam distance a
$E_{sm} I_{sm1}$	uncracked bending stiffness of the soil mix wall, per effective width b_{cl}
$E_{sm} I_{sm2}$	cracked bending stiffness of the soil mix wall, per effective width b_{cl}
EI-1	total uncracked bending stiffness of the composite section, per effective width b_{cl}
EI-2	total cracked bending stiffness of the composite section, per effective width b_{cl}
EI-eff	the effective bending stiffness of the composite section, per effective width b_{cl}
n	the ratio of the E-moduli: E_a / E_{sm}
E_a	the E-modulus of the steel: 210000 N/mm²
I_a	the moment of inertia of the steel beam
I_{sm1}	the moment of inertia of the uncracked cross-section of the soil mix wall
b_{cl}	the effective width of the bending stiffness according to §6.6.2
$h_{sm(eq)}$	the (equivalent) thickness of the soil mix wall
$f_{sm,t}$	the tensile strength of the soil mix material
L_e	the distance between the zero moment points

6.6.2 Effective width b_{cl} for the bending stiffness

The following rules may be applied for the wall width that may be applied with regards to the bending stiffness in the wall calculations:

$$b_{cl} = \frac{L_e}{4} \quad and \quad b_{cl} < the\ average\ axis\text{-}to\text{-}axis\ distance\ a\ of\ the\ beams \tag{6.41}$$

Except in case of large axis-to-axis distances of the steel beams or limited L_e values, the effective width will be able to be maintained from beam to beam. Concretely, this also means that a continuous wall can be implemented in the wall calculations.

6.6.3   Calculation methodology – general

The bending tests of the BBRI (see §2.3.13 in part 2 of the handbook) have always demonstrated:

- that the bending stiffness EI of a reinforced soil mix element is significantly larger than the bending stiffness of only the steel beam;
- that on the other hand – once the moment corresponding to the onset of the cracks in the soil mix is exceeded – the global bending stiffness quickly falls back to a value that is significantly smaller than the bending stiffness of a uncracked soil mix section.

Below are two proposed approaches for estimating the bending stiffness of the soil mix wall.

Method 1
This method is the most appropriate to make a fairly realistic assessment of the bending stiffness of the cracked soil mix wall. It is based on the methodology in §5.4.2.3 of EN 1992-1-1.

Here, the effective bending stiffness EI-eff is calculated as the average of the uncracked stiffness EI-1 and the cracked stiffness EI-2 of the composite section:

$$EI\text{-}eff \ = \ \frac{EI\text{-}1 + EI\text{-}2}{2} \qquad\qquad (6.42)$$

For simplification, the height of the compressed zone in the calculations of the cracked stiffness is determined on the basis of a plastic stress distribution; for the calculation of I_{sm2}, the height x_e may be equal to the height of the compressed soil mix zone. Details for the calculation of EI-1 and EI-2 are described in §6.6.4 and §6.6.5.

Method 2
Here, the effective bending stiffness is calculated as the sum of the stiffness of the reinforcement and the stiffness of the compressed soil mix zone, assuming that the neutral line runs through the middle of the beam. In this way, the stiffness can easily be calculated on the basis of the following formula:

$$EI\text{-}eff \ = \ E_a I_a + E_{sm} \left[\frac{b_{cl} \cdot \left(\frac{h_{sm}}{2} \right)^3}{3} \right] \qquad\qquad (6.43)$$

Above formulas give the bending stiffness per effective width b. These formulas can then easily be converted into bending stiffnesses per 'm wall' by dividing the computed values of EI-eff by the effective width b_{cl}, or in the common situation wherein b_{cl} = a, by dividing by the (average) distance between the beams a. In the latter case, this shows:

for method 1:

$$EI\text{-}eff\ /'m= \frac{EI\text{-}1+EI\text{-}2}{2a} \qquad (6.44)$$

for method 2:

$$EI\text{-}eff\ /'m= E_aI_a/a+E_{sm}\left[\frac{\left(\frac{h_{sm}}{2}\right)^3}{3}\right] \qquad (6.45)$$

6.6.4 Details of the calculation method 1

6.6.4.1 Uncracked bending stiffness EI-1

To calculate the uncracked stiffness, the moment of inertia of the composite cross-section must be determined. For a unit element (with width = the effective width b_{cl}) the uncracked bending stiffness EI-1 is:

$$EI\text{-}1= E_aI_a+E_{sm}(I_{sm}-I_a)=E_{sm}\big[(n-1)I_a+I_{sm}\big] \qquad (6.46)$$

in which:
n: the ratio of the E-moduli: E_a/E_{sm}
E_a: the E-modulus of the steel: $210000\ N/mm^2$
E_{sm}:   the E-modulus of the soil mix material
I_a: the moment of inertia of the steel beam
I_{sm}:   the moment of inertia of the cracked cross-section of the soil mix material.

The moment of inertia of the soil mix material for this unit element is:

$$I_{sm}=\frac{b_{cl}\ h_{sm}^3}{12} \qquad (6.47)$$

in which:
b_{cl} :   the effective width for the bending stiffness according to §6.6.2
h_{sm} :   the equivalent height of the soil mix section

If the tensile strength of the soil mix is known, then the moment leading to the emergence of cracks in the soil mix can be calculated by:

$$M_{cr} = \frac{f_{sm,t}\, I_{sml}}{h_{sm}\,/\,2}$$
(6.48)

in which:

$f_{sm,t}$: the tensile strength of the soil mix material – see §6.3.4.

6.6.4.2 Cracked bending stiffness EI-2

Due to the low tensile strength of the soil mix, the soil mix material will quickly crack when subjected to tensile stress. For the calculation of the cracked stiffness, the tensile zone of the soil mix must be ignored. This will ensure that the stiffness will decrease. However, one must first determine the height of the compressed soil mix material.

By writing out the equilibrium of the normal forces and the equilibrium of the moments, the stiffness of every section can be established. As simplified assumption, the stiffness of the web of the steel beam can be ignored. This way the stiffness can be determined by analogy with a double reinforced rectangular section.

The following variables are defined as follows:

$c_1 = c_2$:	the net cover above and below the beam:	$c_1 = c_2 = \dfrac{h_{sm} - h_a}{2}$	(6.49)
d:	the efficient height of the lower flange	$d = h_{eq} - c_2 - t_f/2$	(6.50)
$c_{1,b}$:	the gross cover on the top flange (up to halve of the thickness of flange)	$c_{1,b} = c_1 + t_f/2$	(6.51)
h_w :	the height of the web of the beam	$h_w = (h_a - 2\,t_f)$	(6.52)

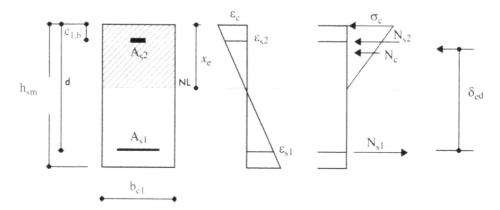

I - Fig. 6.11. Principle calculation for the cracked cross-section.

By writing out the moments balance, one ends up with the following equation (with the E-modulus E_{sm} of the soil mix material as reference):

$$\frac{1}{2} b_{cl} x_e^2 + (n\text{-}1) A_f \left(x_e\text{-} c_{1,b}\right) = n A_f \left(d\text{-}x_e\right) \tag{6.53}$$

The solution of this equation gives

$$\xi_e = -(2 n\text{-}1) \rho + \sqrt{(2 n\text{-}1)^2 \rho^2 + 2 \left[(n\text{-}1)\frac{h_{sm}}{d} + 1\right] \rho} \tag{6.54}$$

in which:

$$\xi_e = \frac{x_e}{d} \tag{6.55}$$

$$\rho = \frac{A_f}{d \, b_{cl}} \tag{6.56}$$

$$A_f (= A_{s1} = A_{s2}) = t_f b_f \tag{6.57}$$

Now that we know x_e, we can calculate the moment of inertia I_2:

$$I_2 = \frac{b_{cl} x_e^3}{3} + (n\text{-}1) A_f \left(x_e\text{-}c_{1,b}\right)^2 + n A_f \left(d\text{-}x_e\right)^2 + n \, t_w \left(\frac{\left(x_e\text{-}c_1\text{-}t_f\right)^3}{3} + \frac{\left(h_w\text{-}x_e+c_1+t_f\right)^3}{3}\right) \tag{6.58}$$

The last term indicates the influence of the web of the beam on the moment of inertia. For the calculation of x_e the web of the beam was not taken into account; the error made here is marginal (1 to 3%). The web of the beam is included in the calculations of the moment of inertia; if not, this would cause an error (10% to 15%) with regards to the bending stiffness.

The cracked bending stiffness (per effective width b_{cl}) of the composite section is thus:

$$EI\text{-}2 = E_{sm}I_2 \tag{6.59}$$

Now that we know I_2, we can calculate the stresses.
Compressive zone of the soil mix:

$$\sigma_c = \frac{M\,x_e}{I_2} \tag{6.60}$$

Tensile zone of the flange:

$$\sigma_{s1} = n\,\frac{M\,(d\text{-}x_e)}{I_2} \tag{6.61}$$

Compressive zone of the flange:

$$\sigma_{s2} = n\,\frac{M\,(x_e\text{-}d)}{I_2} \tag{6.62}$$

6.6.5 Comparison of the calculation methodologies 1 and 2

The I - Tables 6.11 to 6.13 provide a comparison of the detailed calculations of EI (in Nmm²) according to method 1 and the simplified method 2. For information, the bending stiffness $E_a\,I_a$ and the relative input of the steel beam only are included in the last columns.

The illustrative calculations are made considering the following assumptions:
- h_{sm} = 550 mm
- b_{cl} = 1100 mm
- E_{sm} of respectively 2000 MPa, 5000 MPa and 8000 MPa - no creep reduction.

The simplified method 2 gives bending stiffnesses that are approx. 10 to 20% lower than those calculated with the more detailed calculation method 1.

The relative input of the steel beams in the composite bending stiffness of the wall naturally increases with increased beam weight and with decreased E-modulus of the soil mix material.

I - Table 6.11. Determination of bending stiffness EI of a reinforced soil mix element according to methods 1 and 2 for a modulus of elasticity E_{sm} equal to 2000 MPa.

E_{sm} =2000 MPa	Method 1			Method 2		$E_a I_a$ – steel	Input steel
	EI-1	EI-2	(EI-1+EI-2)/2	EI-eff			
IPE200	3.45E+13	1.34E+13	2.40E+13	1.93E+13	81%	4.08E+12	17%
IPE270	4.25E+13	2.26E+13	3.26E+13	2.75E+13	84%	1.22E+13	37%
IPE360	6.43E+13	4.49E+13	5.46E+13	4.95E+13	91%	3.42E+13	63%
HEA240	4.66E+13	2.76E+13	3.71E+13	3.16E+13	85%	1.63E+13	44%
HEA300	6.85E+13	4.91E+13	5.88E+13	5.36E+13	91%	3.83E+13	65%
HEA360	9.93E+13	7.96E+13	8.94E+13	8.48E+13	95%	6.95E+13	78%

I - Table 6.12. Determination of bending stiffness EI of a reinforced soil mix element according to methods 1 and 2 for a modulus of elasticity E_{sm} equal to 5000 MPa.

E_{sm} =5000 MPa	Method 1			Method 2		$E_a I_a$ – steel	Input steel
	EI-1	EI-2	(EI-1+EI-2)/2	EI-eff			
IPE200	8.02E+13	1.96E+13	4.99E+13	4.22E+13	85%	4.08E+12	8%
IPE270	8.81E+13	3.15E+13	5.98E+13	5.03E+13	84%	1.22E+13	20%
IPE360	1.10E+14	5.66E+13	8.31E+13	7.23E+13	87%	3.42E+13	41%
HEA240	9.22E+13	3.91E+13	6.56E+13	5.44E+13	83%	1.63E+13	25%
HEA300	1.14E+14	6.27E+13	8.82E+13	7.64E+13	87%	3.83E+13	43%
HEA360	1.44E+14	9.45E+13	1.19E+14	1.08E+14	90%	6.95E+13	58%

I - Table 6.13. Determination of bending stiffness EI of a reinforced soil mix element according to methods 1 and 2 for a modulus of elasticity E_{sm} equal to 8000 MPa.

E_{sm} =8000 MPa	Method 1			Method 2		$E_a I_a$ – steel	Input steel
	EI-1	EI-2	(EI-1+EI-2)/2	EI-eff			
IPE200	1.26E+14	2.30E+13	7.45E+13	6.51E+13	87%	4.08E+12	5%
IPE270	1.34E+14	3.68E+13	8.53E+13	7.32E+13	86%	1.22E+13	14%
IPE360	1.55E+14	6.44E+13	1.10E+14	9.52E+13	87%	3.42E+13	31%
HEA240	1.38E+14	4.68E+13	9.23E+13	7.73E+13	84%	1.63E+13	18%
HEA300	1.59E+14	7.26E+13	1.16E+14	9.93E+13	86%	3.83E+13	33%
HEA360	1.89E+14	1.06E+14	1.47E+14	1.31E+14	89%	6.95E+13	47%

6.6.6   Assessment and evaluation of the EI value in the design calculations

For illustrative and comparative calculations please refer to Appendix 2.

The following decisions and proposals can be formulated on the basis of this:

1. The bending tests conducted by the BBRI (see §2.3.13 in part 2 of the handbook)indicate that the bending stiffness of the reinforced soil mix wall is significantly larger than those of the beams alone, but smaller than the total bending stiffness of the uncracked wall.
2. For an unanchored wall, the designated EI value has virtually no influence on the internal forces. The use of an effective cracked bending stiffness logically results in larger head displacements than when the uncracked bending stiffness is used. If one assumes an 'average' E value of the soil mix material (and not a safe 'low' value), the theoretical wall displacements are not overly pessimistic.
3. For single anchored or braced walls too, the influence of the bending stiffness on the internal forces is very limited, as well as the anchor forces/strut forces. Here too, the bending stiffness mainly has an effect on the calculated wall displacements, especially on the displacements in the first construction phase (excavation to anchor level).
    For this reason, for such wall configurations, it is often recommended to apply an 'average' E value of the soil mix material (and not a low value).
4. For multiple-supported (and therefore hyper-static) wall configurations, the bending stiffness have a larger effect on the internal forces. But other factors too, including the stiffness of the ground anchors, the (theoretically) indicated pre-stress of the anchors, the phasing, and the applied calculation software have just as much (or even more) influence on the calculated internal forces and deformations.

On the basis of the above finding, the following recommendations are formulated for the EI value to be used in the design calculations (also see §6.3.7):

1. Calculations are made with a middle effective bending stiffness of the soil mix wall, which can be determined by applying the calculation methodologies as described in §6.6.3 and 6.6.4.
2. For the modulus of elasticity of the soil mix material, the characteristic value is taken as equal to the average value. This average value should be consistent with the observed compressive strengths and, if necessary, with the (young) age of the wall in the considered phase.
3. The realisation of the calculations in two phases with, respectively, a 'high' and 'low' estimated E-modulus of the soil mix material, is of little use in most cases. An approach with high or low estimated E-moduli may only be useful for walls with multiple supports and in case of important wall displacements.
4. Optionally, a distinction can be made with the E-modulus as a function of the depth, and the wall may be fragmented into multiple parts in depth with a different bending stiffness.
5. For the long-term, it may be suitable to take the influence of the creep on the deformations and the internal forces into account.

6.6.7   Example calculations

For a number of example calculations, please refer to Appendix 2.

6.7   Determination of the structural strength (M, N, D) of the soil mix wall

6.7.1   General considerations

If the conditions allow to do so, the consideration of the moment capacity of the composite section 'steel soil mix' is completely justifiable. The calculation method can be based on the regulations in Eurocode 4. In case one disregards the cooperating effect of the soil mix material, the moment capacity is only determined by the reinforcement and can be calculated on the basis of the regulations in Eurocode 3. For the calculation of the moment capacity, a number of values must be determined in advance:

1. the effective width b_{c2} for the structural verification;
2. the characteristic value of the compressive strength $f_{sm,k}$ of the soil mix material;
3. the safety factors (material factors) on the steel and soil mix material;
4. any potential uncertainty factors on the geometry;
5. any potential factors for long-term effects.

- First of all, the calculation of the moment capacity of only the steel beam and of the composite section is explained in §6.7.3 and §6.7.4.
- Subsequently, §6.7.5 describes the structural calculation in bending and combined bending of the composite section and related formulas in more detail.
- Finally, §6.7.6 includes design tables and graphs for a number of common soil mix configurations, that can be applied for the design within the prescribed conditions.

6.7.2   Effective width for the control of the internal forces

6.7.2.1   Generalities

In order to determine the effective width b_{c2} for the verification of the internal forces, one may refer to the regulations in EN 1994-1-1 §5.4.1.2. The following principles are applied below:

1. For panel walls, the effective width – in accordance with Eurocode 4 – is limited to:
   - the axis-to-axis distance of the steel beams;
   - ¼ of the distance of the zero moment points;
   - the critical width that is determined on the basis of the maximum adhesion between the soil mix material and the beam.

2. For column walls, it is assumed that no shear forces can be transferred from the steel beams to the adjacent non-reinforced columns. The effective width is consequently limited to the axis-to-axis distance d_2 of the columns (see figure 6.5).

3. Both for the panel walls and the column walls, the effective width is limited to 2 x the flange width (for safety purposes). Provided that this is properly demonstrated (e.g. via a structural FEM analysis), a larger effective width may be applied.

6.7.2.2   In detail

a) Limitations for b_{c2} for panel walls
The effective width is initially determined (=bounded) by:

$$b_{c2} = \frac{L_e}{4} \quad \text{and} \quad b_{c2} < a \tag{6.63}$$

in which:
L_e :   the approximate distance between the zero moment points
a: the axis-to-axis distance of the steel beams

b) Limitations for b_{c2} for column walls
For column walls, the calculations of the moment capacity can best only be based on the reinforced columns. The non-reinforced columns are ignored. The effective width is then set equal to the side of the equivalent square (see figure 6.5) or equal to the axis-to-axis distance d_2 of the columns.

c) Limitations for b_{c2} due to the adhesion

In order to allow the two materials to work together, the shearing forces should be transferred between the steel beam and soil mix material. For the maximum value of the adhesion to be applied, please refer to §6.3.6.

The greater the effective width, the greater the shear strength that must be transferred. By equalising the compressive strength of the compressed soil mix section and the one-sided adhesion between the beam and the soil mix on the flange width b_f and the length L_s, the following boundary value b_{c2} for the zone in compression can be derived:

$$b_{c2} = \frac{b_f L_s f_{bd}}{c_1 f_{sm,d}} \qquad (6.64)$$

With:

c_1 : the cover above the beam; the following applies for an eccentrically placed beam:

$$c_1 = \frac{h_{sm} - 2e - h_a}{2} \qquad (6.65)$$

b_f : the flange width of the beam
e: the eccentricity of the beam (towards the compressed side)
f_{bd} : the design value of the adhesion between soil mix and reinforcement
$f_{sm,d}$: the design value of the compressive strength of the soil mix
L_s : the length over which the shear strength should be spread; this is equal to the distance from the maximum moment to the zero moment point and can be regarded as approximate to $L_e/2$; here the unreinforced part at the bottom the soil mix wall is NOT calculated.

d) Limitations for b_{c2} up to 2 times the flange width

The verification of the adhesion will in most cases (except for large beams and for high compressive strengths of the soil mix material) be decisive for the calculation of the effective width. The greater the width of the beam, the greater the effective width of the soil mix material. In order to somewhat limit the effective compressive zone in case of medium and small beams, especially in combination with limited compressive strengths, in the simplified rule the effective width is generally (alongside the three previous criteria) limited to **2 times the flange width**.

Because of this limitation, a too optimistic calculation is avoided when it comes to large axis-to-axis distance of the beams or large distances between the zero moment points.

The moment capacity of panel walls and column walls are also closer to each other due to this boundary.

6.7.3   Moment capacity of the steel beams only

According to Eurocode 3, the moment capacity of a beam can be calculated in 2 ways: elastic or plastic.

The **elastic moment of resistance** is achieved when the yield stress on the extreme fibres is reached. The beam will deform elastically and will return to its original position if the stress disappears.

The **plastic moment of resistance** is achieved if all fibres flow above and below the neutral line. In other words, all the fibres work at the yield stress. It goes without saying that the plastic moment of resistance is always higher than the elastic moment of resistance. If the plastic moment is achieved, then the beam has been plastically deformed and will thus not return to its original position. However, a beam can only be plastically calculated if it belongs to beam class 1 or 2 when it is subject to bending stress. Most beams that are used for soil mix walls comply with this.

Exceptions include the beams HEA260/280/300 of S355 quality. Beams that belong to class 3 must be calculated elastically. To determine the classification, please refer to Table 5.2 from EN 1993-1-1.

The class is dependent on the geometry of the beam, the type of load and type of steel. The product information of rolled beams often list the classes for each beam. An informative I - Table 6.14 for the most common beams and for steel class S235 and S355 is provided on the next page. However, potential section loss due to corrosion is not taken into account; for a number of beams such a corrosion reduction may result in an increase of class.

The gain on the moment capacity by calculating plastically instead of elastically for the beams most often used for soil mix walls is approx. 10 to 14%. For combined bending the regulations in EN 1993-1-1 apply.

1 - Table 6.14. Beam class in function of the type of beam and steel class (excluding corrosion reduction).

	S235	S355
IPE160	1	1
IPE180	1	1
IPE200	1	1
IPE220	1	1
IPE240	1	1
IPE270	1	1
IPE300	1	1
IPE330	1	1
IPE360	1	1
IPE400	1	1

	S235	S355
IPN160	1	1
IPN180	1	1
IPN200	1	1
IPN220	1	1
IPN240	1	1
IPN270	1	1
IPE300	1	1
IPN320	1	1
IPN360	1	1
IPN400	1	1

	S235	S355
HEA160	1	1
HEA180	1	2
HEA200	1	2
HEA220	1	2
HEA240	1	2
HEA260	1	3
HEA280	1	3
HEA300	1	3
HEA320	1	2
HEA340	1	1

	S235	S355
HEB160	1	1
HEB180	1	1
HEB200	1	1
HEB220	1	1
HEB240	1	1
HEB260	1	1
HEB280	1	1
HEB300	1	1
HEB320	1	1
HEB340	1	1

	S235	S355
HEM160	1	1
HEM180	1	1
HEM200	1	1
HEM220	1	1
HEM240	1	1
HEM260	1	1
HEM280	1	1
HEM300	1	1
HEM320	1	1
HEM340	1	1

6.7.4   Moment capacity of the composite section steel-soil mix

Eurocode 4 includes two ways to calculate the strength of the composite steel concrete beam: elastically or plastically.

The **elastic method** means that the maximum strength of the composite section is achieved if 1 of the 2 materials reaches the maximum elastic stress in 1 of its fibres. The stress-strain diagram is elastic for both materials. Due to the low compressive strength of the soil mix material and because the furthest fibre of the soil mix material is often located much further away from the neutral line than the furthest fibre of the steel, it will always be the soil mix material that determines the elastic moment capacity. Calculations of soil mix walls with realistic parameters show that the moment capacity, calculated in this way, is often lower than the elastic moment capacity of the steel beam alone. Naturally, this is not possible!

In the bending tests conducted on soil mix elements by the BBRI (see §2.3.13 in part 2 of the handbook), it has been determined that the maximum bending moment is always greater than the elastic and plastic bending moment of the steel beam (see Appendix 2). It also shows that the steel stress at the bottom (in the tensile zone) is greater than the steel stress at the top (in the compressive zone). In events of failure, it is also observed that the upper flange starts flowing later than the lower flange. Therefore, there is no other way than the soil mix and steel beam being plastically deformed before failure occurs. That is to say that one should compare the calculations of the moment capacity of the composite section with the methods listed in EN 1994-1-1 § 6.7.3.2 instead. Here a rectangular stress block is applied at the height of the compressed concrete (or soil mix material).

The methods specified in EN 1994-1-1 are **plastic methods**. These methods may only be applied for beams that fall within **class 1 or 2** under the classification system from EN 1993-1-1. Because the flange is generally completely encapsulated with soil mix material, one could possibly assign a more favourable class for class 3/4 beams. §5.5.3 in EN 1994-1-1 lists a number of preconditions for this: *the encapsulating concrete must be reinforced, the concrete must be anchored in the beam by means of dowels or welded reinforcement,...*

However, these are measures that are usually not possible with soil mix walls or would be too expensive. As a result, one cannot help but fall back on the classification in EN 1993-1-1. It should be noted that this could be detrimental to HEA beams with steel quality S355. For IPE and HEB beams (quality S235 and S355) and HEA beams (S235) this shouldn't be a problem. The following parameters should be available for the calculation in accordance with EN 1994-1-1 :

- the yield strength of the steel f_y : S235, S355,...; in accordance with (2) from §3.3 (EN 1994-1-1) f_y may not be greater than 460 N/mm²;
- the safety coefficient for the steel γ_{M0}; this may be taken as equal to the value from EN 1993-1-1, namely 1.0;

- the characteristic compressive strength of the soil mix material $f_{sm,k}$, and the design value of the compressive strength $f_{sm,d}$;
- the E moduli of the 2 materials do not have to be determined for the calculations of the moment capacity or the combined capacity of normal and moment forces, because this is a plastic calculation;
- the coefficient for long-term effects; by analogy with EN 1994-1-1 this could be taken as 0.85. (see §6.3.3); currently little is known about the long-term behaviour of the soil mix material; for safety reasons and considering the currently available knowledge, for the moment capacity it is recommended to only rely on the steel beams (see §6.3.6) and only to deviate from these under specific conditions (see §6.3.6).

6.7.5   Combined bending on the composite section steel-soil mix

The following calculation methodologies and formulas are based on a centrally placed steel beam, with equal cover on both sides of the beam (see figure 6.12). The calculation of the combination of bending and normal force can be carried out in this symmetric situation in accordance with (2) from §6.7.3.2 in EN 1994-1-1.

Given the fact that there is no additional reinforcement (rods) present, the forces f_{sd} from figure 6.12 are not applicable. This figure can be numerically calculated by using widely available software. EN 1994-1-1 also gives a simplified method for creating this interaction diagram by means of a polygonal diagram (see (5) from §6.7.3.2 in EN 1994-1-1).

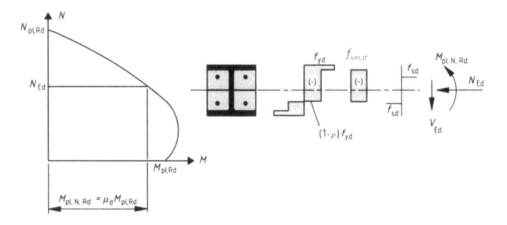

1 - Fig. 6.12. Principle interaction diagram from EN 1994-1-1:
Interaction diagram for a combination of normal compressive force and single bending.

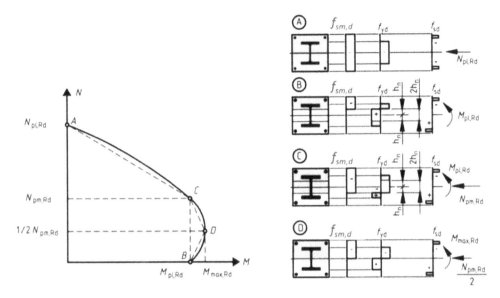

1 - Fig. 6.13. Simplified interaction diagram and corresponding stress distributions.

To create this polygonal diagram (see figure 6.13), four points must be calculated. Simultaneously, the following simplifications are also applied:

- the rolled beam is simplified to 3 rectangular steel plates (2 flanges and 1 web); the extra steel at the height of the connection between the web and the flange as a result of the rolled rounding is thus not taken into account;
- for the calculation of the compression forces in the soil mix material, the section of the steel components within the compressed zone is not removed; given that the beam is always much smaller than the amount of soil mix, this will only lead to a minor error.

Applied notations

A_a	the surface area of the cross section of the steel beam
A_{sm}	the surface area of the wall cross section minus the surface area of the steel beam
f_{yd}	the design value of the yield strength of the steel = f_{yk} / γ_{M0},
$f_{sm,d}$	the design value of the compressive strength of the soil mix material = $\alpha_{sm} \, f_{sm,k} \, \beta / \gamma_{SM}$ (see §6.3.3)
b_{c2}	the effective width of the soil mix material
h_{sm}	the effective (or equivalent) thickness of the soil mix wall
b_r	the effective width of the steel beam
h_a	the theoretical height of the steel beam
t_w	the thickness of the web of the steel beam
t_r	the thickness of the flange of the steel beam

Point A

Only a normal force is present in point A. During the plastic calculation, the design value $N_{pl,Rd}$ of the admissible normal force is:

$$N_{pl,Rd} = A_a f_{yd} + A_{sm} f_{sm,d}$$ (6.66)

Point B

For the calculation of point B, one must determine the location of the neutral line. In most cases, the plastic neutral line (PNL) will be located in the web or the upper flange. For cross-sections with a large cover, with a very large effective width or with high compression strength of the soil mix material, the PNL may be located above the flange. If this is the case, this means that the beam is completely in a tensile state and in ULS that a bit of cracked soil mix is present in between the upper flange of the steel beam and the compressed soil mix zone (top).

Please note: if there are no dowels or other adhesion materials present between the beam and the soil mix material, there is a chance of debonding occurring. In other words, there is the risk of overestimating the moment capacity. As a pre-condition, one can assume that the neutral line should go through the beam to compensate for the lack of dowels. If the PNL does not go through the beam, one can theoretically reduce the cover, reduce the theoretical effective width, choose a different steel beam or reduce the assumed compressive strength of the soil mix material (see below).

First and foremost, one must determine the position of the PNL.
As a first step, we must verify whether the PNL runs through the beam:

Condition 1: if $F_{sm,1} < F_a$ then PNL through beam

with:
$F_{sm,1}$: the maximum compressive force in the soil mix material above the steel beam.

$$F_{sm,1} = b_{c2} \frac{(h_{sm} - h_a)}{2} f_{sm,d}$$ (6.67)

F_a : the maximum tensile force in the steel.

$$F_a = A_a f_{yd}$$ (6.68)

If the above condition is not met, then, theoretically, the height h_{sm} must be reduced. Second, one must determine whether the PNL runs through the web or trough the flange.

Condition 2: if $F_{sm,2} < F_{a,w}$ then the PNL runs through the web

with:

$F_{sm,2}$: the maximum compressive force in the soil mix material above the bottom of the upper flange of the steel beam.

$$F_{sm,2} = b_{c2} \frac{(h_{sm} - h_a + 2\, t_f)}{2} f_{sm,d} \qquad (6.69)$$

$F_{a,w}$: the maximum tensile force in the web of the steel beam.

$$F_{a,w} = t_w (h_a - 2\, t_f) f_{yd} \qquad (6.70)$$

If condition 1 is met but condition 2 is not, then the PNL runs through the upper flange of the beam.

If the PNL runs through the web, we can determine the height of the compressed zone and the maximum moment as follows. Due to the horizontal balance, the compression in the soil mix material must be compensated for by a tensile force in the web (the compressive force in the upper flange is compensated by the tensile force in the bottom flange). The solution to the following equation gives x, the height of the compressed zone:

$$x_w\, b_{c2}\, f_{sm,d} = 2 \left(\frac{h_{sm}}{2} - x_w \right) t_w f_{yd} \quad and \quad x_w > c_1 + t_f \qquad (6.71)$$

Or x is thus:

$$x_w = \frac{t_w\, h_{sm}\, f_{yd}}{f_{sm,d}\, b_{c2} + 2\, t_w f_{yd}} \qquad (6.72)$$

in which:

x_w: is the height of the compressed soil mix material in case the PNL runs through the web.

c_1: is the net cover above the beam:

$$c_1 = \frac{h_{sm} - h_a}{2} \qquad (6.73)$$

Now that x_w has been determined, $M_{pl,Rd}$ can be calculated:

$$M_{pl,Rd,w} = b_{c2} \frac{x_w^{\,2}}{2} f_{smd} + b_f t_f (h_a - t_f) f_{yd} + \left[\left(\frac{h_{sm}}{2} - x_w \right)^2 + \left(\frac{h_a}{2} - t_f \right)^2 \right] t_w f_{yd} \qquad (6.74)$$

$$and \qquad x_w > c_1 + t_f$$

In case the PNL runs through the flange, x can be determined as follows:

$$x_f\, b_{c2}\, f_{smd} + \left(x_f - \frac{h_{sm}-h_a}{2}\right)\, b_f\, f_{yd} = (h_a - 2\, t_f)\, t_w\, f_{yd} + \left(\frac{h_{sm}-h_a}{2} + t_f - x_f\right)\, b_a\, f_{yd} + t_f\, b_f\, f_{yd}$$

(6.75)

$$\text{and} \quad c_1 + t_f > x_w > c_1$$

x_f is thus

$$x_f = \frac{\left(2\, c_1\, b_f + 2\, t_f\, b_f + h_w\, t_w\right) f_{yd}}{f_{sm,d}\, b_{c2} + 2\, b_f f_{yd}}$$

(6.76)

$$\text{and} \quad c_1 + t_f > x_w > d_1$$

in which

x_f : is the height of the compressed soil mix material in case the PNL runs through the flange.

h_w : is the height of the web of the beam.

$$h_w = \left(h_a - 2\, t_f\right)$$

(6.77)

Now that x_f has been determined, $M_{pl,Rd}$ can be calculated:[2]

$$M_{pl,Rd,f} = b_{c2}\,\frac{x_f^2}{2}\, f_{sm,d} + b_f\, t_f\,\left(h_{sm} - x_f - c_1 - \frac{t_f}{2}\right) f_{yd} + h_w\, t_w\, f_{yd}\,\left(\frac{h_{sm}}{2} - x_f\right)$$

(6.78)

$$\text{and} \quad c_1 + t_f > x_w > c_1$$

Point C

For point C, at the same moment as point B, $M_{pl,Rd}$, one must calculate which maximum amount of normal force can be absorbed, $N_{pm,Rd}$.

If the PNL runs through the web:

$$N_{pm,Rd,w} = (h_{sm} - 2\, x)\, b_{c2}\, f_{sm,d} + 2\, (h_{sm} - 2\, x)\, t_w\, f_{yd}$$

(6.79)

If the PNL runs through the flange:

$$N_{pm,Rd,f} = (h_{sm} - 2\, x)\, b_{c2}\, f_{sm,d} + 2\, h_w\, t_w\, f_{yd} + 4\,\left(\frac{h_{sm}}{2} - x - \frac{h_w}{2}\right) b_f\, f_{yd}$$

(6.80)

[2] In the calculation of $M_{pl,Rd}$, the bending around the upper flange is ignored. Due to the limited thickness of the flange and the limited distance from the flange to the PNL, the error made here is very limited.

116

Point D

Point D is determined on the basis of half of the calculated normal force $N_{pm,Rd}$ of point C. The corresponding moment capacity is then:

$$M_{max,Rd} = W_{pl}f_{yd} + b_{c2}\frac{h_{sm}^{2}}{8}f_{sm,d}$$

(6.81)

The plastic moment of resistance W_{pl} of the beam can be found in the product information of the rolled beams. More simply, this can also be calculated as follows (the steel in the rounding radii is ignored here):

$$W_{pl} = \frac{t_w\left(h_a-2\,t_f\right)^2}{4} + b_f\,t_f\left(h_a-t_f\right)$$

(6.82)

6.7.6 Example calculations of the moment capacity

Appendix 2 includes the moment capacities for the wall configurations that have been investigated in the bending tests of the BBRI (see §2.3.13 in part 2 of the handbook) on the one hand, and for a number of cases studies, on the other hand.

The summarising I - Tables 6.15 to 6.20 include for illustration the moment capacity of a number of typical configurations of panel walls and column walls. In addition, the following situations and characteristics were considered:
- type of walls: panel wall height 550 mm and columns diameter 550 mm and 650 mm
- steel beams: from IPE200 to IPE400 and from HEA200 to HEA400, steel S235;
- characteristic compressive strengths $f_{sm,k}$ of the soil mix of resp. 2 MPa, 4 MPa and 6 MPa;
- factors: $\alpha_{sm} = 1.0$, $\beta = 1.0$, $\gamma_{M0} = 1.0$, $\gamma_{SM} = 1.5$;
- eccentricity of the beam: 0 and 50 mm;
- length L_e between the zero moment points: 5000 mm ($L_s = 2500$ mm);
- design value of the adhesion f_{bd} between soil mix and reinforcement: determined by means of the 3 criteria in §6.3.6, including the boundary at 0.30 MPa;
- effective width b_{c2} determined by means of the 4 criteria in §6.7.2.

It should be noted that the moment capacities M(Rd,a,el) and M(Rd,a,pl) (elastic resp. plastic) of the steel beam alone is determined by application of the moments of resistance W_{el} and W_{pl} that are listed in the product catalogues of the beams. On the other hand, for the combined moment, the beam has been replaced by 3 steel plates, which results in a somewhat lower moment of resistance due to the neglect of the rounded beam corners. The total calculated combined moment capacity is therefore slightly underestimated (probably about 3 to 4%), which is on the safe side.

The listed design values of the moment capacities are **applicable per axis-to-axis distance between the beams,** resp. for a characteristic compressive strength of the soil mix material of 2 MPa, 4 MPa and 6 MPa. For the moment capacity per metre wall, these values still need to be divided by the axis-to-axis distance, a, of the beams.

The colouring of the cells gives an indication of the determining factor for the moment capacity. The following colour code is used:

Determining factor:

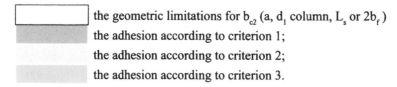

the geometric limitations for b_{c2} (a, d_1 column, L_s or $2b_f$)
the adhesion according to criterion 1;
the adhesion according to criterion 2;
the adhesion according to criterion 3.

1 - Table 6.15. Moment capacities of a panel wall with a thickness of 550 mm reinforced with IPE beams.

PANEL WALL 550 MM – IPE beams

Panel h_{sm} = 550 mm, axis-to-axis reinforcement = 1100 mm, L_e = 5000 mm, eccentricity e = 0 mm, Steel S235, γ_{MO} = 1.00, Soil mix $f_{sm,k}$ = 2, 4 and 6 MPa, α_{sm} = 1.0, β = 1.0, γ_{SM} =1.5

	IPE200	IPE220	IPE240	IPE270	IPE300	IPE330	IPE360	IPE400	IPE450
M(Rd,a,el)	45.7	59.2	76.2	100.8	130.9	167.6	212.3	271.7	352.5
M(Rd,a,pl)	51.8	67.1	86.2	113.7	147.7	189.0	239.5	307.1	400.0
M(Rd,2)	56.0	72.0	90.3	119.2	155.0	193.6	244.1	307.2	400.0
M(Rd,4)	61.9	78.7	97.9	128.4	165.9	205.2	256.5	320.3	412.8
M(Rd,6)	63.2	80.2	99.6	130.4	168.3	210.7	266.2	331.2	424.4

	IPE200	IPE220	IPE240	IPE270	IPE300	IPE330	IPE360	IPE400	IPE450
M(Rd,a,el)	100%	100%	100%	100%	100%	100%	100%	100%	100%
M(Rd,a,pl)	114%	113%	113%	113%	113%	113%	113%	113%	113%
M(Rd,2)	123%	122%	118%	118%	118%	116%	115%	113%	113%
M(Rd,4)	135%	133%	128%	127%	127%	122%	121%	118%	117%
M(Rd,6)	138%	135%	131%	129%	129%	126%	125%	122%	120%

Panel h_{sm} = 550 mm, axis-to-axis reinforcement = 1100 mm, L_e = 5000 mm, eccentricity
e = 50 mm, Steel S235, γ_{MO} = 1.00, Soil mix $f_{sm,k}$ = 2, 4 and 6 MPa, α_{sm} = 1.0, β = 1.0, γ_{SM} = 1.5

	IPE200	IPE220	IPE240	IPE270	IPE300	IPE330	IPE360	IPE400	IPE450
M(Rd,a,el)	45.7	59.2	76.2	100.8	130.9	167.6	212.3	271.7	352.5
M(Rd,a,pl)	51.8	67.1	86.2	113.7	147.7	189.0	239.5	307.1	400.0
M(Rd,2)	55.4	70.9	88.6	116.4	150.5	189.0	239.5	307.1	400.0
M(Rd,4)	60.5	76.5	94.6	123.0	157.8	196.7	247.3	310.7	402.5
M(Rd,6)	61.6	78.6	98.1	128.5	163.9	203.0	254.1	317.9	410.3

	IPE200	IPE220	IPE240	IPE270	IPE300	IPE330	IPE360	IPE400	IPE450
M(Rd,a,el)	100%	100%	100%	100%	100%	100%	100%	100%	100%
M(Rd,a,pl)	114%	113%	113%	113%	113%	113%	113%	113%	113%
M(Rd,2)	121%	120%	116%	115%	115%	113%	113%	113%	113%
M(Rd,4)	132%	129%	124%	122%	121%	117%	116%	114%	114%
M(Rd,6)	135%	133%	129%	128%	125%	121%	120%	117%	116%

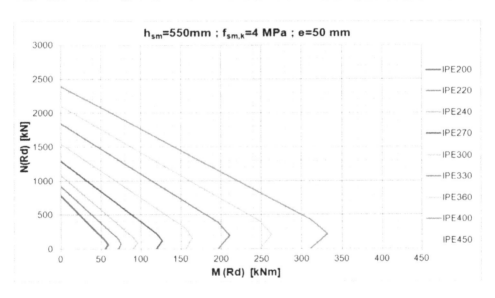

PANEL WALL 550 MM – HEA beams

Panel h_{sm} = 550 mm, axis-to-axis reinforcement = 1100 mm, L_e = 5000 mm, eccentricity
e = 0 mm, Steel S235, γ_{MO} = 1.00, Soil mix $f_{sm,k}$ = 2, 4 and 6 MPa, α_{sm} = 1.0, β = 1.0, γ_{SM} =1.5

	HEA220	HEA240	HEA260	HEA280	HEA300	HEA320	HEA340	HEA360	HEA400
M(Rd,a,el)	121.1	158.6	196.6	238.1	296.1	347.6	394.3	444.4	543.1
M(Rd,a,pl)	133.6	175.0	216.2	261.3	325.0	382.6	434.8	490.7	602.1
M(Rd,2)	142.1	182.8	222.5	269.2	331.1	388.4	439.6	494.4	603.2
M(Rd,4)	153.6	195.7	236.6	285.0	348.7	406.9	458.5	513.7	623.2
M(Rd,6)	156.0	198.5	239.6	288.4	352.4	412.1	467.0	525.4	639.0
M(Rd,a,el)	100%	100%	100%	100%	100%	100%	100%	100%	100%
M(Rd,a,pl)	110%	110%	110%	110%	110%	110%	110%	110%	111%
M(Rd,2)	117%	115%	113%	113%	112%	112%	111%	111%	111%
M(Rd,4)	127%	123%	120%	120%	118%	117%	116%	116%	115%
M(Rd,6)	129%	125%	122%	121%	119%	119%	118%	118%	118%

Panel h_{sm} = 550 mm, axis-to-axis reinforcement = 1100 mm, L_e = 5000 mm, eccentricity
e = 50 mm, Steel S235, γ_{MO} = 1.00, Soil mix $f_{sm,k}$ = 2, 4 and 6 MPa, α_{sm} = 1.0, β = 1.0, γ_{SM} =1.5

	HEA220	HEA240	HEA260	HEA280	HEA300	HEA320	HEA340	HEA360	HEA400
M(Rd,a,el)	121.1	158.6	196.6	238.1	296.1	347.6	394.3	444.4	543.1
M(Rd,a,pl)	133.6	175.0	216.2	261.3	325.0	382.6	434.8	490.7	602.1
M(Rd,2)	140.3	180.0	218.4	263.6	325.0	382.6	434.8	490.7	602.1
M(Rd,4)	149.6	190.0	228.9	274.9	335.6	392.4	443.7	498.7	608.0
M(Rd,6)	152.3	194.5	235.5	283.4	344.6	401.7	453.4	508.7	618.5
M(Rd,a,el)	100%	100%	100%	100%	100%	100%	100%	100%	100%
M(Rd,a,pl)	110%	110%	110%	110%	110%	110%	110%	110%	111%
M(Rd,2)	116%	113%	111%	111%	110%	110%	110%	110%	111%
M(Rd,4)	124%	120%	116%	115%	113%	113%	113%	112%	112%
M(Rd,6)	126%	123%	120%	119%	116%	116%	115%	114%	114%

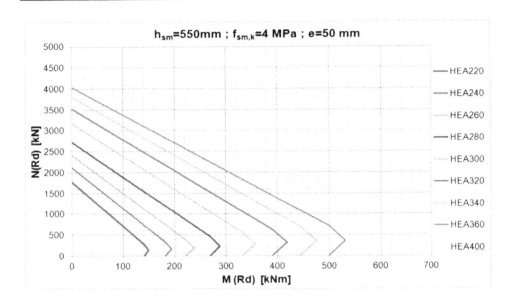

COLUMN WALL Ø 530/460 mm – IPE beams

Column Ø 530 mm, axis-to-axis reinforcement = 920 mm, L_e = 5000 mm, eccentricity e = 0 mm, Steel S235, γ_{MO} = 1.00, Soil mix $f_{sm,k}$ = 2, 4 and 6 MPa, α_{sm} = 1.0, β = 1.0, γ_{SM} =1.5

	IPE200	IPE220	IPE240	IPE270	IPE300	IPE330	IPE360	IPE400
M(Rd,a,el)	45.7	59.2	76.2	100.8	130.9	167.6	212.3	271.7
M(Rd,a,pl)	51.8	67.1	86.2	113.7	147.7	189.0	239.5	307.1
M(Rd,2)	55.4	71.0	88.7	116.5	150.6	189.0	239.5	307.1
M(Rd,4)	60.5	76.6	94.7	123.2	158.0	196.8	247.5	310.9
M(Rd,6)	61.7	78.7	98.1	128.7	164.1	203.3	254.4	318.2
M(Rd,a,el)	100%	100%	100%	100%	100%	100%	100%	100%
M(Rd,a,pl)	114%	113%	113%	113%	113%	113%	113%	113%
M(Rd,2)	121%	120%	116%	116%	115%	113%	113%	113%
M(Rd,4)	133%	129%	124%	122%	121%	117%	117%	114%
M(Rd,6)	135%	133%	129%	128%	125%	121%	120%	117%

Column Ø 530 mm, axis-to-axis reinforcement = 920 mm, L_e = 5000 mm, eccentricity e = 50 mm, Steel S235, γ_{MO} = 1.00, Soil mix $f_{sm,k}$ = 2, 4 and 6 MPa, α_{sm} = 1.0, β = 1.0, γ_{SM} =1.5

	IPE200	IPE220	IPE240	IPE270	IPE300	IPE330
M(Rd,a,el)	45.7	59.2	76.2	100.8	130.9	167.6
M(Rd,a,pl)	51.8	67.1	86.2	113.7	147.7	189.0
M(Rd,2)	53.0	68.3	86.2	113.7	147.7	189.0
M(Rd,4)	56.2	71.7	89.5	117.3	151.5	189.9
M(Rd,6)	58.8	74.6	92.5	120.7	155.2	193.8
	IPE200	IPE220	IPE240	IPE270	IPE300	IPE330
M(Rd,a,el)	100%	100%	100%	100%	100%	100%
M(Rd,a,pl)	114%	113%	113%	113%	113%	113%
M(Rd,2)	116%	115%	113%	113%	113%	113%
M(Rd,4)	123%	121%	117%	116%	116%	113%
M(Rd,6)	129%	126%	121%	120%	119%	116%

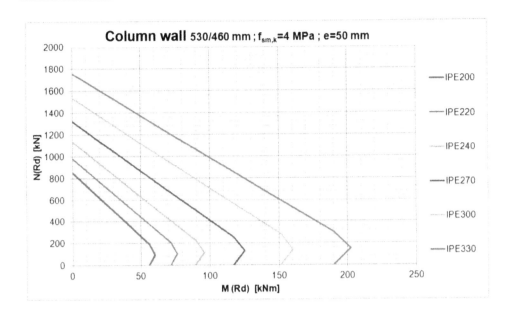

COLUMN WALL Ø 530/460 mm – HEA beams

Column Ø 530 mm, axis-to-axis reinforcement = 920 mm, L_e = 5000 mm, eccentricity e = 0 mm, Steel S235, γ_{MO} = 1.00, Soil mix $f_{sm,k}$ = 2, 4 and 6 MPa, α_{sm} = 1.0, β = 1.0, γ_{SM} =1.5

	HEA200	HEA220	HEA240	HEA260	HEA280	HEA300	HEA320	HEA340	HEA360
M(Rd,a,el)	91.3	121.1	158.6	196.6	238.1	296.1	347.6	394.3	444.4
M(Rd,a,pl)	100.9	133.6	175.0	216.2	261.3	325.0	382.6	434.8	490.7
M(Rd,2)	106.8	140.5	179.6	217.1	261.3	325.0	382.6	434.8	490.7
M(Rd,4)	115.1	149.8	189.5	227.0	271.5	330.7	387.3	438.5	493.4
M(Rd,6)	116.9	152.3	194.6	234.6	279.4	338.9	395.8	447.3	502.4
M(Rd,a,el)	100%	100%	100%	100%	100%	100%	100%	100%	100%
M(Rd,a,pl)	111%	110%	110%	110%	110%	110%	110%	110%	110%
M(Rd,2)	117%	116%	113%	110%	110%	110%	110%	110%	110%
M(Rd,4)	126%	124%	119%	115%	114%	112%	111%	111%	111%
M(Rd,6)	128%	126%	123%	119%	117%	114%	114%	113%	113%

Column Ø 530 mm, axis-to-axis reinforcement = 920 mm, L_e = 5000 mm, eccentricity e = 50 mm, Steel S235, γ_{MO} = 1.00, Soil mix $f_{sm,k}$ = 2, 4 and 6 MPa, α_{sm} = 1.0, β = 1.0, γ_{SM} =1.5

	HEA200	HEA220	HEA240	HEA260	HEA280	HEA300	HEA320	HEA340	HEA360
M(Rd,a,el)	91.3	121.1	158.6	196.6	238.1	296.1	347.6	394.3	444.4
M(Rd,a,pl)	100.9	133.6	175.0	216.2	261.3	325.0	382.6	434.8	490.7
M(Rd,2)	102.7	135.4	175.0	216.2	261.3	325.0	382.6	434.8	490.7
M(Rd,4)	107.9	141.1	180.4	217.8	262.2	325.0	382.6	434.8	490.7
M(Rd,6)	111.9	145.5	185.0	222.5	267.0	326.3	382.9	434.8	490.7
M(Rd,a,el)	100%	100%	100%	100%	100%	100%	100%	100%	100%
M(Rd,a,pl)	111%	110%	110%	110%	110%	110%	110%	110%	110%
M(Rd,2)	112%	112%	110%	110%	110%	110%	110%	110%	110%
M(Rd,4)	118%	117%	114%	111%	110%	110%	110%	110%	110%
M(Rd,6)	123%	120%	117%	113%	112%	110%	110%	110%	110%

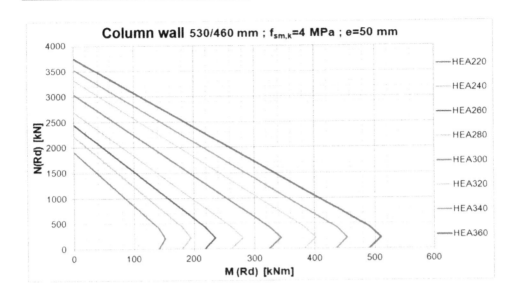

COLUMN WALL Ø 630/560 mm – IPE beams

Column Ø 630 mm, axis-to-axis reinforcement = 1120 mm, L_e = 5000 mm, eccentricity e = 0 mm, Steel S235, γ_{MO} = 1.00, Soil mix $f_{sm,k}$ = 2, 4 and 6 MPa, α_{sm} = 1.0, β = 1.0, γ_{SM} =1.5

	IPE200	IPE220	IPE240	IPE270	IPE300	IPE330	IPE360	IPE400
M(Rd,a,el)	45.7	59.2	76.2	100.8	130.9	167.6	212.3	271.7
M(Rd,a,pl)	51.8	67.1	86.2	113.7	147.7	189.0	239.5	307.1
M(Rd,2)	55.9	71.9	90.2	119.2	154.1	192.7	243.2	307.1
M(Rd,4)	61.6	78.4	97.6	128.1	164.4	203.6	254.7	318.5
M(Rd,6)	62.9	79.9	99.3	130.1	168.1	210.7	264.2	328.6
M(Rd,a,el)	100%	100%	100%	100%	100%	100%	100%	100%
M(Rd,a,pl)	114%	113%	113%	113%	113%	113%	113%	113%
M(Rd,2)	122%	121%	118%	118%	118%	115%	115%	113%
M(Rd,4)	135%	132%	128%	127%	126%	121%	120%	117%
M(Rd,6)	138%	135%	130%	129%	128%	126%	124%	121%

Column Ø 630 mm, axis-to-axis reinforcement = 1120 mm, L_e = 5000 mm, eccentricity e = 50 mm, Steel S235, γ_{MO} = 1.00, Soil mix $f_{sm,k}$ = 2, 4 and 6 MPa, α_{sm} = 1.0, β = 1.0, γ_{SM} = 1.5

	IPE200	IPE220	IPE240	IPE270	IPE300	IPE330	IPE360	IPE400
M(Rd,a,el)	45.7	59.2	76.2	100.8	130.9	167.6	212.3	271.7
M(Rd,a,pl)	51.8	67.1	86.2	113.7	147.7	189.0	239.5	307.1
M(Rd,2)	52.5	68.0	86.2	114.1	149.2	189.0	239.5	307.1
M(Rd,4)	55.4	71.5	89.8	119.1	155.6	196.7	247.3	310.7
M(Rd,6)	56.1	72.3	90.8	120.3	157.0	199.0	254.1	317.9
M(Rd,a,el)	100%	100%	100%	100%	100%	100%	100%	100%
M(Rd,a,pl)	114%	113%	113%	113%	113%	113%	113%	113%
M(Rd,2)	115%	115%	113%	113%	114%	113%	113%	113%
M(Rd,4)	121%	121%	118%	118%	119%	117%	116%	114%
M(Rd,6)	123%	122%	119%	119%	120%	119%	120%	117%

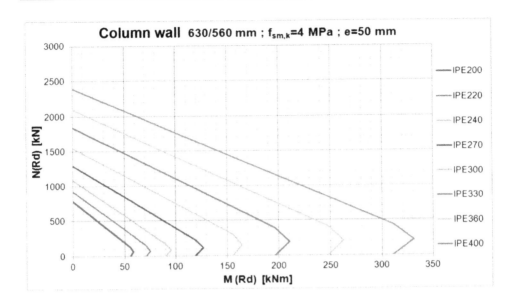

COLUMN WALL Ø 630/560 mm – HEA beams

Column Ø 630 mm, axis-to-axis reinforcement = 1120 mm, L_e = 5000 mm, eccentricity
e = 0 mm, Steel S235, γ_{MO} = 1.00, Soil mix $f_{sm,k}$ = 2, 4 and 6 MPa, α_{sm} = 1.0, β = 1.0, γ_{SM} =1.5

	HEA220	HEA240	HEA260	HEA280	HEA300	HEA320	HEA340	HEA360	HEA400
M(Rd,a,el)	121.1	158.6	196.6	238.1	296.1	347.6	394.3	444.4	543.1
M(Rd,a,pl)	133.6	175.0	216.2	261.3	325.0	382.6	434.8	490.7	602.1
M(Rd,2)	141.9	182.5	222.2	269.0	329.0	385.5	436.6	491.3	602.1
M(Rd,4)	153.0	195.1	236.0	284.4	345.2	402.1	453.6	508.7	618.1
M(Rd,6)	155.4	197.8	238.9	287.6	351.6	411.5	466.5	522.3	632.4
M(Rd,a,el)	100%	100%	100%	100%	100%	100%	100%	100%	100%
M(Rd,a,pl)	110%	110%	110%	110%	110%	110%	110%	110%	111%
M(Rd,2)	117%	115%	113%	113%	111%	111%	111%	111%	111%
M(Rd,4)	126%	123%	120%	119%	117%	116%	115%	114%	114%
M(Rd,6)	128%	125%	122%	121%	119%	118%	118%	118%	116%

Column Ø 630 mm, axis-to-axis reinforcement = 1120 mm, L_e = 5000 mm, eccentricity
e = 50 mm, Steel S235, γ_{MO} = 1.00, Soil mix $f_{sm,k}$ = 2, 4 and 6 MPa, α_{sm} = 1.0, β = 1.0, γ_{SM} = 1.5

	HEA220	HEA240	HEA260	HEA280	HEA300	HEA320	HEA340	HEA360	HEA400
M(Rd,a,el)	121.1	158.6	196.6	238.1	296.1	347.6	394.3	444.4	543.1
M(Rd,a,pl)	133.6	175.0	216.2	261.3	325.0	382.6	434.8	490.7	602.1
M(Rd,2)	139.3	178.9	217.2	262.4	325.0	382.6	434.8	490.7	602.1
M(Rd,4)	147.9	188.2	226.9	272.8	332.1	388.8	440.1	495.0	604.1
M(Rd,6)	151.6	193.9	234.3	280.6	340.2	397.2	448.8	504.0	613.5
M(Rd,a,el)	100%	100%	100%	100%	100%	100%	100%	100%	100%
M(Rd,a,pl)	110%	110%	110%	110%	110%	110%	110%	110%	111%
M(Rd,2)	115%	113%	111%	110%	110%	110%	110%	110%	111%
M(Rd,4)	122%	119%	115%	115%	112%	112%	112%	111%	111%
M(Rd,6)	125%	122%	119%	118%	115%	114%	114%	113%	113%

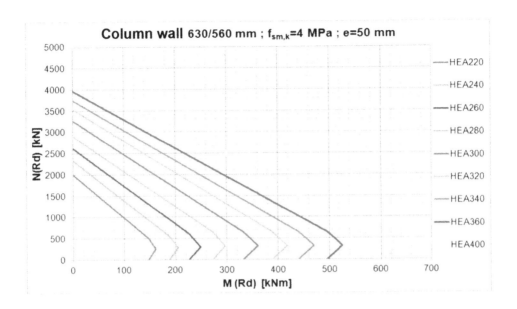

6.7.7 Calculation of the shear resistance

For the calculation of the shear resistance, one may refer to EN 1993-1-1. The shear force should be fully absorbed by the web of the steel beam.

$$V_{Ed} < 0.5\ V_{pl,Rd} \tag{6.83}$$

with

$$V_{pl,Rd} = \frac{A_v f_{yd}}{\sqrt{3}} \tag{6.84}$$

A_v is the active plastic shear surface and can be found in the product information of the rolled beams.

If

$$0.5\ V_{pl,Rd} < V_{Ed} < V_{pl,Rd} \tag{6.85}$$

a reduction should be applied for the calculation of the moment capacity at the yield stress for the shear surface (in accordance with the requirements in EN 1993-1-1). This process can also be simplified by theoretically reducing the web thickness.

6.7.8 Verification of the strength of the soil mix wall under vertical loads

Vertical loads can affect the soil mix wall due to load transfer on the top of the wall, through intermediate floors and the base plate or through the vertical component of anchor forces or strut forces. These forces may occur in the temporary phase (short term) or the permanent phase (long term). The vertical forces on the wall due to the soil pressures (inclined soil surface) and due to the weight of the wall can often be ignored.

The constructive methods that can be used for the transfer of the construction loads to the wall are succinctly described in §6.11.3. For the anchor loads, please refer to §6.8.7 to 6.8.10.

The verification of the strength of the soil mix wall involves two aspects:

1. on the one hand, the vertical stress on the soil mix material and reinforcement must be limited to the admissible values, and that in combination with the acting bending moments and shear forces;

2. additionally, one must verify whether there is sufficient adhesion to avoid any debonding or punching through the beams under the vertical loads that affect them.

Distribution of the vertical loads between soil mix and reinforcement
The following is recommended when applying a stiff concrete tranfer beam for transferring the vertical loads on the head of the soil mix wall:
- the transfer of force on the head of the wall can be estimated in proportion to the compressibility of both ' soil mix' and 'steel' materials, or in proportion to their respective E-modulus; here the lowest value over the entire height of the wall must be taken into account for the soil mix material;
- in the event of highly concentrated loads, the load distribution must take into account the stiffness of the transfer beam;
- for the use phase (long term), the E-modulus should be divided by the creep factor $(1+\Psi) = 2.0$;
- for the long term, the interaction with the soil mix material will only be considered if its characteristic compressive strength is at least 3 MPa, the material is protected against climatic influences and in the absence of contaminants that over time may affect the integrity of the soil mix material;
- for the transmission of the vertical loads in function of the depth, starting from the excavation level, the acting forces can be reduced with a reliably estimated value of the positive skin friction on the soil mix wall;
- under the base of the steel beams one must verify that the acting vertical forces are only absorbed by the soil mix material.

For the loads caused by intermediate floors and the bottom plate, which are transferred to the reinforcement by means of the mechanical connections, it is logically assumed that these loads are 100% transferred to the steel beams.

For the loads caused by the vertical component of the anchor forces, the load transfer is dependent on the applied methodology (steel anchor tube in the soil mix, purlins welded on the reinforcement, …).

The above-mentioned loads should be determined in ULS with application of the necessary load factors.

Verification of the permissible stresses
For the temporary phase, the soil mix wall can be calculated considered the acting combination of bending moments M and normal forces N in accordance with the methodology described in the previous paragraphs.

For the permanent phase, in which – aside from exception – the moment capacity may only be calculated on the basis of the reinforcement, the assessment of the structural resistance of the soil mix wall may take place as follows:

- the steel stresses are calculated on the basis of the most critical combination of the bending moments M and the part N_1 of the normal force N that, by means of the force transfer analysis, directly affects the reinforcement; these steel stresses must be limited to the design value of the admissible steel stresses, in accordance with EN 1993-1 and EN 1993-5;
- the stresses on the soil mix material that follow from the part N_2 of the normal force N that, by means of the force transfer analysis, directly affects the soil mix, must be limited to the design value of the compressive strength of the soil mix material $f_{sm,d}$ (including a long-term factor of 0.85).

Verification of the adhesion between the steel beam and soil mix material
There should be sufficient adhesion in order to prevent the maximum vertical load that affects the steel beams resulting in punching between the steel beam and soil mix.

Here, the following rules apply:
- the design value of the adhesion f_{bd} is calculated as described in §6.3.6;
- for long term situations, the adhesion may only be calculated from a characteristic compressive strength of 3 MPa and for the part above the excavation level only if the reinforcement is protected against corrosion.

6.8   Horizontal support

Frequently, the soil mix wall realised without horizontal support does not meet the specifications in terms of deformations, forces and/or economic design. A supported soil mix wall may be realised in several ways.

Hereinafter the different ways of supporting a soil mix wall will be explained. This concerns the transfer of the forces from the supporting points to the soil mix wall.

6.8.1   Different anchors and strut frame constructions

- Strut frame against the top of the steel beams; (a)
- Strut frame against soil mix wall; (b)
- Horizontal tensile bar against the top of the steel beams; (c)
- Horizontal tensile bar against soil mix wall; (d)
- Horizontal tensile bar deeply installed against the soil mix wall; (e)
- Inclined anchor against the top of the steel beams; (f)
- Inclined anchor against the soil mix wall; (g)
- Inclined anchor deeply installed against the soil mix wall; (h)

Note: in case a concrete beam is used over the soil mix wall, the beam should be properly calculated.

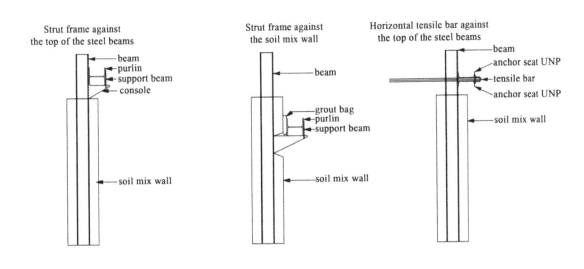

Strut frame against
the top of the steel beams
— beam
— purlin
— support beam
— console
— soil mix wall

Strut frame against
the soil mix wall
— beam
— grout bag
— purlin
— support beam
— soil mix wall

Horizontal tensile bar against
the top of the steel beams
— beam
— anchor seat UNP
— tensile bar
— anchor seat UNP
— soil mix wall

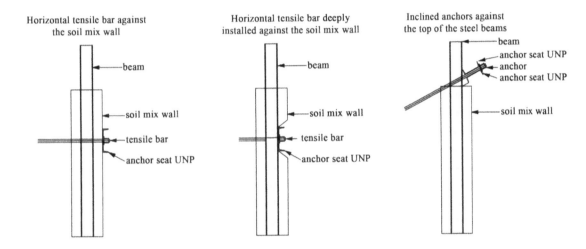

Horizontal tensile bar against
the soil mix wall
— beam
— soil mix wall
— tensile bar
— anchor seat UNP

Horizontal tensile bar deeply
installed against the soil mix wall
— beam
— soil mix wall
— tensile bar
— anchor seat UNP

Inclined anchors against
the top of the steel beams
— beam
— anchor seat UNP
— anchor
— anchor seat UNP
— soil mix wall

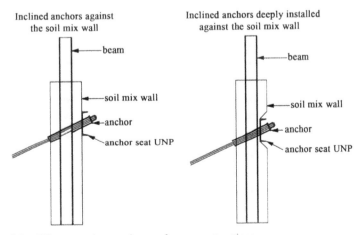

Inclined anchors against
the soil mix wall
— beam
— soil mix wall
— anchor
— anchor seat UNP

Inclined anchors deeply installed
against the soil mix wall
— beam
— soil mix wall
— anchor
— anchor seat UNP

1 - Fig. 6.14. Overview of the different anchors and strut frame constructions.

133

6.8.2 Strut frame against the top of the steel beams

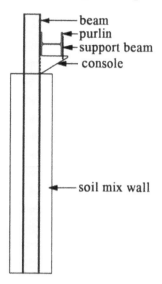

I - Fig. 6.15. Strut frame against the top of the steel beams (vertical cross-section).

The purlin is placed against the top of the steel beams, where the area in between the beam and purlin should be sufficiently stuffed.

The load transfer runs from the soil mix wall via the beams as concentrated loads in the purlin. If the axis-to-axis distance between the beams is approx. 1.0 m, the load may also be modelled as a q-load on the purlin beam.

To be controlled/calculated:
- the stability of the top of the steel beams;
- the consoles for the purlins;
- optionally the to be transferred forces parallel to the soil mix wall.

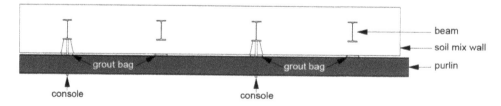

I - Fig. 6.16. Strut frame against the top of the steel beams (top view).

6.8.3   Strut frame against the soil mix wall

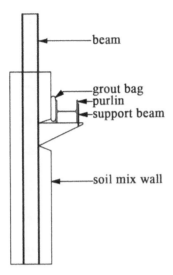

I - Fig. 6.17. Strut frame against the soil mix wall (vertical cross-section).

Distinction between 2 connections between soil mix wall and purlin:

Grout bags
The purlin is placed against the soil mix wall, with the space between the purlin and soil mix wall stuffed with grout bags at the location of the steel beams. Sufficient room in between the purlin and the soil mix wall must be maintained in order to properly place the grout bags.

To be controlled/calculated:
- the compressive stresses at the location of the connection between soil mix wall and grout bag;
- the consoles for the purlins;
- optionally the to be transferred forces parallel to the soil mix wall.

Casting
the purlin is placed against the soil mix wall, with the space between the purlin and soil mix wall stuffed with the cast in situ mortar along the entire length of the soil mix wall. Sufficient room in between the purlin and the soil mix wall must be maintained in order to properly apply the cast in situ mortar. The cast in situ mortar must always be present next to the web of the purlin.

To be controlled/calculated:

- the compressive stresses at the location of the connection between soil mix wall and cast in situ mortar;
- the consoles for the purlins;
- optionally the to be transferred forces parallel to the soil mix wall.

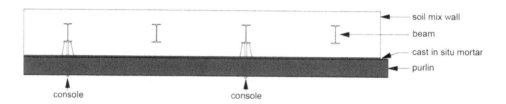

I - Fig. 6.18. Strut frame against soil mix wall (top view).

6.8.4 Horizontal tensile bar against the top of the steel beams

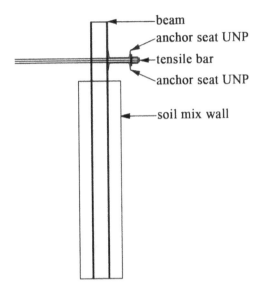

I - Fig. 6.19. Horizontal tensile bar against the top of the steel beams (vertical cross-section).

The purlin is placed against the top of the steel beams, where the area in between the beam and purlin should be sufficiently stuffed. The load transfer runs from the soil mix wall via the beams as concentrated loads in the purlin. If the axis-to-axis distance between the beams is approx. 1.0 m, the load may also be modelled as a q-load on the purlin. The purlin can continuously run across multiple beams (as in the figure below), but short purlins across two beams at a time and the anchor may also be applied.

To be controlled/calculated:
- the stability of the top of the steel beams;
- the purlin;
- the consoles for the purlins.

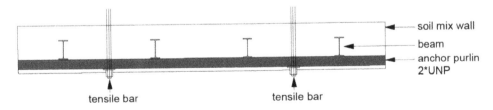

I - Fig. 6.20. Horizontal tensile bar against the top of the steel beams (top view).

6.8.5 Horizontal tensile bar against the soil mix wall

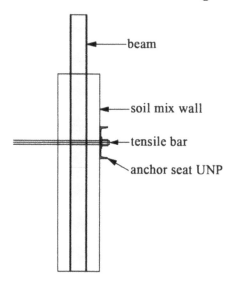

I - Fig. 6.21. Horizontal tensile bar against the soil mix wall (vertical cross-section).

The anchor seat is placed against the soil mix wall, in the middle, between the two steel beams. A typical axis-to-axis distance between the tensile bars is 2x the axis-to-axis distance between the steel beams. A UNP beam is a common beam for an anchor seat.

To be controlled/calculated:
- transfer of the anchor force to the soil mix material via the anchor seat;
- transfer of anchor force by arch effect to the steel beams.

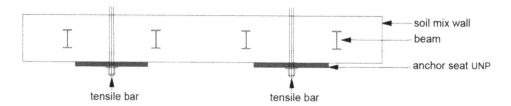

I - Fig. 6.22. Horizontal tensile bar against the soil mix wall (top view).

6.8.6 *Horizontal tensile bar deeply installed against the soil mix wall*

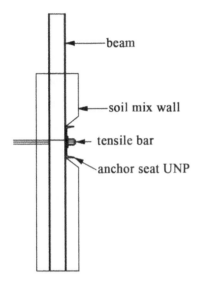

1 - Fig. 6.23. Horizontal tensile bar deeply installed against the soil mix wall (vertical cross-section).

The anchor seat is placed against the soil mix wall, in the middle, between the two steel beams after a recess is made in the soil mix wall. A typical axis-to-axis distance between the tensile bar is 2x the axis-to-axis distance between the steel beams. A UNP beam is a common beam for an anchor seat.

To be controlled/calculated:
- transfer of the anchor force to the soil mix material via the anchor seat;
- transfer of anchor force by arch effect to the steel beams.

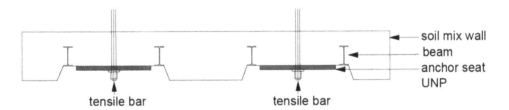

1 - Fig. 6.24. Horizontal tensile bar deeply installed against the soil mix wall (top view).

6.8.7 *Inclined anchors against the top of the steel beams*

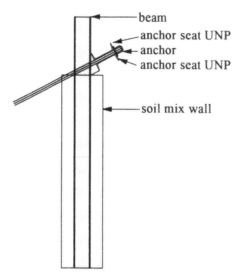

I - Fig. 6.25. Inclined anchors against the top of the steel beams (vertical cross-section).

The purlin is placed against the top of the steel beams, where the area in between the beam and purlin should be sufficiently stuffed. The load transfer runs from the soil mix wall via the beams as concentrated loads in the purlin. If the axis-to-axis distance between the steel beams is approx. 1.0 m, the load may also be modelled as a q-load on the purlin. The purlin can continuously run across multiple beams (as in figure 6.26), but short purlins across two beams at a time and the anchor may also be applied. The vertical component of the anchor force must be transferred via a shear-resistant connection.

To be controlled/calculated:
- the stability of the top of the steel beams;
- the purlin;
- the absorption of horizontal and vertical force from the anchor;
- the consoles for the purlins.

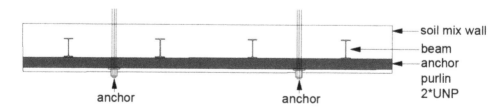

I - Fig. 6.26. Inclined anchors against the top of the steel beams (top view).

140

6.8.8 Inclined anchors against the soil mix wall

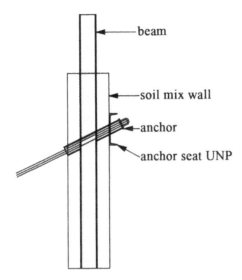

I - Fig. 6.27. Inclined anchors against the soil mix wall (vertical cross-section).

The anchor seat is placed against the soil mix wall, in the middle, between the two steel beams. A typical axis-to-axis distance between the anchors is 2x the axis-to-axis distance between the steel beams. A UNP beam is a common beam for an anchor seat. The anchor seat is equipped with a tube that enters the soil mix wall in order to absorb the vertical component from the anchor.

To be controlled/calculated:
*   transfer of the anchor force to the soil mix material via the anchor seat;
*   transfer of anchor force by arch effect to the steel beams.
*   transfer of the vertical anchor force in the soil mix wall.

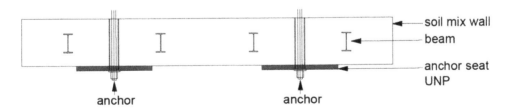

I - Fig. 6.28. Inclined anchor against the soil mix wall (top view).

6.8.9   Inclined anchor deeply installed against the soil mix wall

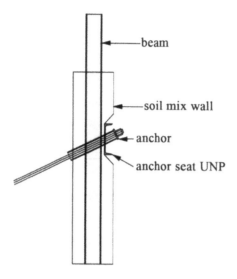

The anchor seat is placed against the soil mix wall, in the middle, between the two steel beams after a recess is made in the soil mix wall. A typical axis-to-axis distance between the anchors is 2x the axis-to-axis distance between the steel beams. A UNP beam is a common beam for an anchor seat. The anchor seat is equipped with a tube that enters the soil mix wall in order to absorb the vertical component from the anchor. Other ways of transferring the vertical anchor force are also possible.

To be controlled/calculated:

- transfer of the anchor force to the soil mix material via the anchor seat;
- transfer of anchor force by arch effect to the steel beams;
- transfer of the vertical anchor force in the soil mix wall.

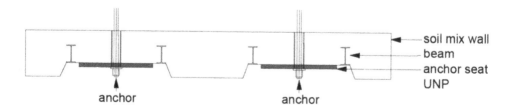

I - Fig. 6.30. Inclined anchors deeply installed against the soil mix wall (top view).

142

6.8.10 Important points for the verification of the anchor seat construction

A common solution is one in which the soil mix wall is equipped with an anchor or tensile bar. Here, a rectangular cross-section of a plate or steel beam is used (e.g. UNP), allowing the anchor force to be transferred to the soil mix wall.

In the verification of the anchor seat construction there are a number of points one must consider in terms of the transfer of force to the soil mix material.

A. Transfer of the horizontal component of the anchor force to the soil mix material:
The following verifications must be carried out for this:

1. Verification of the compression stresses underneath the plate or beam.

2. Verification of the additional compression stresses, which occur during the transfer to the steel beams.

3. Verification of the resistance of the soil mix material against punching failure.

A1. Design load:
The anchor seat should be verified on the basis of the acquired design value of the horizontal component of the anchor force, in accordance with Step 8 in paragraph 6.2.1 (Belgium, including factor 1.35) and Step 9 in paragraph 6.2.2 (the Netherlands, including factor 1.25).

A2. Verification of the contact stress directly underneath the plate or the beam:
The maximum contact stress underneath the anchor plate or steel beam is:

$$\sigma_{sm;contact} = P_{hor;d} / A_{load} \tag{6.86}$$

with:
A_{load} : Loaded surface area underneath the anchor plate or steel beam, taking into account the deduction for the presence of a potential hole for the anchor.

To transfer the load into the soil mix wall as evenly as possible, a certain deformation of the anchor plate or steel beam is allowed. Compression stress is present in a vertical direction on the side of the contact surface. In accordance with Eurocode 2, the admissible compression stress here is:

$$\sigma_{sm;adm;d} = f_{sm,d} \tag{6.87}$$

One should comply with :

$$\sigma_{sm;contact} \leq \sigma_{sm;adm;d} \tag{6.88}$$

143

A3. Verification of the additional compression stresses, which occur during transfer to the steel beams:

A finite element model has been used for determining the occurrence of compression stress trajectories, the results of which are shown in figure 6.31.

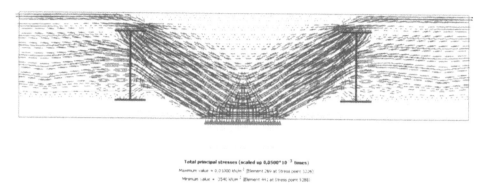

Total principal stresses (scaled up 0,0500* 10 ³ times)
Maximum value = 0,0 1000 kN/m ² (Element 269 at Stress point 3226)
Minimum value = 3540 kN/m ² (Element 441 at Stress point 5288)

1 - Fig. 6.31. Compression stress trajectories at the height of the anchor seat (FEM analysis).

The FEM stress distribution is consistent with the load distribution in accordance with the strut and tie model of Thürlimann as indicated below.

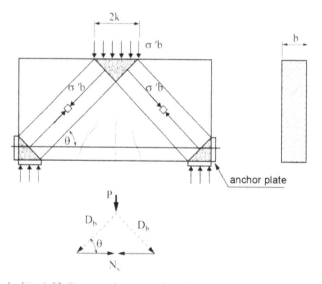

anchor plate

1 - Fig. 6.32. Compression stress distribution according to the strut and tie model of Thürlimann.

The length of the contact surface in this case is 2k. The width is b. The stress directly underneath the contact surface is therefore $\sigma'_{sm} = P / (2bk)$.

The stress in the compression stress trajectories $\sigma'_{sm} = P / (4bk \sin\theta^2)$

144

In the vertical direction, one may also take a stress distribution into account for determining the stress at the location of the connection area with the beam. With a vertical distribution under 45° this results in:

$$\sigma'_{sm,contact} = P_{hor;d} / (2L_y (L_z + 2 h_{bg}) \sin\theta^2) \text{ with } \tan\theta = h_{bg} / (0.5 \text{ } a)$$

with:

a: axis-to-axis distance between the beams [mm]

h_{bg} :   maximum effective height available for the pressure arch
       = height to exterior of wall [mm].

L_y :   horizontal dimension of the anchor plate or the steel beam

L_z :   vertical dimension of the anchor plate or the steel beam

Tensile stress is present on the side of the connection area with the steel beam, in the vertical direction. In accordance with Eurocode 2, the admissible compression stress here is:

$$\sigma_{sm;adm;d} = 0.6 \text{ } x \text{ } (1 - f_{sm,k}/250) \text{ } x f_{sm,d} \tag{6.90}$$

One should comply with :

$$\sigma_{sm;contact} \leq \sigma_{sm;adm;d} \tag{6.91}$$

A4. Verification of the resistance of the soil mix material against punching failure

With the use of an anchor in between the steel beams of the soil mix wall, a punching failure cone may develop from the anchor connection and that between the steel beams.

One must therefore verify whether the resistance of the soil mix wall against punching failure is sufficient. This verification must be carried out in accordance with EN 1992-1-1.

As a testing perimeter, the contour present at a distance of $2h_{sm}$ around the anchor plate can be considered. However, in transverse direction this distance is limited to size D_{uy}.

The following applies for D_{uy} :

$$D_{uy} = 0.5 * (a - L_y - B) * (h_{sm}/(h_a + c_v)) \tag{6.92}$$

After all, the angle at which the punching failure cone may arise, is limited in the transverse direction by the presence of the steel beams. See figure 6.33 for the associated shape of the punching failure cone in the transverse direction.

Incremental deviatoric strain $\Delta\gamma_s$

ud   Maximum value = 0,9370*10³ (Element 268 at Node 9681)
      Minimum value = 0,000 (Element 6339 at Node 61398)

I - Fig. 6.33. Deformation path of the punching failure cone at the height of the anchor seat.

From a front view, the shape shown in figure 6.34 (top) results from the testing perimeter.

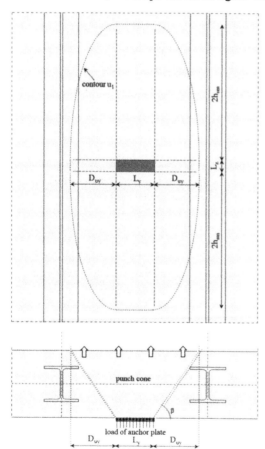

I - Fig. 6.34. Shape and perimeter of the punching failure cone.

If the anchor plate is located near the top of the wall, the testing perimeter must be determined as a straight line to the top, if this results in a contour of the perimeter that is smaller than the indicated value for u_1.

The contour (u_1) of the oval-shaped testing perimeter can be described as follows:

$$u_1 = 2L_y + 2L_z + 4,17 D_{uy}^2 / (2h_{sm}) + 8 h_{sm} \quad if \quad D_{uy} \leq 2h_{sm}/3 \tag{6.93}$$

$$u_1 = 2L_y + 2L_z + \pi * [1.5 * (D_{uy} + 2h_{sm}) - \sqrt{1(2h_{sm} * D_{uy})}] \quad if \quad 2h_{sm}/3 < D_{uy} \leq 2h_{sm} \tag{6.94}$$

The verification value for the punching stress is:

$$V_{Ed} = \beta * P_{hor;d} / (u_1 * h_{sm}).$$

The value β can be determined in accordance with 1992-1-1 and is generally $\beta = 1.15$. If the test perimeter is determined by the presence of the top side of the wall, the value is $\beta = 1.4$.

The limit value of the punching stress $V_{Rd,sm}$ is equal to the design value of the shear strength of the soil mix material τ_{Rsm}, in accordance with §6.3.5.

The positive influence of the underlying soil pressure within the control perimeter may be taken into account when determining the resistance against punching failure.

The following should apply:

$$V_{Ed} \leq V_{Rd,sm} \tag{6.95}$$

B. Transfer of the vertical anchor force in the soil mix wall:
For anchors that are vertically placed with a specific anchor angle, a vertical anchor force component will be applicable, which should be transferred onto the soil mix wall.

Generally speaking, an extra steel anchor tube is used here and is installed throughout the hole in the wall.

This steel anchor tube is fixed to the steel plate of the steel beam (UNP) and transfers load onto the soil mix wall by means of contact pressure.

As a maximum admissible contact pressure, one employs:

$$\sigma_{sm;adm;d} = f_{sm,d} \tag{6.96}$$

One should comply with :

$$\sigma_{sm;contact} \leq \sigma_{sm;adm;d} \tag{6.97}$$

6.9   Determination of the vertical bearing capacity

6.9.1   *Introduction – Principles*

In addition to the earth or water retaining functions, soil mix elements may also be used for the development of vertical bearing capacity. Soil mix elements can be used to transfer loads from an upper structure (e.g. a building) onto the bearing layers of the subsoil. Additionally, a vertical force may also exert itself onto a soil mix wall from an anchor. Generally speaking, the development of vertical bearing capacity will involve deep embedment of soil mix elements in the soil. Analogous to the approach for pile foundations, with deep embedment, the development of the geotechnical bearing capacity consists of a base resistance and a shaft friction. The following paragraphs include calculation methods for this purpose, elaborated for deep embedment with calculation methods contiguous to both Dutch and Belgian situations. In addition, §6.7.8 includes a number of recommendations for the structural verification of the soil mix wall subjected to vertical forces.

In case of shallow embedment of soil mix elements or block stabilisation, the vertical bearing capacity must be calculated as a shallow foundation or a caisson (drilled piers), in accordance with the applicable standards and guidelines.

Soil mix elements are considered suitable for the development of compression bearing capacity. The verification and calculation of the tensile bearing capacity of soil mix elements should be done with caution.

6.9.2   *Vertical bearing capacity in the event of axial compression load – calculation method NL*

As a calculation method for the bearing capacity, the Koppejan method must be applied for deep embedment. This is extensively shown in NEN 9997-1+C1. For deriving a vertical bearing capacity, soil mix elements are classified into the pile category 'concrete pile, formed in the ground with the aid of supporting liquid, bored or drilled'.

In line with the Deep Walls Handbook (CUR 231), the soil mix elements are thereby classified into the same category as 'diaphragm walls'. The to be deduced pile class factors are shown for (gravelly) sand in I - Table 6.21. For the values for clay, loam and peat, please refer to I - Table 7.d in NEN 9997-1+C1. The listed values can be maintained, unless specified otherwise on the basis of load tests (see §6.9.5).

1 - Table 6.21. Pile class factors for the bearing capacity of soil mix walls in sand and sandy/gravelly soil (in compression).

Description	index	factor
Pile class factor for the shaft resistance	α_s	0.006
Pile class factor for the base	α_p	0.5*
Load settlement line in accordance with figure 7.n and 7.o (NEN 9997-1)	-	3
Pile base form factor	β	1.0
Pile base form factor (ratio cross-section)	s	dependent on the shape of the soil mix element

*From 1-1-2017 the pile class factors α_p for the base will be adjusted in the Netherlands; factor 0.5 will be reduced to a factor of 0.35.

The pile base form factor is dependent on the ratio between the largest and smallest side of the cross-section of the soil mix element or wall and must be determined in accordance with NEN 9997-1 paragraph 7.6.2.3. This factor is important in the case of rectangular elements or continuous walls. As a lower limit, one may maintain a value of s = 0.62 for embedment in a stable sand layer, analogous to those for dam walls in accordance with CUR 166. For individual panels or any other base, the value must be determined separately.

In addition, for the calculation of the vertical bearing capacity of soil mix elements, one must always take into account all aspects which are normally considered for foundations. For example, think of the influence of excavations (reduction of the cone resistance), the presence of negative skin friction, etc.

Considering the installation process of the soil mix element, one must take into account some disturbance and relaxation of the soil at the base and along the shaft. In a normal procedure of the mixing process and with a permanent presence of sufficient internal pressure of the fresh cementitious mixture, the degree of disturbance is usually fairly limited. As a result of any disturbance at the base as a result of the mixing process and in the absence of any displacement or compaction effects, the base may be quite weak. For this reason, a fairly large vertical deformation may be necessary in order to mobilise a significant bearing capacity at the base. This weak behaviour is reflected in the applicable pile class factor. On the basis of a load settlement line 3 (in accordance with NEN 9997-1+C1, figure 7.n), for safety reasons it is assumed that the base resistance has only fully developed at a settlement equal to 20% of the equivalent pile diameter. In case of a round soil mix element with a diameter of 0.55 m, the base of the column must be displaced by 0.11 m to realise the full development of the base resistance. On the other hand, the full shaft friction will be fully developed at an earlier stage, namely at 25 mm of absolute displacement (NEN 9997-1+C1, figure 7.0).

6.9.3 Vertical bearing capacity in the event of axial compression load-calculation method BE

The geotechnical design of pile foundations in Belgium should be carried out in accordance with NBN EN 1997-1 and ANB. In Belgium people have opted for design approach 1 (DA1). The Belgian national annex includes additional information on the semi-empirical calculation methodology on the basis of the CPT as described in (BBRI, 2009). De Beer method is applied for the unit limit bearing capacity at the pile base. The shaft friction is derived from the cone resistance and soil type.

Since there are currently few pile load test results with regards to soil mix elements available, on the basis of the execution principles of soil mix elements it would seem reasonable to include these in "Category III - Piles with soil removal – Bored piles executed with supporting fluid".

For this category, the following empirical installation factors for pile base (α_b) and shaft friction (α_s) apply:

In tertiary over-consolidated clay: $\alpha_b = 0.8$ en $\alpha_s = 0.5$

In other soil types: $\alpha_b = 0.5$ and $\alpha_s = 0.5$.

The calculation methodology as described in (BBRI 2009) also includes the following important points with regards to the vertical geotechnical bearing capacity of soil mix elements:

- determination of the equivalent base diameter D_{eq} of rectangular elements, that must be included in the De Beer method and in the formula for determining the factor ε_b, which calculates the crack degree of tertiary clay;
- determination of the shape factor β (for rectangular elements); for a wall $\beta = 1/1.3 = 0.77$;
- determination of the correlation factors ξ_3 and ξ_4 that are dependent on the penetration test and on the stiffness of the structure; for soil mix walls a constant redistribution of forces is possible (rigid construction) so that one does not have to apply reductions to the values listed in (BBRI, 2009);
- application of the model factor γ_{rd} : for bored piles the model factor γ_{rd} is equal to 1.15 and independent of whether pile load tests have been conducted on the soil mix system or not;
- the application of the partial safety coefficients γ_b and γ_s. For bored piles the following values apply for DA1/1 and DA1/2: $\gamma_b = 1.2$ and $\gamma_s = 1.0$ (DA1/1) and $\gamma_b = 1.65$ and $\gamma_s = 1.35$ (DA1/2). The factor γ_b may be reduced up to 1.0 (DA1/1) and 1.35 (DA1/2) in case the execution of the piles is done in accordance with a 'quality guarantee system'. Guidelines with regard to a 'quality guarantee system' are included in Chapter 8.

Annex 2 of (BBRI, 2009) also includes procedures and conditions that on the basis of pile load tests (also see §6.9.5) allow one to apply appropriate (improved) installation factors in the proposed calculation methodology.

On the other hand, for a specific project it is also possible to base the geotechnical design on prior pile load tests. More details about the method and factors to be used are included in the ANB of the NBN EN 1997-1 (see §6.9.5).

It is important to note for the calculation methodology in (BBRI, 2009) that this only applies for the geotechnical bearing capacity in ULS and that the calculation method is calibrated on the basis of a conventional failure criterion, namely that the failure load of a pile foundation corresponds to the load under which a pile base settlement of 10% of the pile base diameter occurs. This applies to all pile categories.

In addition to the ULS, the SLS must also be tested. However, (BBRI, 2009) does not yet include a methodology for the SLS. If relevant, the group effect should also be tested; similarly, no methodology is yet included in (BBRI, 2009).

Finally, it should be noted that for the verification of the structural bearing capacity of vertically-loaded soil mix elements, combination 1 (DA1/1) is decisive. §6.7.8 includes a number of guidelines for the structural verification of the soil mix wall subjected to vertical forces.

6.9.4   *Vertical bearing capacity in combination with a retaining function*

In case of a combination of vertical bearing capacity and a earth and water-retaining function, the design of the wall must be verified in greater detail. If the vertical load is relatively small, an approach in which the bending and vertical balance can be geotechnically considered separately is sufficient. In line with the approach to sheet pile walls, a separate calculation of the horizontal and vertical stability may be sufficient in case the average shear stress as a result of the external vertical load is ≤ 12.5 kN/m², for more information see the more detailed descriptions in §4.10.10 of CUR 166, part 2. This average shear stress τ is calculated from $\tau = V_s /(2L - l)$, with V_s = the external vertical load per 'm wall, L = the total height of the wall and l = the soil-retaining height of the wall.

A vertical compression load on a soil mix wall may be introduced from external loads (upper construction), but also from the anchoring on the wall (vertical load component). Depending on the interaction with the upper construction, it is also possible that a head moment and horizontal force acts on the upper section of the soil mix wall. In all cases, an extensive constructive verification must be carried out, in which one must demonstrate that the forces are geotechnically and structurally absorbable by the soil mix wall and the corresponding steel beams.

In most cases, with earth-retaining walls, the wall friction on the active side will be directed downwards and on the passive side upwards. If a vertical load is exerted on the wall, then the direction of the wall friction may change. This may influence the horizontal earth pressures on the wall. Due to the weak behaviour of the base resistance, the wall friction will occur earlier with soil mix elements.

As a result, a wall that is subject to heavy vertical loads will mainly transfer the load onto the upper section, making the wall friction angle on the active side change. Consequently, the horizontal load on the wall will increase significantly, making the bending moment, stamp and/or anchor forces greater (CUR 231). If a so-called 'spring model' is used for the calculations, please refer to CUR 166 for more information on how to use this model.

For more heavily loaded walls, introduction of loads from the superstructure or critical cross-sections, the interaction between the vertical and horizontal behaviour should be assessed in more detail. In such situations, a calculation with a finite element model (FEM) must be conducted, given that the direction, height and distribution of the wall friction becomes clear from the calculations, without these having to be calculated for each individual load phase. When using an FEM, the safety consideration of the development of force in the wall must be in accordance with what has been described and is required on the basis of NEN 9997-1+C1 and NBN EN 1997-1 + ANB. Here, the minimum variations in retaining height, water levels, terrain loads, design values with upper and lower limit values for the stiffness of the soil parameters, etc. must be taken into account.

For more information about the design and influence of vertical loads in combination with bending loads, please refer to CUR 166 and the Diaphragm Walls Handbook (CUR 231).

6.9.5   Pile load tests on soil mix elements

For the more detailed calculation of the vertical bearing capacity of soil mix elements, it is possible to conduct on-site load tests. For this purpose, one can realise test elements on the site, on which – with the assistance of jacks – loads are applied. The displacements are consequently measured with high accuracy. Load tests must be conducted in situations described in par. 7.5.1, paragraph 1 of NEN 9997-1+C1 and NBN EN 1997-1 + ANB. Given the limit experience with vertical bearing capacity of soil mix panels, conducting pile load tests may be advantageous in optimising design and execution. Tests are especially important in situations with a large-scale use of soil mix as foundation element, where there are uncertainties about the soil mix's behaviour or with a high risk profile.

Part 1 of Eurocode 7, par. 7.5 includes the approach and procedures with regards to pile load tests (NEN 9997-1+C1 and NBN EN 1997-1 + ANB). This procedure may also be applied for soil mix elements. For the load procedure it is important to carefully consider the size of the load steps, so that the collapse/failure of an element can be examined as accurately as possible. After obtaining the results of the test loads, one must determine the failure bearing capacity. On the basis of the method described in part 1 of Eurocode 7, the results of various test loads must be assessed for acquiring the characteristic values ($R_{c;k}$) by taking into account factors on the minimum and average obtained collapse/failure values such as ξ_1 and ξ_2 (in statical pile load tests). Because of the need to take into account ξ factors, which become more favourable with an increasing number of tests, it is recommended to conduct multiple tests. The design value of the resistance $R_{c;d}$ (bearing capacity) must then be calculated by taking into account a partial resistance factor γ_R.

On the basis of loads tests, this factor for 'augered piles' is set to 1.2 in the Dutch Annex. The different factors to be taken into account with loads test are different for the Netherlands and Belgium.

6.10   Influence of the wall installation on adjacent foundations

When executing a soil mix wall directly adjacent to a foundation, one must take into account the influence of the wall on the foundation. The bearing capacity and stability of the existing foundation may temporarily be lowered and deformations of the foundation may occur. The wall realisation is characterised by alternately executing panels. During the intensive mixing of the soil with cement, the panel will act as a heavy liquid with a unit weight of order 19 kN/m³ and a negligible shear strength, that is to say an angle of internal friction and cohesion almost equal to zero. After hardening of the panels, the strength and stiffness will quickly increase and the adjacent panel is executed.

In order to determine the order of magnitude and negative influence, two preliminary calculations have been made for the realisation of a soil mix wall next to a shallow foundation. PLAXIS 3D allows for excellent modelling and calculation of this problem. The calculation includes a wall with a thickness of 0.55 m and a panel width of 2.4 m. Every panel allows for two units of HEA400 beams with a spacing of 1.1 m. The distance to the beam in the adjacent panel is also 1.1. The steel beams are not included in the calculation model. Next to the soil mix wall is a strip foundation with a width of 1.0 m. The distance between the side of the strip foundation and the side of the wall has been designated as 1.0 m. The strip foundation itself is uniformly loaded with 120 kPa. The phasing of the calculation is as follows :

1. Initial stress state in the soil;
2. applying load at the location of the foundation;
3. excavating/mixing centre panel (undrained); with the chosen geometry, the excavation of the middle panel includes a simultaneous excavation of panels with a centre-to-centre distance of 2.2 m;
4. middle panel is 1 day old and excavation/mixing adjacent panels (undrained).

The preliminary calculations are executed for a homogeneous subsurface consisting of moderately dense sands and moderately hard clay. The selected soil parameters are indicated in I - Table 6.22.

I - Table 6.22. Parameters for the orienting calculation.

Material model		Sand, moderately firm Hardening Soil	Sand, moderately firm Hardening Soil	Soil Mix fresh Mohr-Coulomb	Soil Mix 1 day old Mohr-Coulomb
		Drained	Undrained	Undrained	Undrained
γ_{unsat}	kN/m³	17	20	19	19
γ_{sat}	kN/m³	20	20	19	19
c'	kN/m²	0.1	1.0	0.1	[1]Sand: 200 Loam 150
φ'	°	32.5	27.5	10	0
ψ	°	2.5	0	0	0
Tensile strength	kN/m²	0	0	0	0
E	kN/m²	-	-	200	[2]Sand 712000 Loam 566000
E_{50}^{ref}	kN/m²	45000	30000	-	-
E_{oed}^{ref}	kN/m²	45000	15000	-	-
E_{ur}^{rewf}	kN/m²	135000	90000	-	-

[1] Based on UCS$_{1 day}$: Sand 0.4 MPa and Loam 0.3 MPa
[2] Based on E = 1482 UCS$^{0.5}$

154

Figure 6.35 includes the geometry of the phases 2, 3 and 4.

Phase	Description	
2	Load at the location of the foundation • strip width is 1.0 m • distance to panel is 1.0 m • load is 120 kPa	
3	Excavating/mixing mid-panel • unit weight is 19 kN/m³ • angle of internal friction is 10° • cohesion is 0 kPa	
4	Excavating/mixing side-panels • unit weight is 19 kN/m³ • angle of internal friction is 10° • cohesion is 0 kPa • mid-panel is 1 day old • angle of internal friction is 0° • cohesion is 200 kPa, 150 kPa for Sand and Loam respectively	

I - Fig. 6.35. Plaxis 3D calculation influence adjacent foundations by installation of soil mix wall geometry of the phases 2, 3 and 4.

The calculation results for the moderately firm sand package result in the following :

- When executing soil mix process, limited vertical deformation occurs at the location of the panel as a result of the greater weight of soil mix compared to the original subsurface. The vertical deformation also causes limited vertical deformation next to the panel, including at the foundation. In addition, a horizontal displacement occurs, perpendicular to the wall towards the outside. This is caused by too much horizontal stress in the panel compared to the soil (higher unit weight and K_0 of almost 1.0).
- The settlement of the foundation is about 1.2 mm after creating the mid-panel and a total of approx. 1.6 after realising the complete wall.
- If all panels are excavated/mixed in one phase, vertical deformation of approx. 4.4 mm will occur at the location of the foundation.

In the figure 6.36, typical results are presented.

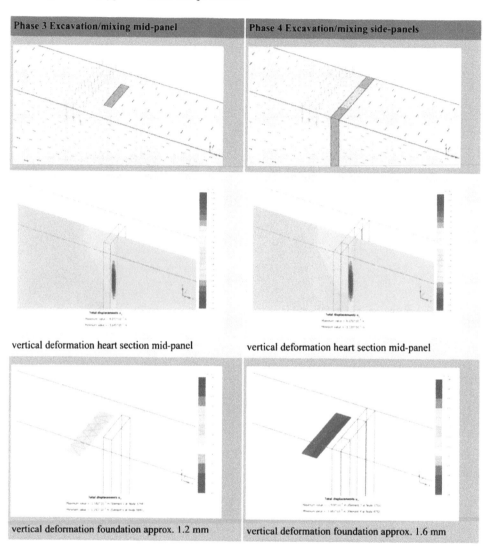

vertical deformation heart section mid-panel   vertical deformation heart section mid-panel

vertical deformation foundation approx. 1.2 mm   vertical deformation foundation approx. 1.6 mm

I - Fig. 6.36. Plaxis 3D calculation of the influence on adjacent foundations by installation of soil mix wall – characteristic results of the phases 3 and 4.

The results in the moderately firm loam subsurface show greater deformations:
- The settlement of the foundation is about 2.9 mm after creating the mid-panel and a total of approx. 3.8 after realising the complete wall.
- If all panels are excavated/mixed in one phase, vertical deformation of approx. 9.5 mm will occur at the location of the foundation.

156

It should be noted that when applying the foundation load on the surface in question, a vertical deformation of approx. 44 mm of the foundation occurs.

It is concluded that the influence of executing the soil mix wall must be analysed further if shallow foundations, wells or piles are located near the wall. The construction phases can fairly easily be modelled in a phased 3D finite element calculation.

6.11   Constructive design aspects

6.11.1 Corner solutions

Constructive attention is drawn to the realisation of a corner solution.
The following aspects are at play:
- the formation of a pressure arch in the horizontal surface of the cross-section;
- single-sided stresses affecting the steel beam;
- the absorption of a load in the surface of the soil mix wall in case of a stamped solution.

This paragraph describes the relevant aspects and possible measures. It includes several examples of solutions, but of course, other well-researched solutions are also possible.

A distinction is made between an internal and external corner (see figure 6.37).

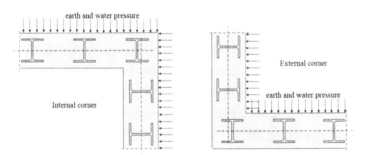

1 - Fig.6.37. Internal and external corners in a soil mix wall.

a) The formation of a pressure arch in the corners
The corners lack a flange on 1 side of the corner beam for the base of a pressure arch. Nevertheless, pressure arching may occur, as shown in 6.38.

In particular, the emergence of the displayed half-pressure arch in the external corner is dependent on the maximum friction between the steel of the beam and the soil mix material.

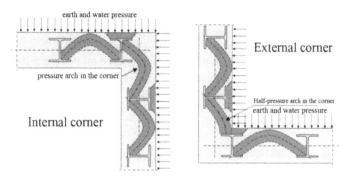

I - Fig. 6.38. Possible pressure arches in the corners.

b) Single-sided stresses affecting the steel beam

In case of a continuous wall, both sides of the web of the beam will engage a pressure arch. This results in an equilibrium of forces, because this load is two-sided and is assumed to be equal in magnitude.

With a corner solution this is not the case. On the one side there is no pressure arch present, or there is a more limited pressure arch with much lower stresses. Here, under influence of the pressure arch, the beam will be pressed outwards. This is especially the case with external corners, see figure 6.39.

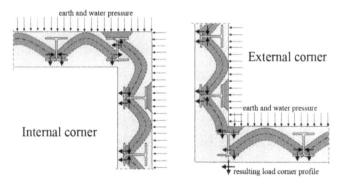

I - Fig. 6.39. Affecting loads from the pressure arches at the location of an internal and external corner.

The steel beam of a continuous wall (intermediate beam) should be able to absorb 2x the load of a pressure arch around the strong axis. In this situation, it is embedded into the soil and, if necessary, supported by a strut frame and/or anchoring at the top.

For the corner solutions given, the beam placed in the corner must always be able to absorb a pressure arch load around the strong axis. This does not differ much from the beam of the continuous wall.

In the other, weaker, direction a resultant pressure arch load is also present in an external corner.

158

In order to limit the governing internal forces in this weak axis, besides the embedment at the lower side in the ground, additional support is considered on the top in the form of a welded steel frame or application of an anchor, which absorbs part of the load and limits the mutual displacements, see figure 6.40.

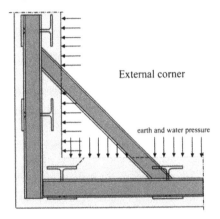

External corner

earth and water pressure

1 - Fig. 6.40. Welded steel frame on the top at the location of an external corner.

However, for the design of the steel, the corner beam must be assessed with respect to the corresponding force effect around the weak axis. With an external corner, one may take into account a reduced earth pressure, because the stresses spread occurs in two directions here.

c) The absorption of a load in the surface of the soil mix wall in case of a stamped solution
If the soil mix wall can act as a free-standing (cantilever) wall, each steel beam absorbs the loads from its own zones.

For corner solutions with strut frames, the situation is different. The load on the strut frame from the longitudinal wall is guided to both ends of the strut frame. Here, the load is transferred onto the connecting purlins. These connecting purlins can subsequently transfer this load onto the soil mix wall, creating a wall horizontally loaded onto the surface, see figure 6.41.

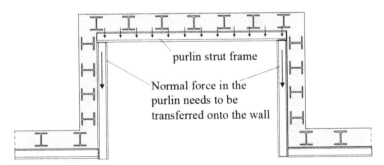

purlin strut frame

Normal force in the purlin needs to be transferred onto the wall

1 - Fig. 6.41. Principle transfer of load from the strut frame onto the cross-walls.

As a result of the the load transfer, the cross-walls will be loaded onto the surface. CUR 166 describes how this load can be absorbed with regards to sheet pile walls (sheet pile horizontally loaded onto the surface). Because it is assumed that sheet pile wall locks have little shear resistance, portal elements are created for sheet pile walls, by rigidly fixing the foundation elements on the top side. Consequently, a slice is created that allows for more efficient absorption of the load in question.

This same principle can also be applied for the soil mix wall. In this situation, the beams must be attached to each other rigidly on the top side. This can be done, for example, by placing the beams rigidly in a concrete covering cap. Additionally, a horizontal continuous steel beam can also be placed above the heads of the steel beams of the soil mix wall, which is rigidly fixed/welded onto each beam. In order to transfer the load between the purlin and the steel beams of the cross-wall, the purlin must also be attached to the vertical steel beams of this cross-wall. This can be done by means of intermediate welding of the steel plates. A solution with the use of anchors in order to absorb the resultant horizontal force is of course also possible.

d) Potential cracking of external corner
As a result of the two-sided force effects and displacement, cracking may occur in the soil mix material of an external corner. One must therefore consider this aspect when positioning the beams and take measures to prevent/control this situation.

One solution may be a welded steel frame, as shown in 6.40, or anchoring. Another crack-limiting solution is a reinforcement cage hoisted over the retaining height in the correspondingcorner. An example of an application of this solution is shown in figure 6.42. A focus point here is that, simultaneously, two panels should be diametrically opposed to each other.

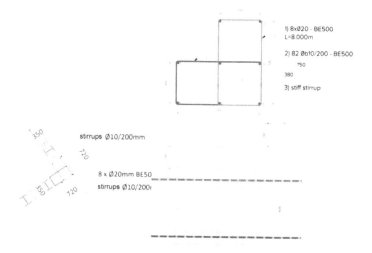

I - Fig. 6.42. Example application of a reinforcement cage as part of the corner solution.

6.11.2 Lower side of the beam

The steel beam does not need to continue up until the underside of the soil mix wall.

On the contrary, analogous to the application of reinforcements in a diaphragm wall, it is recommended to apply at least a 0.2 m distance between the bottom of the soil mix material and the level of the end of the steel beam (in accordance with EN 1538 en CUR 231). Here, the reinforcement must hang downwards, so that it is applied upright under its own weight. When placing on the bottom of the trench, an initial compression force can be introduced into the reinforcement, as a result of which it may undergo buckling. That can also be the result of an uneven bottom surface.

The reinforcement can be stopped earlier if, for the bottom end of the wall, the soil mix material is able to absorb loads without cracking. At any rate, an anchorage length should always be used in these cases, see figure 6.43. An axial tensile strength occurs in the flange of the steel beam in the cross-section of the bottom side, where there is slight cracking in the tensile zone. This must be transferred onto the soil mix material over a requisite anchorage length l_v outside the cracked zone.

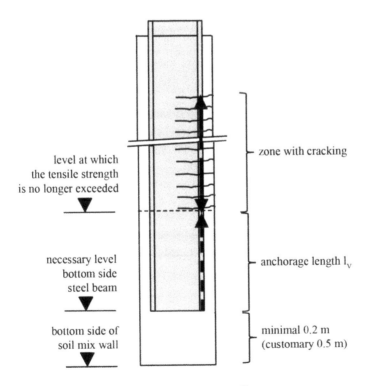

1 - Fig. 6.43. Lower side of the beam in a soil mix wall.

Determination of the lower side of the cracked zone:
This is the cross-section, in which the tensile stress in the soil mix material is limited to $f_{sm,td}$.

$f_{sm,td}$: design value of the tensile strength of the soil mix material in accordance with §6.3.4.

The corresponding moment at which cracks arise is:

$$M_{r;sm} = 1/6 * b_{c2} * h_{sm}^2 * f_{sm,td} \tag{6.98}$$

Determination of the anchorage length:
Under the cross-section where $M_d = M_{r;sm}$ applies, an anchorage length of l_v must be applied. Here, a minimum of $l_v = 500$ mm should be applied.

The hereby resulting shear stress around the circumference of the flange with a common steel beam is, for a maximum admissible tensile load, smaller than, for example, an equivalent round rod with minimum anchorage length $l_{v;min} = 10\ D_{eq}$ in accordance with Eurocode 2. The length of 500 mm is therefore considered as practically applicable.

Determination of the minimum required length of the steel beam:
For determining the height of the wall, which should be equipped with reinforcement, the following calculation methodology can be used:
- On the basis of a constructive calculation, one must determine what the minimum required embedment depth is, in which one complies with the requirements with regards to stability, strength and deformation.
- On the basis of this calculation, it is determined at what level on the bottom side of the wall the moment at which cracks arise $M_{r;sm}$ is reached, (see figure 6.44).
- Below this level, an anchorage length $l_v = $ of 500 mm should be calculated for determining the minimum required steel beam length.
- The height between the bottom side of the wall and bottom side of the steel beam should be at least 0.2 m. If necessary, the insertion depth should be increased for this purpose (recalculation not necessary).
- Subsequently, if applicable, the wall are milled even deeper, for example, to reach a "impervious" layer. This over-length is not included in the constructive calculations.

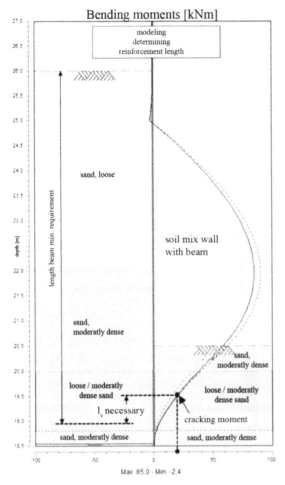

I - Fig. 6.44. Overview modelling and minimum required steel beam length.

6.11.3 Constructive finishing aspects for permanent retaining function

For soil mix walls with a long-term retaining function, a number of constructive finishing aspects should be considered for the benefit of the long-term behaviour of the wall. It is particularly important to note:

1. Any corrective measures, required as a result of material or execution deviations in the wall;
2. Measures for long-term transfer of the applied loads on the wall;
3. Applying a finishing layer or protection barrier/wall for aesthetic reasons and/or for ensuring the desired earth or water retaining functions in the long term.

1. Corrective measures

Corrective measures (if necessary also for temporary functions) may be required if the wall properties or geometry exhibit characteristics due to which the proper functioning or stability of the wall is compromised. Think of the following: major leaks, visible presence of large soil inclusions or poor hardening, deviations in the reinforcement, protrusions of the wall within the excavation section etc. The required corrective measures must be considered and analysed for each specific case.

2. Transfer of forces

When introducing vertical loads (and potential head moments or horizontal forces), it is of great importance that the loads are transferred onto the soil mix wall in a staggered fashion. Here, one must not forget that the properties of the soil mix material are more limited and heterogeneous compared to a concrete diaphragm wall or secant pile wall.

For the transfer of **vertical loads due to the superstructure,** usually a rigid reinforced concrete head beam is placed on the head of the soil mix wall. This head beam is applied after optional sweeping and cleaning of the head of the soil mix wall and clearing the steel beams. Under appropriate conditions (please see the guidelines in § 6.7.8), for the design of the head beam one may assume that the top loads are partly transferred onto the soil mix material and partly transferred onto the steel beams. The load distribution and consequent bending moments in the head beam can be based on the ratio of the stiffness of both elements (elastic approach) or on the ratio of the compressive strengths of both elements (plastic approach). At the location of the beams, necessary constructional precautions must be taken (free beams, welded reinforcement rods or dowels) to ensure that the load transfers onto the beams are adequately safe (also see principle sketches in figure 6.45).

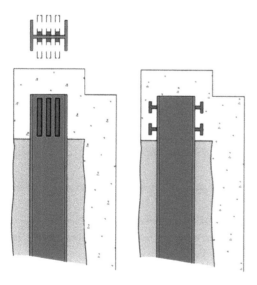

I - Fig. 6.45. Principles for applying a head beam with welded rods (left) or dowels (right).

164

When introducing the external loads onto the wall (N and potentially also M and H), one must assess – after distribution of the loads and in combination with the internal forces due to the earth-retaining function – whether the compressive strength of the soil mix material and steel stress are not exceeded at any level. Here one may calculate using a composite cross-section, where in suitable conditions (see §6.3.6 and §6.7.4 to §6.7.8) one may assume an interaction between the soil mix material and the steel beams. In the other case, one must only calculate using the reinforcement.

The transfer of **vertical loads from the underground floors** and the bottom plate of the soil mix wall can be done by welding steel connecting elements (e.g. reinforcement rods or steel beams) in between the floors and the steel beams of the soil mix wall (see example sketches in figure 6.46).

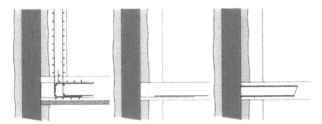

1 - Fig. 6.46. Principle examples of the constructive detailing of the connection between the base floor (left) and the intermediate floors by means of welded reinforcement rods (middle) and welded beams e.g. HEM100 (right).

The **long term horizontal stability** can be ensured by means of definitive ground anchors. However, in most cases the temporary supports (anchors, struts, support beams…) are replaced with the supporting function of the underground floors.

These floors either directly rest on the reinforcement of the soil mix wall, or indirectly on the protection barrier/wall which itself rests on the reinforcement of the soil mix wall.

3. Earth and/or water retaining function – finishing visible front area - covering
The finishing of the visible side area of the soil mix wall has one or more functions:
- creating an aesthetic touch
- preventing pulverisation of the front area
- protecting the soil mix wall against climatic influences: freeze/thaw effects, drying out
- protecting the wall against accidental damages, e.g. caused by collisions, fire etc.
- the transfer of horizontal (earth pressures and/or water pressures) and vertical loads.

Potential finishings include:
- aesthetics: milling front area, shotcrete, protection barrier/wall of concrete or masonry
- pulverisation: coating, shotcrete
- climate protection: coating, shotcrete, protection barrier/wall (with or without cavity)
- accidental damage: concrete protection barrier/wall
- water-retaining: impervious protection barrier/wall or protection barrier/wall with drainage cavity
- earth-retaining: cooperating protection barrier/wall, with or without lateral force connections in between the soil mix wall and protection barrier/wall.

Figure 6.47 includes a number of example sketches of the constructive detailing of the connection between a protection barrier/wall with the soil mix wall.

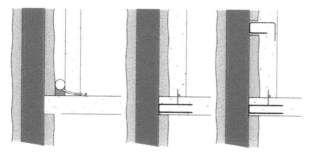

1 - Fig. 6.47. Principle examples of the constructive detailing of protection barrier/wall.

6.12 Watertightness design aspects

If soil mix walls are applied as cut-off walls, there are a number of important design aspects that should be taken into account. In this case, the wall has a primary cut-off function, but is supported on both sides by soil (initially ground level, embankments or support banks).

If the wall only has an earth-water cut-off function, the compressive strength of the soil mix can remain low (e.g. 0.5 - 1.5 MPa). In certain conditions it is also possible to use walls with an even lower compressive strength. In such circumstances, one must adequately examine the lifetime, risks of deformation in the context of the environment and macro-stability. This may be a possibility for soil mix walls with a temporary function in the construction phase (< 2 years) and a low risk profile.

The cut-off soil mix wall is then executed without reinforcement on both sides due to the soil support. At the same time, this means that the embankment stability of excavations near the soil mix wall is of great importance. If the stability of excavations near the soil mix walls in the construction or final phase is not sufficient, significant damage to the wall and function loss may occur by sliding. For this reason it is important that the stability of (temporary) excavations is computed in respect to execution.

In addition to stability of the embankments, deformation of the wall is also a major factor. Due to differences in earth and water pressures on the wall, the wall will start to deform. It is important that the interaction between construction phasing, water pressures, any elements in the wall (e.g. a geomembrane) and stiffness differences in corners and connections to other structures, are adequately analysed for each (construction) phase.

After all, it is important that the cut-off wall deforms gradually across all construction and use phases. Major curvature in the wall results in unacceptably large bending and tensile stresses and in potential crack formation. Curvature of the soil mix wall may occur as a result of local excavations (e.g. trenches), loads caused by material or traffic, or large stiffness differences in the transition from natural soil layers. Suitable software for comprehensive consideration of the stability of present embankments and the deformation behaviour is a finite element model (FEM), in which the soil behaviour and the walls can be adequately modelled. An example of an FEM model of a water-retaining soil mix wall with Plaxis is included in I - Fig. 6.48. The calculated deformations are given in I - Fig. 6.49.

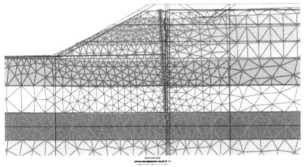

I - Fig. 6.48. Example model of a cut-off soil mix seal wall with the assistance of a finite element model (Plaxis).

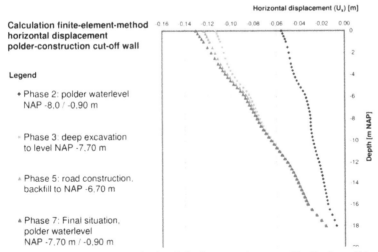

Horizontal displacement (U$_x$) [m]

Calculation finite-element-method horizontal displacement polder-construction cut-off wall

Legend

◆ Phase 2: polder waterlevel NAP -8,0 / -0,90 m

▪ Phase 3: deep excavation to level NAP -7,70 m

▲ Phase 5: road construction, backfill to NAP -6,70 m

▲ Phase 7: Final situation, polder waterlevel NAP -7,70 m / -0,90 m

I - Fig. 6.49. Example of a calculated displacement in a cut-off soil mix wall in the different construction phases (Plaxis).

If deformations across the wall during the design are determined per construction phase, feedback can be established with the actual deformation behaviour during execution. This guarantees that the deformation behaviour, in practice, corresponds to what is determined in the design. For monitoring techniques that include, for example, placement of inclinometers in the wall, please refer to paragraph 8.7.

For useable results of the modelling, the characteristics of the soil mix material and soil layers must be determined as accurately as possible (see paragraph §2.3 in part 2 of the handbook and §6.3). For determining the sensitivity on the deformations, it is recommended to make calculations with both upper and lower values of the characteristics of the soil mix wall (strength and stiffness parameters). For insight into relations between strength and stiffness parameters, please refer to paragraph 2.3 in part 2 of the handbook. The stiffness of the soil mix wall is furthermore influenced by the thickness of the wall. The choice of wall thickness and strength characteristics determines the stiffness and therefore the deformation behaviour. In addition to the characteristics of the soil mix material, the strength and stiffness parameters of natural soil layers are also of major importance. Depending on the strength and stiffness of the soil mix wall, these will have major influences on the predicted deformation behaviour.

The calculated moments can be verified at the cracking moment in accordance with paragraph 6.11.2. If the cracking moment is exceeded, the wall may be equipped with a geomembrane or the thickness and the stiffness parameters of the wall may be varied. It should be noted that potential cracking as a result of bending/deformation will only occur on one side (= on tensile side of the wall => outside the excavation). Crack formation, under the influence of bending-compressive stresses of the deformation and the compressive stress resulting from the settlement of the soil and negative friction on the wall due to water lowering, will not occur on the excavation side (=compressive side). Any one-sided (micro) crack formations on the exterior of the wall should not cause any problem in terms of the functionality of the wall.

If soil mix walls are used as water-retaining cut-off walls, it should be apparent that the permeability of the soil mix material is a crucial aspect in the design. For example, the soil mix walls are used as vertical screen with placement in a horizontal "impervious" soil layer (creating a natural polder). From a D&C contract, the client may place a ceiling on the leakage rate of the overall structure (e.g. 20 to 100 m^3/day). In determining the admissible leakage rate for the construction, one must take into account permit issues (e.g. water permits), environmental factors and the possibility of draining the leakage rate or returning it to the soil. With the aid of hydrogeological calculations, in the design one must demonstrate that the maximum leakage rate is realised (verification) with the applicable geometry and wall properties. The total rate includes a component related to 'leakage' through the wall and through a horizontal "impervious" soil layer.

It is therefore highly important that the characteristics and potential deviations in a naturally "impervious" soil layer are meticulously taken into account in the design process. The total leakage rate can be determined on the basis of data on the permeability of the walls and the "impervious" soil layer. Depending on the project, these will mainly consist of sophisticated hydrogeological calculations in a 2D or 3D groundwater flow model (e.g. Microfem, Plaxis, etc). Additionally, when performing the design, it is recommended to consider the 'realistic' permeability of the wall, also see I - Table 3.2 in paragraph 3.2.6. The range of permeability (k value) may be located between 1×10^{-7} and 10^{-10} m/s. It it also recommended to include calculations with upper and lower limit parameters, so that one can gain an insight into the sensitivity of the total leakage rate. Experiences with calculations with regards to the permeability of various projects and authors are included in paragraph 2.3.9 in part 2 of the handbook. It should be noted that the permeability of the walls in the field will be greater than the permeability values determined on small samples in a laboratory. Important aspects influencing this may be the homogeneity of the soil mix wall (presence of inclusions) and the proper overlap of soil mix panels. If panels, dependent on verticality/execution method, do not properly connect, it should be apparent that large leaks may arise locally. In such a case, mitigating measures are required. With the aid of full-scale pumping tests (see paragraph 8.8), the leakage rate of compartments and total structure can be determined and compared with the calculated leakage rate from the design process.

In case of embankments against the soil mix wall, one should be alert to the effect of bulging of the groundwater level. On the excavation side of the soil mix wall, a polder level will generally be set.

Depending on the base and position of the dewatering system, a bulge of the ground-water may occur in the embankment. This bulge may arise as a result of input from above (rain), potential leakage through the wall or leakage above the wall. This latter phenomenon may occur in the event that the groundwater level is greater outside the pit than the top of the wall. A bulge of the groundwater level in the embankment will have a major impact on the embankment stability, and hence the safety of the structure. For maximum deformations of the wall, the situation is governing when considering a flat polder level. For the design of the wall, both a situation with and without a bulge of the ground water level should be taken into account.

If geomembranes are used in the soil mix wall (see paragraph 2.2.6 in part 2 of the handbook), the effect thereof should be included in the FEM calculations. The material of the geomembrane used is very smooth, as a result of which a surface with a highly reduced strength will appear on the boundary area between the geomembrane and soil mix material. Subsequently, the adhesion between the materials will significantly decrease (decreased interface strength).

In case of greater bending of the wall, the wall may act as a 'sandwich construction', in which, on both sides of the geomembrane, the wall elements will theoretically be able to shift independently in relation to each other (two walls with half the thickness of the total wall). One should consider whether or not this will lead to significant shear stresses in the geomembrane or undesired behaviour of the wall. No values are currently available for the value of the lowered interface strength. The most conservative starting point in the calculations is to allow for no adhesion on the surface between the geomembrane and the soil mix wall. If the calculations show that this will result in undesirable behaviour, the value may be updated by conducting lab tests (shear test on adhesion between geomembrane surface – soil mix material).

Depending on the on-site surface and groundwater levels, a finishing level for the soil mix wall should be determined. If the critical groundwater level in the area is high, the finishing level of the soil mix wall should be slightly beneath the surface level. In such a case, one should pay close attention to the protection of the top zone of the soil mix as well as protection against meteorological influences (freezing/thaw effects up to a depth of approx. 0.8 m). This may affect the cut-off function of the wall. For this purpose, certain measures can be taken such as raising the surface level above the walls, applying a short geomembrane, or wrapping the head of the soil mix wall in a clayey form. Additionally, for walls located slightly beneath the surface level there is the possibility of mechanical damage of the wall during nearby excavation or drilling activities. In such a situation, for the protection of the head of the wall one may choose for finishing with a concrete block mattress (concrete tiles on a geotextile).

Such mattresses have a great resistance to mechanical damage, and function as a 'warning' indicating the presence of the underneath wall during future excavation activities.

6.13   Fire

When soil mix walls are built in a situation where the soil mix material may be affected by fire, the design must be adequately verified for fire safety. For conducting these calculations, certain principles must be determined with regards to:
* the verification framework
* the fire load
* the critical cross-sections
* the material characteristics.

Subsequently, one can calculate the fire safety.

6.13.1 Verification framework

The verification framework is defined in the Eurocodes.
* Eurocode 1: EN 1991-1-2 (+NA): Actions on structures – Part 1-2 General actions – Actions on structures exposed to fire.
* Eurocode 2: EN 1992-1-2 (+NA): Design of concrete structures – Part 1-2 General rules – Structural fire design.
* Eurocode 3: EN 1993-1-2 (+NA): Design of steel structures – Part 1-2 General rules – Structural fire design.
* Eurocode 4: EN 1994-1-2 (+NA): Design of composite steel and concrete structures – Part 1-2 General rules – Structural fire design.

Since these design standards for fire are not established for soil mix, one has to make a decision. If steel beams are included in the soil mix walls, applying Eurocode 4 seems the most suitable. Naturally, when assessing fire safety one must take into account the differences in material properties between soil mix and concrete.

If the walls are used with a structural function, the following aspects should also be considered:

- For structures in the Netherlands, the Guideline for the Design of structures (Richtlijnen Ontwerp Kunstwerken – ROK) by Rijkswaterstaat (RWS) applies. This ROK does not include any additional requirements with respect to EN 1994-1-2.
- The ROK is written for concrete structures and, in addition to Eurocode 2, lists (unless specified otherwise in the query specification) the following requirements during 120 minutes for tunnels:
  - for the closed part of tunnels: the RWS fire curve (EN 1991-1-2/NA, art. 3.2.4);
  - for the non-closed section (ramps): the hydrocarbon curve (EN 1991-1-2, art. 3.2.3).

6.13.2 Fire load

Different curves have been established (see figure 6.50) for the temperature development during a fire situation, such as:

- Standard fire curve (ISO 834)
- Hydrocarbon curve
- Rijkswaterstaat curve

Standard fire curve
The standard fire curve in accordance with ISO 834 is the most common scenario for assessing the fire resistance of buildings. This is a ventilation controlled natural fire, with a cellulose-containing fuel (wood, paper, etc.). The temperature increase after 30 minutes is 842 °C.

Hydrocarbon curve
This fire curve is a ventilated gasoline fire with a rapid rise in temperature of 900°C after 5 minutes and a peak of temperature of 1110 °C. This fire scenario represents a fire load of 200 kW/m² (heat flow per m²).

Rijkswaterstaat curve
This fire curve established by Rijkswaterstaat is the most severe fire load for use in tunnels. This represents a fire load of 300 MW (total heat flow), in which a rapid temperature rise of 1200 °C and a peak of temperature of 1350 °C is achieved.

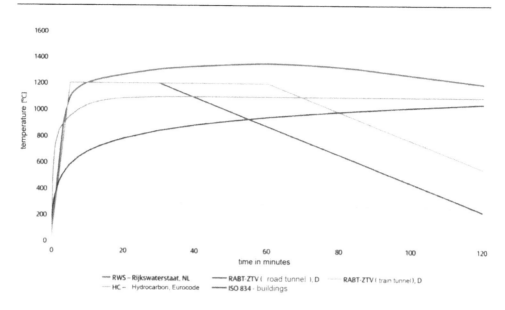

1 - Fig. 6.50. Fire curves.

Research on concrete structures has demonstrated that in order to meet the Rijkswaterstaat (RWS) curve, a cover of 75 mm is required for a fire that lasts 60 minutes. Taking into account the effect of the fire on the concrete, the minimum required cover should be 100 mm. For a 120 minute fire, the minimum required concrete cover should be 125 mm.

Depending on the application of the soil mix walls, one of these fire curves should be taken into account during the considerations.

6.13.3 Cross-sections

Steel beams are often placed in soil mix walls for absorbing the bending moment. For example, as a result of earth pressures. The height of these moments may vary along the length of the wall. When assessing the fire safety, the most critical cross-sections must be accurately analysed.

During the analysis, specific attention should be given to the positioning of the steel beams in the wall (centric / eccentric), the cover on the steel beams, and measures taken to protect the wall during the use phase against environmental influences (protection barrier/wall, shotcrete, coating, ...). In relation to this cover, one must also take into account the positioning tolerance of the steel beams in the soil mix walls, in which tolerances of approx. 50 mm are not uncommon (see §8.2).

Depending on the measures taken to protect the wall against environmental influences and/or potential presence of local contaminations in the use phase, it may also be appropriate to take into account an extra loss of soil mix cover.

172

6.13.4 Material characteristics

For analysis of the behaviour of the soil mix wall during fire loads, there is currently little information available. However, it is clear that soil mix material behaves differently than concrete. The focus point here is that the material characteristics differ per location, because the (soil) composition of the soil mix varies. In terms of the fire resistance, there are two factors at play:
- The thermal conductivity of the soil mix material
  (in relation to the cover on the steel beams).
- The burst behaviour or material loss as a result of fire.

Thermal conductivity
The thermal conductivity of the soil mix material is mainly determined by the soil present at the location of the wall and the amount of injected binder. A common value for the thermal conductivity coefficient of soil mix is therefore not readily available. In comparison with concrete, soil mix is characterised by a relatively low strength and high porosity. For example, the presence of stones and/or specific soil types (such as peat, loam, etc.) may have a relevant influence. This means that, for each project, the soil mix material should be considered as an isolated cementitious material. In case of a sandy soil, one may be able to create an indicative comparison with a cement plaster or cement-based floor (see I - Table 6.23).

I - Table 6.23. Material characteristics.

Material	Specific weight (kg/m³)	Thermal conductivity coefficient λ (W/mK)	Heat capacity c (J/kgK)
Concrete	2400	1.70 – 2.20	840
Cement plaster	1900	0.93 – 1.50	840
Aerated concrete	1300	0.52 – 1.20	840

Bursting behaviour
With concrete, the moisture inside the material causes internal material stresses when heated, with a loss of the concrete cohesion as a result. The bursting of concrete mainly occurs in concrete with a high moisture content and high density (strength).

From this understanding, it can be reasoned that the relatively porous soil mix material will be less likely to burst compared to concrete. However, due to its low strength, the material will burst faster in the event of fire loads. Research data with regards to burst behaviour is currently not yet available. Additionally, one must take into account the moisture content and presence of organic (possibly flammable) contaminants in the material, which can stimulate bursting.

6.13.5 Fire safety

When the above-described points are established, the fire-retardant properties of the soil mix and fire safety of the construction can be evaluated. First of all, the heating of the structure by fire is calculated, in order to determine the decrease in strength of the steel beams over time.

Decrease of the yield strength of the steel
For this assessment, the decrease in the yield strength of steel as a result of a rise in temperature is described in EN 1993-1-2. A table, included in this standard, shows that the effective yield strength of steel decreases from a temperature of 400°C. For protection of the steel in the as-built situation, effective management of the positioning tolerance is of essential importance.

Decrease of the compressive strength of the soil mix material
The decrease in compressive strength of concrete as a result of a rise in temperature is described in EN 1992-1-2. For concrete, the compressive strength does not significantly decrease until a temperature of 400°C. No similar temperature is known for soil mix material. This requires further research. Recent results concerning the behaviour of soil mix at high temperature are available in Helson (2017).

Furthermore, for the soil mix material it can be reasoned that when the yield strength of the steel is not reached during fire, the pressure arch in the soil mix in between the steel beams will not be affected by the fire.

Soil mix material burst
The bursting behaviour of soil mix material during fire will differ per construction site. A higher porosity compared to concrete will have a positive influence. In contrast, the relatively lower strength, moisture content (positioning with respect to the groundwater level) and the presence of contaminants will have a negative effect. Research into these influencing factors is desired.

The risk of bursting of the soil mix material during fire is also determined by the measures adopted for protecting the wall against environmental influences in the definitive phase (protection barrier/wall, shotcrete, coating, ...). Depending on the chosen protection measures, in relation to the risk of burst of the soil mix wall surface, for the fire safety of the construction it is important that the cover by the soil mix material on the steel beams calculated in the design, is realised during execution.

6.14 Durability

Long-term data with regards to the durability of soil mix walls is currently not yet available. For information on potential damage mechanisms, please refer to Chapter 9, which includes an overview on the basis of available insights in the area of concrete technology. Further information concerning the durability of the soil mix material is also given in section 2.4 in part 2 of the handbook.

This paragraph shows how one may conceptually approach the aspect of durability in the design and the execution of soil mix walls with the currently available knowledge.

6.14.1 Soil mix walls with a temporary function

For walls with a temporary function, no special measures should be taken under normal circumstances with regards to both the durability of the soil mix material or the corrosion of the steel beam.

In order to avoid leaching (erosion) of the soil mix material, the characteristic compressive strength of the soil mix material should be at least 0.3 MPa. When executing soil mix walls in soils with aggressive components, a prior evaluation of the effect on the setting reaction (curing time) and the strength development of the soil mix material should be conducted (see §5.3). Insofar as the minimum requirements with regards to the mechanical characteristics of the soil mix material are met, no additional measures are required.

At the same time, the impact of the aggressive components/contaminants on the corrosion of the steel beams should be evaluated (see Chapter 9). EN 1993-5 suggests that for steel pile or sheet pile elements with a lifetime of less than 4 years, the loss of steel section as a result of corrosion is in most cases negligible.

6.14.2 Soil mix walls with permanent function

For soil mix walls with a permanent function, the following measures are taken to guarantee that the soil mix functionality is ensured during its lifetime:

Soil mix material
Exposure of soil mix material to ambient air should be prevented in order to avoid evaporation shrinkage and the resulting decrease in strength and stiffness over time. This is especially important in (strongly) ventilated zones, in which progressive degradation of the soil mix material could occur. The soil mix wall may be protected with a protection barrier/ wall, shotcrete, specific coating, ... (also see §6.11.3).

One should avoid (parts of) the soil mix wall being exposed to freeze-thaw cycles. If one deviates from this point, one should be aware that, on the basis of current knowledge, a risk is being taken and that it is necessary to take specific measures (minimum cement content, minimum compressive strength, max. vol % unmixed soft soil inclusions, creep tests, tests with freeze-thaw cycles, ...) and regularly conduct tests during the lifetime of the wall.

In order to avoid leaching (erosion) of the soil mix material, the characteristic compressive strength of the soil mix material should always be at least 0.5 MPa. If one wishes to take into account the structural strength of the soil mix and the adhesion in between the soil mix and the steel beams for the vertical bearing capacity in the permanent solution, the characteristic compressive strength of the soil mix should be at least 3 MPa.

In the presence of aggressive components in the soil, one should conduct an evaluation in advance on the effect on the setting reaction (curing time) and the strength development of the soil mix material (see §5.3). At the same time, the impact of the aggressive components/contaminants on the steel beams should be evaluated (see below).

The design value of the compressive strength of the soil mix material that is included in the design phase is based on the 28-day compressive strength. In this way a 'hidden safety' is introduced. This because the literature shows that after 28 days the compressive strength can still significantly increase.

For definitive applications a number of reduction factors are applied to determine the design value of the compressive strength (factor α_{sm}) and the characteristic value of the stiffness of the soil mix material E_{sm} (reduction factor $1+\Psi$).

Corrosion of the steel beams
EN 1993-5 includes the following measures for the protection of steel piles or sheet piles against corrosion:
- providing protective coating (paint, galvanisation...);
- cathodic protection;
- a minimum cover with grout, mortar or concrete;
- an additional steel section that may corrode during the lifetime.

In principle, soil mix material has an alkaline character and will to some extent protect the steel beam against corrosion (see §9). The alkalinity of the soil mix material, and thus the way the soil mix material protects the steel against corrosion is determined by several factors: the binder type and the binder content, soil type(s), groundwater (whether or not brackish), potential contaminants etc. Because there is currently little concrete information available on this aspect, it is recommended to not consider the soil mix material in itself as a corrosion protective measure, but to instead take different protection measures as described in EN 1993-5. If one opts for permissible corrosion on the steel beams, one may apply the corrosion tables from EN 1993-5 and the national annex (ANB) (see §9.4.1).

Below the groundwater level – and in the presence of polluted and aggressive soils (including peat) – in the long term a corrosion allowance of 0.3 mm per 25 years will be considered.

The use of corrosion allowance as a mitigating measure is only applicable if the effective occurrence of corrosion does not affect the functionality of the soil mix wall during its lifetime.

Interaction between steel and soil mix

For definitive applications of soil mix walls, a number of limitations are imposed on the interaction between the soil mix wall, steel beam and protection barrier/wall.

In terms of the bending stiffness of the soil mix walls: the composite bending stiffness as suggested in § 6.7 may be considered, but only in combination with the long-term E_{sm} modulus of the soil mix material (reduction factor of $1+\Psi$).

The soil mix material may be taken into account for the permanent phase in the transfer of soil pressures onto the steel beams via a pressure arch (only if applying the factor α_{sm}, zie § 6.3.3). For the verification of the internal forces in the permanent phase, the basic principle is that the interaction between soil mix and steel beam is not taken into account and only the steel beam is used for calculations. Only under 'suitable' circumstances may the client / contracting authority provide permission to take into account a certain adhesion in between the steel beam and soil mix in the long term in accordance with the criteria in § 6.3.6.

In terms of the vertical bearing capacity in the permanent phase, one may calculate using the interaction between the steel beams and soil mix, in accordance with the criteria in §6.3.6 and the guidelines of §6.7.8.

Interaction between soil mix wall and protection barrier/wall

For long-term applications, the soil mix wall should be protected against exposure to ambient air, drying, freeze-thaw cycles, etc. on the excavation side.

If for this purpose a protection barrier/wall is cast against the soil mix wall, one should assume that the leakage through the soil mix wall over time will be sufficient to ensure the transfer of the water pressures on the protection barrier/wall. This also means that the protection barrier/wall should be designed with the purpose of retaining the complete water pressure.

Naturally, this is not the case if a protecion barrier/wall is provided with cavity and drainage. In those cases, the soil mix wall in the permanent phase should transfer the earth and water pressures onto the steel beams via a pressure arch.

The interaction between the steel beams/soil mix wall and the protection barrier/wall depends on the connections applied between the two.

Chapter 7 Execution

7.1   Execution standard

As its title indicates, the EN 14679 'Execution of special geotechnical works - Deep Mixing' standard relates to the execution of 'deep mixing' of soil with a binding agent. Parts of this standard are applicable to soil mix walls.

7.2   Execution methods

7.2.1   Introduction

The construction principles and execution processes for soil mix walls that are applicable in Belgium and the Netherlands were previously mentioned in §2.1 in part 1 of the handbook. The main important principles regarding the execution process that affect the final product are repeated below.

- One can distinguish three methods of executing soil mix walls, namely: column walls, panel walls and walls realised by means of a chain cutter device (trench mixing system). The way in which these walls are executed in practice (execution order of the various soil mix elements) is described in further detail in the following paragraphs.
- The execution parameters (binder type, w/c factor, ...) and the way in which the mixing procedure is carried out (e.g. the number of upward and downward movements, movements in accordance with the sawtooth pattern, penetration and withdrawal speeds, ...) are mostly determined on the basis of the amount of binder one wishes to mix into the soil. In most cases, taking into account the experience of the deep mixing contractor, these parameters are optimised in situ in the early stage of the work activities.
- Generally speaking, the more mixing energy per unit volume soil mix material, the more homogeneous the end result will be. In the international literature, this mixing energy is expressed in terms of the 'blade rotation number' or BRN, and, depending on the type of soil, people refer to a minimum value of 350 to 430 rotations per metre for obtaining sufficient homogeneous mixture. However, the definition and criterion is dependent on the type of mixing installation, its configuration and the type of soil. They can thus not be unambiguously determined for all types of augers and cutters that are used for the soil mix process in Belgium and the Netherlands. §2.3.2 in part 2 of the handbook includes more information on this matter.
- Depending on the soil and execution procedure, some amount of mixed material will flow back to the surface. In unsaturated and highly permeable soils, the volume of the spoil will remain limited. In saturated soils and especially in cohesive soils (clay), this volume may be up to 30% or more of the volume of the realised soil mix element.

179

7.2.2   Column walls

The soil is mechanically mixed in place with a binding agent with the aid of a special mixing auger (shaft configuration). This mixing procedure results in soil mix columns. By edging such columns into each other, it becomes possible to create a continuous wall with a earth/water-retaining function (see I - Fig. 7.1). In this system one may also choose to install the primary columns less deep than the secondary columns (staggered installation).

For a wall with soil mix columns, a guiding frame is installed in advance that indicates the position of the columns and provides support when inserting the mixing tool at depth.

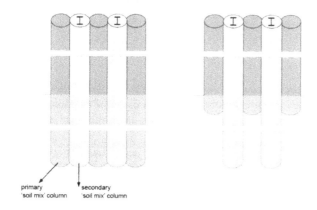

primary   secondary
'soil mix' column   'soil mix' column

I - Fig. 7.1. Column wall, traditional and staggered installation.

The execution takes place in different phases (see figure 7.2):
- in a first phase, a series of primary unreinforced columns are installed at the positions $1 - 5 - 9 - 13 - ...$;
- in a second phase, a second series of primary unreinforced columns are installed at the positions $3 - 7 - 11 - ....$;
- in a third phase, the secondary reinforced columns are installed at the $2 - 4 - 6 - 8 - 10 - 12 - ...$ positions. Here the primary column is partially drilled away. If the soil mix wall is installed in the vicinity of settlement-sensitive structures, the secondary columns are also carried out in two phases $2 - 6 - 10 - ...$ and subsequently $4 - 8 - 12 - ...$;
- the construction pit is subsequently excavated up to the installation level of a possible horizontal support;
- if necessary, the horizontal support (anchors, tensile piles, stamps, ...) is applied;
- the construction pit is further excavated up to the installation level of a possible additional horizontal support or up to the final excavation level.

180

During the installation of a soil mix wall made with columns, the following aspects should
be taken into account:
- At least 6 hours should pass between the execution of the first and second series of
  primary columns.
- The installation of the secondary columns takes place as soon as possible after the
  installation of the primary columns. However, the installation of the secondary columns
  may be started at the earliest 8 hours after concluding the primary column process.
  Depending on the soil characteristics and environmental conditions, a separate phasing
  may be required.
- The overlap between the primary and secondary columns is always at least 60 mm. If
  the soil mix wall is realised as a silo structure and/or with a water-retaining function,
  the overlap is at least 1/8 of the column diameter, but always more than 60 mm. Here
  one must always take into account the placement tolerances.
- Sufficient time should pass between the installation of the columns and the excavation,
  so that the soil mix material reaches the minimum required compressive strength and
  stiffness (which are applied as design principles).

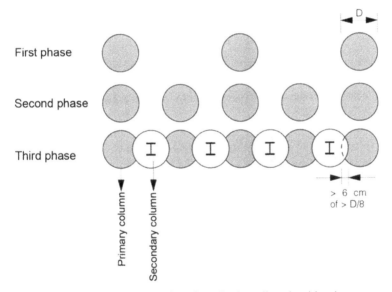

1 - Fig. 7.2. Top view of the execution of a soil mix wall made with columns.

Multiple shaft configuration systems can realise two to three columns simultaneously.
The production phases are shown in figures 7.3 and 7.4. Figure 7.3 indicatively shows the
location of the beam.

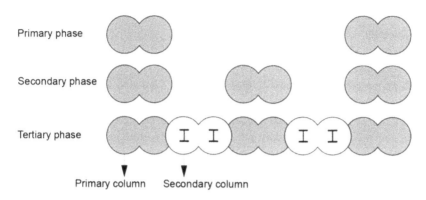

Primary phase

Secondary phase

Tertiary phase

Primary column   Secondary column

1 - Fig. 7.3. Execution process during which two augers (multiple shaft system) are applied at the same time.

If extremely stringent requirements are imposed regarding the permeability of the soil mix wall (e.g. for permanent water-retaining or cut-off walls), it is recommended to construct the wall with either panels or a multiple shaft configuration system with sufficient overlap between the panels or columns, or, with a continuous chain cutter (trench mixing) system.

A sufficient overlap between panels or multiple columns can, for example, be obtained by carrying out the execution procedure in accordance with the so-called 'pilgrim procedure', see figure 7.4, in which the mixing of the material occurs multiple times as part of the production process.

As an extra measure for ensuring an optimal water-retaining function of the wall, a geomembrane may be placed in the wall (see §2.2.6 in part 2 of the handbook).

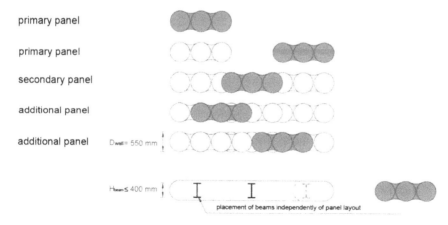

primary panel

primary panel

secondary panel

additional panel

additional panel D_{wall} = 550 mm

H_{beam} ≤ 400 mm   placement of beams independently of panel layout

1 - Fig. 7.4. Execution process during which three augers are applied at the same time – in accordance with the principles of the 'pilgrim procedure'.

7.2.3 Panel walls

The soil will be mechanically mixed in place with a binding agent by means of a cutter device. By installing the panels in an edging fashion, one ends up with a wall that can function as a earth and/or water-retaining wall or cut-off wall (see figure I - 7.5).

For a wall made with soil mix panels a guiding frame can be applied in advance that indicates the positions of the panels and provides support when inserting the cutter device at depth.

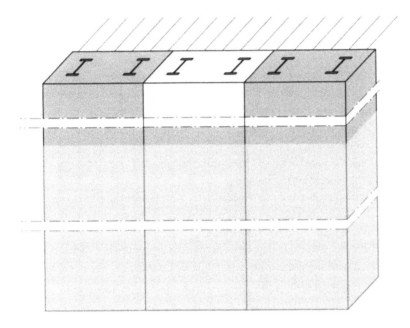

1 - Fig. 7.5. Wall made with soil mix panels.

Execution occurs in different phases (see figure I - 7.6):
- in a first phase, the primary panels are installed and, if necessary, reinforced at positions 1-3-5-7-9-11-...;
- in a second phase, the secondary panels are installed and, if necessary, reinforced at positions 2-4-6-8-12-... Here the primary panels are partially milled away;
- the construction pit is subsequently excavated up to the installation level of a possible horizontal support;
- if required, the horizontal support is applied;
- the construction pit is further excavated up to the next installation level of any additional horizontal support or up to the final excavation level.

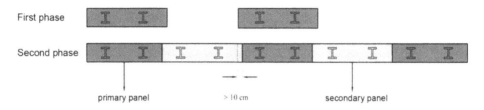

First phase

Second phase

primary panel   > 10 cm   secondary panel

1 - Fig. 7.6. Top view of the execution of a soil mix wall made with panels.

During the installation of a soil mix wall made with panels, the following aspects should be taken into account:

- The installation of the secondary panels takes place as soon as possible after the installation of the primary panels. However, the installation of the secondary panels may be started at the earliest 8 hours after concluding the primary panel process.
- Depending on the soil characteristics and the environmental conditions, a separate phasing may be required;
- The overlap between the panels in a soil mix wall is at least 100 mm. Here one should take into account the placement tolerances;
- sufficient time should pass between the installation of the panels and the excavation, so that the soil mix material reaches the minimum required compressive strength and stiffness (which are applied as design principles).
- An adjusted installation phasing in which the mixture of a panels takes place multiple times in accordance with the principle of the 'pilgrim procedure' is also possible with the cutter technology (see figure 7.4).

7.2.4 Walls made with a chain cutter device (trench mixing)

With the aid of a chain cutter device (trench mixing system), the soil is mechanically mixed in place with a binding agent, thanks to which a continuous wall can be realised in a single pass. It is especially cut-off walls that are made by means of this method.

7.3 Horizontal alignment of the walls

During execution, it is the aim to create walls that are as straight as possible. This can be achieved by using lasers or guide frames. For walls with the single column technology or in the event of strict requirements with regards to the execution tolerances, it is advisable to use guide frames.

7.4   Steel beams

After creating the column or panel wall, steel beams can be placed in the wall as reinforcement. The beams should preferably be installed by their own weight, i.e. without pressure. The upper side of the beams can be placed both above and beneath the working level. It is important to consider the position and orientation of the beams. On the field, a clear drawing must be available, on which the location of the beams is adequately indicated. The length, steel quality and beam type must be determined on the basis of calculations. With steel, a delivery certificate must be provided that lists the steel quality, beam length and beam type.

The placement tolerance of beams is usually approx. 50 mm. If necessary, a broad frame can be used, allowing the accuracy of the position of the beams to be limited to approx 25 mm. An example of a steel frame is given in figure 7.7. Here the vertical beams are vibrodriven into the soil on which steel pieces are welded on which a steel beam is finally attached horizontally. This beam is accurately adjusted to height and location and serves as a reference when executing the wall. When the wall is executed, the frames are attached to the beam, in which the steel beams can then be accurately placed in the correct position.

1 - Fig. 7.7. Guide frame (with the courtesy Hoffmann).

7.5   Reducing the wall thickness

If there is limited room for the soil mix wall, the wall thickness may be reduced. After excavation, this can be carried out by means of a milling cutter and more space becomes available for the construction of, for example, a basement/cellar.

The focus point here is the remaining cover on the beams and the effect of this reduction on the design of the soil mix wall.

7.6   Anchoring/bracing

The transfer of the forces of the horizontal support on the wall is realised with the help of purlins. A local transfer of the force with, e.g. an anchor seat, is also possible. In this case, the ground anchors can be placed between or at the location of the steel beams. The verification of the potential punching failure mechanism has to be performed.

7.7   Execution near existing buildings

In practice, soil mix walls are usually realised in the vicinity of existing buildings. The wall can be installed very close to or against existing foundation elements. The influence of installing the soil mix wall along existing structures should be assessed for each individual case. The execution process must be adjusted accordingly. For this reason, soil mix panels near foundation elements are usually not realised in one continuous process. Instead, they must be applied in separate stages. One focus point is the realisation of a soil mix wall near foundation piles where the wall continues up to below the level of the pile base.

In all cases, an analysis must be made, before the beginning of the works, of the influence of installing the soil mix wall along the existing structure (see §6.10).

The application of foundation elements right next to a soil mix wall may affect the integrity of the wall. This is especially true for installing the foundation elements during and immediately after the execution of the wall due to the limited developed strength of the wall itself. The installation of foundation elements at a distance of less than 5 metres and within one week after the execution of the soil mix panels should always be prevented.

After the soil mix wall has developed sufficient strength, the properties of the foundation elements determine the degree of influence on the wall, such as the degree of soil displacement, relaxation of the soil due to installation, the vibration intensity and the temporary generation of water overpressure during installation. To avoid displacement / pressure on the wall, for displacement piles it is recommended to bear in mind a minimum distance from centre pile to front wall of at least 3.5 times the pile diameter. For non-displacement pile systems there are no specific requirements with regards to the soil mix wall. The determining factor will be what influence the wall exerts on the pile system. If the soil mix walls also have a bearing function and the to be applied adjacent pile system has a deeper base level, there are stricter requirements to prevent negative influence on the bearing capacity of the wall.

With regards to the vibration intensity, no further requirements are established for the foundation system to be applied.

7.8 Execution monitoring

During execution, process parameters should be registered by means of which the deep mixing contractor can prove proper execution of his tasks and adjust the production process during execution. These parameters are not used to demonstrate compliance of the final product. The process parameters must be registered unequivocally. It is often required to provide the client with this information within a certain period of time. A distinction is made between parameters that must be checked once and parameters that must be checked continuously or at a regular interval. I - Table 7.1 provides an overview of the process parameters to be registered. For additional information, one may also refer to the European execution Standard EN 14679.

The equipment and sensors used must be sufficiently accurate and calibrated periodically. The deep mixing contractor must be able to deliver the calibration certificates. The manner and frequency of measurement of the different parameters are recorded in a work plan before starting the work activities.

Figure 7.8 includes an example of a registration form, with graphs in which the different parameters are shown.

Figure 7.9 includes an example of what one can produce with the registered process parameters. The figure shows a cross-section of a column wall. Such a cross-section can be constructed at any depth. In the figure one can visualise the individual circles, caused by drilling with augers. If at a certain depth the wall should not connect, this can be indicated. From this reconstruction, one can deduce that at a slightly meandering connecting wall has been realised at the depth in question.

Process parameter[3] per soil mix element	Unit	one-time registration	continuous registration
Soil mix element reference (ID number)		X	
Ground level and construction level	m NAP	X	
Size mixing tool (auger / cutting head), thickness of the wall and tolerance	m	X	
Weather conditions (extremes in rain, wind, sun)	-	X	
Position machine/mixing tool x/y/z (expansion)	m	X	
Work order (Installation order)	-	X	
Type of binding agent and composition: amount of water / viscosity / water- cement factor / unit weight	litre, -, kg/m³	X	
Verticality machine / angle of the mast during process	% /degrees/m		X
Date, start and end time of the mixing	date/time		X
Level top and bottom of the mixing	m level		X
Flow rate and pressure of binding agent / air pressure	l/min, bar		X
Vertical speed of the mixing tool (during penetration and retrieval)	mm/rev or m/min		X
Rotational speed of the mixing tool (during penetration and retrieval)	rev./min		X
Binder content per metre, m³ or depth (during penetration and retrieval)	kg/m³ or l/min		X
Quantity of surplus/spoil[1]	m³ or l/min	X	X (l/min)[1]
Settlement of the grout mixture in the soil mix element	m³ level	X	
Location, orientation and height of the steel beams[2] or other elements (e.g. geomembrane)	x/y/z	X	
Location and depth of any samples taken during the mixing process	m level	X	
Peculiarities, problems or deviations during execution (e.g. obstacles, interruption of process, deviation flow rate/ pressure, etc.)	-		X

[1] In any case, inspect visually / estimate quantity.

[2] After removing the steel beams under the top side of the wall, the height and orientation can no longer be checked. The procedure in which the beams are placed should be such that proper placement is realised.

[3] Many of the process parameters are dependent on equipment, and are not self-selectable or adjustable during execution.

Herstellungsprotokoll CSM

Baustelle:			
Auftraggeber:	ecovat	Auftrags-Nr.:	uden

Gerätefahrer:	a.klootwijk	Stich: 20
		Datum: 25-jul-2014
Bohrgerät:	Getriebe:	Wanddicke/Durchm.: m
I-Nr.:	I-Nr-1:	Stichbreite: m
	I-Nr-2:	Bohrtiefe: 20.01 m
	I-Nr-3:	Neigung: 0 c

Bewehrung	lt. Plan-Nr.:	
	Art:	
	Länge:	0.00 m

CSM-Rezept...	Zusatzstoffe	
Zement:		
Betonit:	Menge:	
W/Z:		

Beginn:	8:25:42	Soll Susp-Verbr.:	0.00 m³
Ende:	10:36:55	Ist Susp-Verbr.:	17.171 m³
Herstellungsz...	02:11:13	Susp-Verbr. ab:	17.171 m³
		Susp-Verbr. auf:	0.000 m³

Bemerkungen:

Polier/Bauleiter: Auftraggeber:

WINMIXb.v.

1 - Fig. 7.8. Example of registration form.

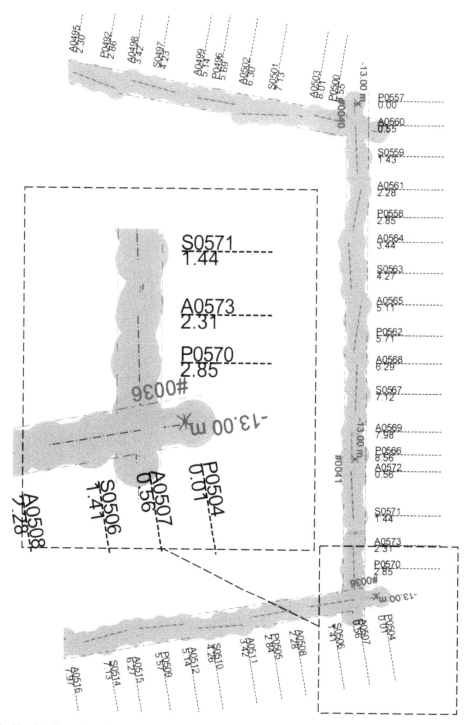

I - Fig. 7.9. Example of a cross-section of a column wall (with the courtesy of Bauer).

Chapter 8 Quality Control

8.1   Introduction

Quality control of the execution process and of the final product is of great importance in the realisation of a soil mix wall. The wall should meet the requirements of the final product. One must consider:

- strength
- stiffness
- permeability
- evenness; length, location etc.

8.2   Quality plan

Before commencing the work activities, a quality plan must be created in which the inspections and frequency of inspections are recorded both during and after the execution. The control of the execution process of the soil mix wall and registration of the execution parameters such as specified in §7.8, should be carried out continuously, independent of the application area as previously defined in I - Table 3.1. The application area of the soil mix wall is important for determining the type and amount of required tests to control the predetermined characteristics of the realised wall. Additionally, the action taken, if a controlled element does not comply, should always be described. The quality plan also includes documents, drawings and procedures that are necessary for the execution of the work.

The permissible tolerances should be specified in advance. These are dependent on the design and are specified by the designer. Under the assumption of careful execution of soil mix walls, one can assume that the following tolerances prevail:

- accuracy of the horizontal position of soil mix columns at the level of the work platform: 25 mm (execution with guide beams) and 75 mm (without guide beams);
- accuracy of the horizontal position of soil mix panels at the level of the work platform: 50 mm;
- accuracy of the vertical position of the bottom side of the soil mix elements: ±100 mm;
- accuracy of the vertical position of the top of the soil mix elements: +/- 50 mm;
- accuracy of the top level of the soil mix elements: is dependent on the soil characteristics and on the possible use of a guide beam;
- accuracy of the slope of soil mix elements: 1.3%;
- for local protrusions, an additional tolerance of 100 mm is customary. In certain conditions (e.g. in presence of local cavities, large hard stones in the soil or weak layers) larger protrusions are unavoidable.

If more stringent tolerances are imposed, for example with regards to the slope in water-retaining soil mix walls or silo structures (e.g. 0.5%), it may be necessary to take additional measures.

With regards to the tolerances on the placement of the steel beams, the following apply:
- accuracy of the horizontal position of the centre of the beam at the level of the work platform compared to the centre of the 'as-built' soil mix element: 50 mm or 25 mm in case of application of a frame;
- maximum deviation on the horizontal position of the centre of the steel beam compared to the centre of the 'as-built' soil mix element; ± 50 mm;
- maximum deviation on the vertical position of the steel beam in a soil mix element: ± 50 mm;
- angular deviation of the main axis of the beam compared to the one perpendicular to the wall < 5°;
- for cases in which the cover of the steel beam is of great importance (e.g. because of fire resistance, cathodic protection...) special attention should be paid to the control of the placement and slope tolerances of both the soil mix elements and the steel beams. The use of guiding frames that allow for the decoupling of tolerances, is highly recommended in such circumstances.

8.3   Suitability tests

In order to estimate the strength of the wall in advance, various approaches are applied in practice. The first approach involves estimating the to be realised strength on the basis of penetration test results and practical experience. In a second approach, a suitability test is conducted and a soil bore is performed in advance during which soil samples are taken and lab mixtures are tested or preliminary test panels or columns are created beforehand (see Chapter 5).

8.4   Sampling

8.4.1   Location and quantity samples

By testing samples, one can assess to what extent the properties of the soil mix wall comply with the values used in the design phase.

After realisation of the wall, samples are taken for analysis in the laboratory. For this purpose, samples should be taken from the representative zones of the work. The representative zones are highly dependent on the individual project details and the application area and should be determined on a case-by-case basis. Generally speaking, the best samples are taken at a depth of at least 1 m from the top of the soil mix wand. For determining the location of the sampling, one may take the guidelines in CUR Recommendation 84 (see figure 8.1) as an indication.

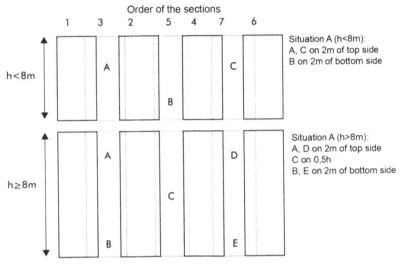

1 - Fig. 8.1. Overview of sampling (A. B, C. D. E are locations of sampling).

The sampling process should be in line with the application area (see table 3.1) and the reliability classes RC1 to RC 3 (NL: EN 1990 / CUR 166) or risk classes RK1 to RK3 (BE: BGGG, 2015b) in which the structure is classified. For this purpose, Table 8.1 includes the basic test programme with the type and amount of control tests on the soil mix material. In addition, depending on the function of the soil mix elements, the selected contractor should always at least comply with the test programme in the execution phase. Naturally, the selected contractor may always conduct more tests if necessary. Moreover, the amount and type of tests should be aligned with the requirements of the client. A contract with the client may include stricter requirements than those listed in the test programme below.

The table provides a basic test programme per representative zone. A 'representative zone' refers to each zone/layer in the wall of the project, of which one can assume that the parameters of the wall may or will vary. Important parameters herein are the potential soil stratification and the mixing method to be applied. For example, in case of soil mix in a soil stratification presenting 2 layers (e.g. first 3 metres of clay, followed by sand), this means there are 2 representative zones. In case of an execution process of soil mix technology in which homogeneous mixing occurs across the entire panel, derivation from 1 zone may be sufficient. Nevertheless, in such a situation one must demonstrate that homogeneous mixing has indeed been conducted. In case there is a strongly heterogeneous soil stratification (successions of various soil layers across the entire construction site) or local presence of contaminants, this should be taken into account in determining the amount of representative zones and thus the test programme. The following table includes an inclusion analysis. A simplified line method may suffice for this analysis, with an estimation of the volume percentage of unmixed soft soil inclusions (see paragraph 8.5).

1 - Table 8.1. Type and amount of control tests on soil mix in function of the application area and risk class (NL) or risk class (BE).

Application area and function	Requirements with regards to quality control of the soil mix material
A. Temporary earth-retaining soil mix wall	**Basic test programme per representative zone** Number of samples: 1 sample per 150 m³ with a minimum of 6 Tests: UCS, density, inclusion analysis[1] Interpretation: UCS_{char}, estimation vol% inclusions, $E^{(2)}$ **Deviations from the basic test programme** - RK 1 – RC 1 → specific empirical data[3] are sufficient - RK 2 – RC 2 with no interaction steel-soil mix taken into account → specific empirical data[3] are sufficient - RK 3 – RC 3 → basic test programme x 2
B. Permanent earth-retaining soil mix walls C. Temporary water-retaining and earth-retaining soil mix walls D1. Permanent earth-retaining and temporary water-retaining	**Basic test programme per representative zone** Number of samples: 1 sample per 75 m³ with a minimum of 12 Tests: UCS, density, inclusion analysis[1], E Interpretation : UCS_{char}, estimation vol% inclusions, E **Deviations from the basic test programme** - RK 1 – RC 1 → Limited number of samples up to 1/150 m³ with a minimum of 6 → Determination of $E^{(2)}$ permitted - RK 2 – RC 2 in which specific empirical data[3] are available → Limited number of samples up to 1/150 m³ with a minimum of 6 → Determination of $E^{(2)}$ permitted - RK 3 – RC 3 → basic test programme x 2

Application area and function	Requirements with regards to quality control of the soil mix material
D2. Permanent earth-retaining and permanent water-retaining D3. Temporary or permanent earth-retaining and permanent water-retaining	**Basic test programme per representative zone** Number of samples: 1 sample per 75 m³ with a minimum of 12 Tests: UCS, density, inclusion analysis[1], E, k Interpretation: UCS$_{char}$, estimation vol% inclusions, E, k **Deviations from the basic test programme** - RK 1 – RC 1 → Limited number of samples up to 1/150 m³ with a minimum of 6 → Determination of E[2] permitted - RK 2 – RC 2 in which specific empirical data[3] are available → Limited number of samples up to 1/159 m³ with a minimum of 6 → Determination of E[2] permitted - RK 3 – RC 3 → basic test programme x 2 → Determining k-wall with the aid of pumping test
E. Soil mix wall with temporary bearing capacity function	**Basic test programme per representative zone** Number of samples: 1 sample per 150 m³ with a minimum of 6 Tests: UCS, density, inclusion analysis[1] Interpretation: UCS$_{char}$, estimation vol% inclusions, E[2] **Deviations from the basic test programme** - RK 2 – RC 2 in which no specific empirical data[3] are available → basic test programme x 2 - RK 3 – RC 3 → basic test programme x 2
F. Soil mix wall with permanent bearing capacity function	**Basic test programme per representative zone** Number of samples: 1 sample per 75 m³ with a minimum of 12 Sampling over the full depth Tests: UCS, density, inclusion analysis[1], E Interpretation: UCS$_{char}$, estimation vol% inclusions, E **Deviations from the basic test programme** - RK 1 – RC 1 → Limited number of samples up to 1/150 m³ with a minimum of 6 → Determination of E[2] permitted - RK 2 – RC 2 in which specific empirical data[3] are available → Limited number of samples up to 1/150 m³ with a minimum of 6 → Determination of E[2] permitted - RK 3 – RC 3 → basic test programme x 2

Application area and function	Requirements with regards to quality control of the soil mix material
G. Temporary (water) cut-off soil mix wall	**Basic test programme per representative zone** Number of samples: 1 sample per 150 m^3 with a minimum of 6 Tests: UCS, density, inclusion analysis[1], k Interpretation : UCS$_{char}$, estimation vol% inclusions, E[2] **Deviations from the basic test programme** - RK 1 – RC 1 → specific empirical data[3] are sufficient - RK 2 – RC 2 in which specific empirical data[3] are available → specific empirical data[3] are sufficient - RK 3 – RC 3 → basic test programme x 2
H. Permanent (water) cut-off soil mix wall	**Basic test programme per representative zone** Number of samples: 1 sample per 75 m^3 with a minimum of 12 Tests: UCS, density, inclusion analysis[1], E, k Interpretation : UCS$_{char}$, estimation vol% inclusions, E, k **Deviations from the basic test programme** - RK 1 – RC 1 → Limited number of samples up to 1/150 m^3 with a minimum of 6 → Determination of E[2] permitted - RK 2 – RC 2 in which specific empirical data[3] are available → Limited number of samples up to 1/150 m^3 with a minimum of 6 → Determination of E[2] permitted - RK 3 – RC 3 → basic test programme x 2 → Determining k-wall with the aid of pumping tests

[1] With simplified line method (see § 8.5).

[2] On the basis of correlation with UCS values (in accordance with § 6.1).

[3] Specific empirical data = at least two applications with the same execution procedure in similar circumstances (same soil type)

8.4.2 Sampling techniques

Wet grab samples
A liquid sample is taken from the freshly realised soil mix wall at a specific depth. It is recommended to take these samples from a depth of at least 1 metre in the recently realised wall. In order to take wet grab samples at a greater depth, specific tools (equipped with a valve that can be mechanically opened and closed above the ground surface) are required.

Once the sample has been removed, above the ground surface, it is cast into a mould and will continue to cure. The benefit of this sampling technique is the low cost price and the fact that one can acquire the samples and test results at an early stage (e.g. strength development before excavating the construction pit or lowering the groundwater). Potential disadvantages of wet grab samples are the representativeness of the samples (especially dependent on the tool used and soil type) and the cure conditions that differ from those of the in situ soil mix material.

Because of the aforementioned reasons, the test results of wet grab samples are not acceptable for the quality assurance of the postulated parameters.

Only if it can be demonstrated for a specific site that the applied wet grab method provides representative samples and test results (e.g. by means of a comparison with test results on cored samples), the technique can be applied on a wider scale at the site in question.

Coring
Cores are drilled in the representative zones of the realised wall. For retaining structures, horizontal cores are usually drilled in the visible surface of the wall after (partial) excavation of the construction pit. Optionally, one may also vertically core from the top of the soil mix wall. If one wishes to core at greater depths, special (expensive) technology is usually required.For more specifications with regards to sampling with the aid of core drilling, please refer to EN 12504-1.

Double-walled vertical liners
An alternative to conventional core drilling in the soil mix material is to opt for double-walled liners. This technology was first successfully applied in the Netherlands with soil mix walls for the Westelijke Invalsweg aqueduct in Leeuwarden (see § 2.2.6 in part 2 of the handbook). The liners consists of a double-walled system, with a steel outer tube and plastic inner tube. The bottom of the steel tube is equipped with a cutting edge. Quality sealing between both tubes is necessary. When installing the tubes into the slurry, through the opening on the bottom of the system, the soil mix material flows into the liner across the full length of the liner. For the aqueduct in Leeuwarden, this approach was used for installing 8 liners up to a depth of approx. 16 m below the ground level (see Figure 8.2).

I - Fig. 8.2. Installation of double-walled liner (plastic inner tube, steel outer tube) in a soil mix wall.

I - Fig. 8.3. Photographed sample material in liners opened in a geotechnical laboratory.

After installation, the soil mix material can harden in the liners installed in the wall. Bear in mind that one should check that the inner liner remains separated from the outer tube during the hardening process. After 28 days, the inner liner can be completely removed with a crane, during which the liner is cut into handleable elements of, for example, 1metre. It is important that the sample material in the liner elements is disrupted as little as possible. To this end, specific measures should be taken for sealing the front ends of the liner pieces, transport and conditioned storage (both on the worksite and in the lab). In the lab, the plastic liner tubes can be cut longitudinally, after which the homogeneity of the soil mix material can be verified and described. The sample material can then also be photographed (see figure 8.3). As a result, one can gain insight into the homogeneity of the mixing process, in which sample material is obtained as in the case of a long boring. The process / result is somewhat comparable to that obtained with a so-called Begemann drill. For the Westelijke Invalsweg project in Leeuwarden, the quality of the samples proved to be excellent. If the liner tubes are treated with care, this technology allows one to obtain sample material of sufficient quality for the application of compression and permeability tests. The benefits of a sampling with liners are:

- obtaining sample material over the entire depth of the wall;
- the possibility of verifying the homogeneity of the soil mix for the client / adviser;
- the possibility to perform sample selection based on observation of the homogeneity or of deviations over the height;
- the hardening of the samples in field conditions. The results of the tests will be more representative of the properties of the wall than with wet grab samples hardened in the laboratory;
- obtaining sample material with as little disturbance as possible, the liner is removed statically, resulting in no or minimal hairline cracks in the sample material (as may occur in a conventional drilling method).

8.4.3 Dimensions of the samples

The strength of a test specimen is partly dependent on its shape and dimensions. For soil mix samples, it is recommended to opt for the following dimensions for the cored samples:
- Diameter = 100 mm ± 15 mm
- Length of the cores preferably greater than 200 mm

For lab and wet grab mixtures, it is also recommended to cast the soil mix material in cylindrical moulds with similar dimensions. In this way, the influence of the shape and dimensions of the sample on the test results is minimised.

8.5 Storage, treatment and preparation of the test specimens and determination of the volume percentage of inclusions

After taking soil mix samples, it is important that the samples are conditioned and stored away. Before the samples are transported to a geotechnical laboratory, they must first remain on the construction site for some time. It is important the samples are sealed and not disturbed. The samples should therefore be placed in a tray of water situated indoors. The samples should then be transported to a laboratory as soon as possible. In the lab, the samples should be stored in a conditioned room with a humidity greater than 95% (or in water bins/trays) and a constant temperature (see figures 8.4 and 8.5). Most laboratories allow for conditioned storage at 10 or 20°C. Samples taken for quality control should be stored at a temperature of 20 °C. If the samples are taken for maturity testing, they should be stored at a temperature that is equal to the average soil temperature (approx. 10 °C). Samples that are stored at a low temperature usually show a lower strength development. For comparison with the field properties, one should take into account that the average soil temperature is approx. 10 ± 2°C. One should also consider the temperature storage in this case.

I - Fig. 8.4. Conditioned storage with constant temperature / humidity.

I - Fig. 8.5. Conditioned storage of soil mix samples.

Before the cored material is cut into useful test specimens, a visual analysis and an estimation of the volume percentage of the unmixed soft soil inclusions is conducted. For example, this can be done with a simplified method (see figure 8.6), in which 4 lines (every 90°) are followed along the cored material in the longitudinal direction of the core. The cumulative length of soft inclusions along these lines is subsequently measured. The line percentage of inclusions is then equal to the ratio of this cumulative length divided by the total length of the lines and provides an estimation of the volume percentage of the unmixed soft soil inclusions in the soil mix core.

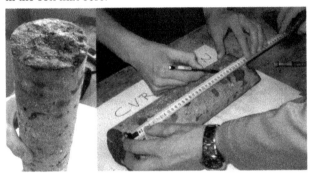

1 - Fig. 8.6. Determination of the line percentage of soft inclusions in cored soil mix material.

Before testing the soil mix material, the cores must first be cut and flattened into useful test specimens. The height-diameter H/D of the test specimen is usually 1 or 2 depending on the type of test. In order to ensure an objective selection of the exploitable part of the sample to be tested (= test specimen), the procedures illustrated in figures 8.7 to 8.9 may be applied.

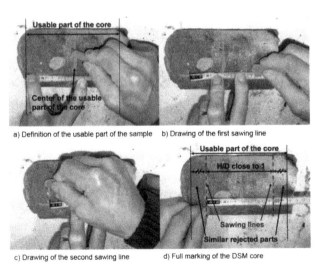

a) Definition of the usable part of the sample   b) Drawing of the first sawing line

c) Drawing of the second sawing line   d) Full marking of the DSM core

1 - Fig. 8.7. Determination of the exploitable part of the core (= test specimen) with H/D = 1 from a core D < H < 2D.

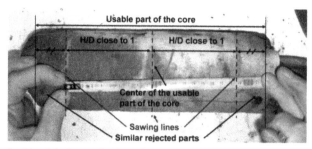

I - Fig. 8.8. Determination of the exploitable part of the core (= test specimen) with H/D = 2 from a core H > 2D.

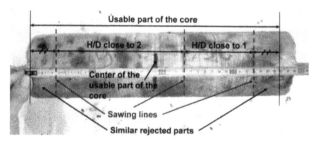

I - Fig. 8.9. Determination of the exploitable part of the core (= test specimen) with H/D = 1 from a core with H > 2D.

8.6   Determination of the soil mix characteristics

Once the test specimens are prepared, the required characteristics of the soil mix material can be determined. I - Table 8.2 provides an overview of the characteristics that can be determined in the laboratory. I - Table 8.2 also refers to the test standards that can be used and shows the recommended dimensions of the test specimens. As there are no specific test standards for soil mix material, test standards for concrete and soil should be used.

In addition to the inclusion analysis conducted on the basis of visual inspection of the intact sample piece, after conducting the destructive tests, it is important to visually inspect the material internally in the core and to record the dimensions of possible (large) inclusions (which may affect the test result).

The most required characteristics are the density and unit weight, the strength (UCS) and the stiffness (elasticity modulus) and the permeability of the soil mix material. In addition, reference is made to the test standards that can be used for determining the porosity, the tensile splitting strength and the wave propagation speed in the soil mix material. This property closely correlates with the strength and stiffness of the soil mix material and is an interesting, non-destructive test method for determining the evolution of strength and stiffness over time.

I - Table 8.2. Test standards and recommendations for the dimensions of the test specimens for determining the characteristics of the soil mix material.

Density	• Standard: EN 12390-7 • Minimum volume of the test specimens: 0.785 litre • For D = 100 mm, H/D ≥ 1
UCS	• Standard: EN 12390-3 and EN 12504-1 • Test specimens with H/D = 1 and H/D = 2 are permitted • H/D = 1 is recommended because of the maximum exploitation of the cored material and equivalence with cube-shaped test specimens (in Belgium)
Modulus of elasticity	• Standard: NBN B 15-203 and ISO/FDIS 1920-10 • Test specimens with H/D = 2 to 4 • H/D = 2 is recommended • For measuring the deformation, it is recommended to measure the deformation directly on the test specimen • The test result provides an E_{tg} (between $\sigma_{10\%ucs}$ and $\sigma_{30\%ucs}$) • The test result also gives a compressive strength (UCS)
Permeability	• Standard: DIN 18130-1 • minimum section of test specimen is 10 cm² for cohesive soil and 20 cm² (cohesionless soil) • for soil mix, D = approx. 100 mm is recommended
Tensile splitting strength	• EN 12390-6 • same dimensions as for UCS tests are recommended
Porosity	• NBN B 15-215 • minimum volume test specimens = 800 cm³ • for D = 100 mm, H/D > 1
Wave propagation speed	• ASTM C597-09 • min[D, H] > wavelength ultrasound vibrations

8.6.1 Compression tests and deduction of the modulus of elasticity

For obtaining insight into the strength and stiffness parameters of the soil mix material, it is common to conduct compression tests in a geotechnical laboratory (see figures 8.10 and 8.11).

These are compression tests with an approach similar to the unconfined compressive strength (UCS) tests of concrete cubes. During the test, in addition to the compressive strength, the deformation is also measured; this is best done on the tests specimen itself. On this basis, a modulus of elasticity is derived from the test graphs.

By taking multiple samples from the same location in the wall, it is possible to determine the development of the compressive strength over time. For this process, one may conduct compression tests at 7, 14, 28 and 56 days. Obtaining insight into the strength development

203

of the soil mix material will be crucial at the beginning of a project and/or in circumstances in which people have limited experience with the to be expected characteristics. The development of strength and stiffness over time can be evaluated on the basis of a collection of tests. An example of such an evaluation is shown in figure 8.12. With the consideration of control values included in the contract, it is clearly visualised that the values obtained after 56 days are sufficient.

I - Fig. 8.10. Execution of lab test – UCS.

I - Fig. 8.11. Deformation of the sample material and initiation of the cracks.

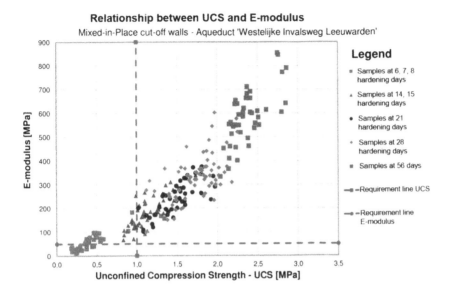

I - Fig. 8.12. Example of an evaluation of the achieved compressive strength and modulus of elasticity with development over time (Gerritsen, 2013).

8.6.2 *Permeability tests*

For testing the permeability of the soil mix samples, permeability tests may be conducted in a geotechnical laboratory. The water permeability of soil mix walls is important for situations in which these act as cut-off walls for polder constructions or other situations in which achievement of a particular hydrogeological resistance is important (permanent walls). For a description of the different aspects with regards to the permeability, please refer to paragraphs 2.3.9 in part 2 of the handbook and 6.3.9 and 8.8 in part 1 of the handbook. With regards to the characteristic values of soil mix material, the 'falling head' test is the most suitable for determining the permeability in a laboratory. This test can be conducted in a triaxal cell. If necessary, the sample can be consolidated on the basis of the field stresses, being the depth at which the sample was taken. Figure 8.13 includes the arrangement of such a test. Permeability tests can be conducted on all types of samples (wet grab/cored/liner).

1 - Fig. 8.13. Permeability test set-up.

When preparing the permeability samples, it is important that the soil mix sample fully aligns with the cell. This is because leakages on the sides can have a major influence on the results. By taking multiple samples from the same position in the wall, it is also possible to determine the development of the permeability over time for these test. Experience shows that the hardening of soil mix material reduces its permeability (for example, see figure 2.45 in part 2 of the handbook). The development of the permeability over time can be evaluated on the basis of a collection of samples.

8.7   Monitoring of the wall deformations

During execution, it is important that the deformation of the walls is monitored. By monitoring deformations over the course of the different construction phases, one gains insight into whether the deformation behaviour of the wall corresponds to the behaviour calculated during the design phase. If the deformation behaviour during execution is significantly different from the calculation, this should be immediately reported to the designers. Moreover, it is recommended to determine limit and alarm values for the deformations.

For measuring the deformations of the soil mix wall, one may choose for the simplest arrangement of applying fixed points at the top of the wall. For this purpose, measuring nails or bolts may be drilled into the soil mix wall. These fixed points can then be measured over time in the x/y/z direction, on the basis of which the movement on the top of the wall can be monitored. The disadvantage of this is that no information can be obtained about the deformation behaviour in depth.

With the installation of inclinometers next to or in the soil mix wall, this information can be obtained. Installation in the soil mix wall can take place by installing inclinometer tubes in the fresh soil mix material immediately after mixing. By using a suitable ballast (e.g. steel weight) and, for example, filling it with water, one can prevent the raising of the inclinometer tubes from the wet slurry during the hardening of the wall. Installation of inclinometers can also be done by drilling these afterwards with conventional drilling techniques. In this case, installation should occur at the back of the wall. However, the disadvantage of such a setup is that the measurements will be less accurate and that doubt may arise whether the measurements follow the movement of the wall or movement of the soil.

I - Fig. 8.14. Installation of inclinometers near a soil mix wall (with the courtesy of Witteveen+Bos).

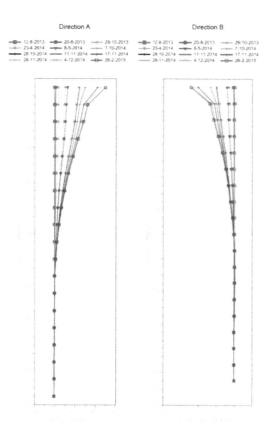

1 - Fig. 8.15. Measurement of the wall deformations of a soil mix wall as function of the depth during the different construction phases (with the courtesy of Witteveen–Bos).

8.8 Determination of the in situ permeability of the soil mix walls: Pumping tests

In addition to the limited measurement on sample material, the permeability of the soil mix wall and of the entire system can be determined by conducting a full-scale pumping test. Before conducting the pumping tests for soil mix walls, one must take the following into account:

- The test should take place before the excavations, in or in the vicinity of the walls. If leakages are observed, corrective measures may be taken.
- Walls should be properly hardened before the pumping test is initiated. Preferably, a minimum waiting period of 28 days after installation of the screens should be maintained.

- The water level at the soil mix walls should be reduced gradually. With a gradual reduction of the water level, the wall may also deform gradually in the soil. With a gradual deformation, the soil mix material is able to follow the deflection of the wall without major tensile stresses occurring in the wall. This aspect is especially important for unreinforced walls.
- Creating a measuring net of monitoring pipes in a sufficiently refined grid, inside and outside the walls. In this situation, it is important that the different filter depths and distances from the monitoring pipes are set such that leaks through the walls or sealing (soil) layers can be observed in the entire geology of the site.
- Consideration of the way in which distinction is made in the leakage rate from the horizontal component (through the walls) and vertical component (through the sealing horizontal soil layer or artificial injection layer). Making a distinction between and accurately quantifying these components appears to be very difficult in practice. For example, it may result in contractual discussions between the contractor and client about whether leaks occur as a result of the realised walls or the horizontal sealing layer. At present, the only method seems to be to place the measurements in a hydro-geological model, in which the permeability of the walls and soil layers are fitted with sufficiently measured reductions and flow rates.

Another option is to perform a pumping test before placing the walls, in order to determine value of the geohydrological parameters of the natural sealing soil layer. Subsequently, after placement of the walls, the flow rate can be monitored during excavation and draining in the construction phase. On the basis of this flow rate and the value of the geohydrological parameters determined earlier for the soil layer, one can calculate what share of the flow rate flows horizontal via the walls and one can calculate the average water-tightness of the wall.

8.9   Leak detection method for soil mix walls

If one wants certainty about the presence of potential leakage and the location thereof, one may decide to use a leak detection method in the soil mix walls.

Since constructive walls are equipped with steel beams, a geo-electrical measurement method is not expected to product quality results. After all, steel foundation elements cause too much disturbance. A method that can be applied in such cases is the spontaneous potential method (Electro Chemical Response - ECR, Electrical Flux Tracking - EFT). With this method, any leaks can be detected. Moreover, depending on the desired accuracy, the grid of the sensors can be adjusted with this method. A major advantage of this method is that any leaks can be detected with a tolerance of approx. +/- 1 metre. As a result, on the basis of the flow line image, one is theoretically able to determine whether the leakage is in the wall or in the floor. When detecting a leakage and finding that it is not permissible, corrective measures may be taken (e.g. additional soil mix wall or panel behind the leakage, or use of injection technology).

When measuring, it is important to note that the method measures whether there are openings in the boundaries (leakages). In the example of a soil mix wall, this will mean that if incomplete mixing has occurred across the wall and large-scale soil inclusions have become part of the wall (e.g. clay inclusions), no warning will follow. In such a situation, the wall is considered 'sealed'. The method is therefore not suitable to asses the homogeneity of the wall or presence of inclusions. A disadvantage of this method is that, during measurement of a leakage, no conclusions can be drawn about the local leakage rate. Instead, a pumping test will be required. Another disadvantage of this method is the considerable cost, especially if measurements take place in an intensive grid of sensors. For a more detailed description of the electrical potential method for use in construction pits, please refer to Brouwer et al. (2011).

8.10   Permanent walls

Permanent walls should also retain their function in the long term. Due to the effect of frost and chemicals, the strength of the soil mix material may reduce over time. For this reason, an extensive analysis of the mixture, the soil and groundwater should be made for permanent walls.

The standard EN 206-1 'Beton - Deel 1: Specificatie, eigenschappen, vervaardiging en conformiteit' ['Concrete - Part 1: Specification, characteristics, manufacturing and conformity'] addresses chemical degradation of concrete by natural soil and groundwater and non-natural chemicals.

Chapter 9 Lifetime considerations, management and maintenance

9.1   Introduction

For soil mix walls with a temporary function (less than two years), a design focusing on strength is generally sufficient. In order to determine if the intended lifetime of permanent walls (on the basis of current technological insights) can be realised, a project-specific verification of the lifetime of the walls is required. This verification should involve the execution process and the exposure circumstances.

Long-term data on the durability of the soil mix material and concerning the degradation mechanisms for soil mix walls are not (yet) available. Because soil mix is a cement-based material with a certain water-tightness, alkaline properties and (mostly) embedded steel, a verification of the lifetime can be made similar to those made for concrete constructions. The big difference compared to concrete is that the constituents of concrete can be consciously chosen, whereas with soil mix walls these are dependent on the (local) soil conditions. In addition, a dispersion of properties in depth and along the length of the walls is also possible.

With these uncertainties caused by the disperion of different parameters, the lifetime should be determined on the basis of the predictable degradation mechanisms of the walls. On the basis of this verification, an economically optimal attunement – in order to maintain the functionality of a soil mix wall during its intended lifetime – may have to be made between:

- the modification (if necessary) of the design, especially for situations in which corrective measures can no longer be applied after realisation;
- the management measures required during the use phase;
- the required maintenance of the soil mix wall during the intended lifetime.

The current and necessary insights for this verification are summarised in this chapter. For the translation of this knowledge into concrete conceptual requirements and a design method that conforms to the structural Eurocodes, please refer to Chapter 6 of this handbook, in particular par. 6.14 related to the durability.

211

9.2 Degradation mechanisms

The verification of the lifetime is focused on the damages that may occur on the soil mix walls during their intended lifetime. These damages can be divided into two main groups (see I - Fig. 9.1):

- damage to the soil mix itself;
- damage as a result of corrosion of integrated steel elements (beams and permanent anchors).

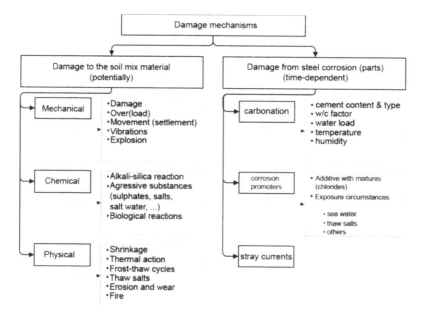

I - Fig. 9.1. Overview of damage mechanisms, based on EN 1504-9.

In addition to this overview, the following degradation mechanisms specifically for soil mix walls apply:

- Cracks may occur as a result of drying (shrinkage) and/or deformations of the wall. In addition to the risk of loss of bearing capacity and water-tightness (leakages), the protection of the steel components against corrosion by the alkaline soil mix material may be partially lost.
- Aggressive constituents/contaminants in the soil are mixed into the soil mix material and may stimulate corrosion of integrated steel (H or I) beams, reinforcements and anchors.

It should be noted that direct experience with degradation mechanisms of soil mix walls are limited. Permanent walls in particular are only applied since a few years.

9.3   Damage to soil mix material

Soil mix is a mixture of soil and cementitious binding agent. Per project, and even within a project itself, the composition of the soil may differ. A preliminary (before design phase) analysis of the soil composition is a prerequisite for recognising potential damage to the soil mix walls in advance and, if necessary, taking preventive measures. The most common damages to the soil mix material are described below.

9.3.1   Mechanical damage to soil mix material

Mechanical damage to the walls often occurs due to cracks, which may be accompanied by inadmissible deformations. In the event of short-term loads (e.g. an impact or explosion), both smaller and larger pieces of soil mix may come loose. In these situations, the relationship between the damage and its cause is fairly obvious.

Damage as a result of long-term limited (over)loads or settlement of support points occurs at a much slower rate, given that the soil mix material creep plays a major role in this. By conducting a stability test, the influence of any overload can be determined in advance. Although vibrations are often unpleasant for users of the structure, they rarely cause damage. If, however, damage occurs, a measurement campaign can be performed on the basis of the German standard DIN 4150.

By assessing the occurring loads, the crack-width can be calculated on the basis of the design. Depending on this crack-width, measures should be taken (for more information see par. 9.3.3 – Shrinkage).

9.3.2   Chemical damage to soil mix material

§5.3 (part 1 of the handbook) and § 2.4 (part 2) already provide an overview of chemical substances that may impact the binding process (slowing or accelerating effect). Additionally, there are a number of chemical substances and reactions that must be considered when assessing the lifetime of the soil mix wall. These are listed below.

Alkali-silica reaction (ASR)
ASR can be described as a reaction between alkaline constituents in the soil mix (originating from the cement, additives and/or adjuvants) and potentially reactive (that is to say alkaline-sensitive) granulates in the soil: silicates (silicic acid) in the form of opal, chalcedon, cristobalite,tridymite and cryptocrystalline quartz. This reaction gives rise to the formation of expansive products, including alkali-silica gel, which attracts water and thus swells.

This causes internal tensile stresses in the soil mix wall, that eventually result in cracking of the material.

The chance of an alkali-silica reaction is considerable if the following conditions are met:
- the reactive granulates fall within the critical area (pessimum). The lower and upper limits of this area are dependent on the mineral composition of the soil;
- the constant or regular presence of sufficient moisture in the soil mix material;
- A sufficiently high alkali concentration as a result of the applied binding agent.

In unreinforced soil mix walls, ASR usually manifests itself in the form of a random cracking pattern (craquelé). However, in reinforced walls, the free expansion of the walls in the direction of the reinforcement is hindered, so that the cracking pattern follows the underlying reinforcement.

For assessing the risk of ASR, one must first determine whether the above conditions can be met.

Aggressive substances: sulphates

Sulphates from the environment (soil, water) can react in the soil mix material and cause ettringite formation. This crystal formation is accompanied by a huge expansion ($\pm$ 300 %) and may occur both during the plastic phase of the soil mix (primary) and the hardened soil mix (secondary). Only secondary ettringite formation is harmful to the soil mix material. This is because the internal pressure caused by expansion can lead to cracking and failure of the structural component. Additionally, sulphates may decalcify the main constituents of the binding in the cement (C-S-H), reducing the strength and potentially causing stability problems.

Aggressive substances: salts

Hardened cement can be affected by aggressive salts such as ammonium and magnesium salts, which can, for example, be present in fertilisers and/or industrial waste water.

Aggressive substances: acids

Because soil mix material is basic, it can be affected by acids, for example from the chemical industry or agricultural sector. These acids react to the calcium compounds in the hardened cement (calcium hydroxide and hydrated calcium silicates and aluminates) and form calcium salts and silicon dioxide. The limestone granulates are also affected by acids. The rate of degradation depends on the following factors:
- acidity and concentration of acid solution: in a stable environment, a pH between 6.5 and 5.5 is considered to be slightly aggressive, a pH between 5.5 and 4.5 moderately aggressive and a pH between 4.5 and 4 as very aggressive;
- stationary or flowing acidic groundwater;
- the solubility of the formed salts;
- the porosity of the soil mix material.

In order to assess the risk of agression against cement stone (porous and loss of strength) by aggressive substances, a (ground)water test according to EN 206-1 is recommended.

Biological aggression

The main form of biological degradation is the biogenic sulfuric acid attack (BAA). This especially occurs in sewers and drainage systems in which the waste water is rich in sulfur compounds from rotting processes:

- an anaerobic environment may arise in slowly flowing or stationary waste water, in which sulfate-reducing bacteria convert sulfur compounds into hydrogen sulfide;
- this gas is then released into an oxygen-rich sewage environment, where it is converted into elementary sulfur;
- this sulfur is then converted into sulfuric acid by aerobic sulfur oxidising bacteria, which in turn breaks down the cement stone (e.g. in gypsum), causing the soil mix material to lose its cohesion.

The number of situations in which the biological degradation of soil mix material must be taken into account is small, but it should nevertheless always be considered.

9.3.3 Physical damage to soil mix material

Shrinkage

Depending on the origin of the shrinkage, a distinction can be made between plastic shrinkage, endogenous shrinkage and dehydration shrinkage. Shrinkage results in cracks if the shrinkage is hindered by the surrounding structure or occurs non-homogeneously. Shrinkage may occur during the various hardening stages of the soil mix material and depends largely on the composition.

During the plastic phase, the soil mix material shrinks mainly due to the loss of unbound water. This plastic shrinkage is greatest in soil mix compositions with a higher water-cement factor and is significantly influenced by sun exposure, wind and/or the presence of a water-absorbing or water-permeable soil. The associated cracks are usually wide and erratic.

During the hardening (hydration), the soil mix material shrinks because the reaction products (hardened soil mix) occupy a smaller volume than the non-hydrated cement and water. This endogenous shrinkage is greatest with a low water-cement factor.

Finally, dehydration shrinkage may occur after further hardening of the soil mix material. One should also consider dehydration shrinkage if the walls are fully or partially excavated. After all, this shrinkage may cause new cracks and widen existing cracks.

The to be expected shrinkage (reduction) and the width of the cracks in the soil mix walls must be taken into account for the verification of the lifetime. One problem here is that empirical data about occurring shrinkage is very limited. The following general aspects should be considered for the verification of the lifetime:

- Shrinkage of the soil mix material is dependent on the grain structure and density of the soil and amount of cement used.
- Chance of shrinkage due to temperature fluctuations and variations in moisture content is substantially zero in cases where the soil mix walls are permanently enveloped by soil and/or groundwater.

When the occurring crack width is too large, CO_2, chlorides, oxygen and water may penetrate and come into contact with steel components in the walls. In parallel with the situation for concrete structures, there are no requirements for a maximum admissible crack width in soil mix walls. In parallel with the situation for concrete structures, it should be noted that the admissible crack width is dependent on the environmental class and that, depending on the cover available, the admissible crack width should be increased by a k-factor.

Thermal action

Temperature fluctuations may occur in soil mix walls caused by:
- the hydration heat of the fresh soil mix material;
- a too quick excavation of the walls;
- the differential surface heating of the hardened walls (by the sun or another heat source).

These temperature fluctuations cause differences in thermal expansion, which in turn cause tensile stresses, to which the soil mix is poorly resistant, with crack formation as a result.

Frost-thaw cycles

If no or insufficient precautions are taken, the execution of the walls during a cold period may result in frost damage. In case of freezing, the water in the fresh soil mix wall will undergo strong expansion.

In the plastic phase of the soil mix material, this expansion may occur unhindered, so that the hardened soil mix will not exhibit any external damage. However, it will be of poor quality.

On the other hand, with young, hardened soil mix, the expansion is prevented. This causes internal stresses that may cause the soil mix surface to peel off (usually in different layers).

It is generally assumed that cement-bound material can withstand these stresses, as long as the compressive strength is greater than 5 N/mm^2. Such a compressive strength should be reached if an air temperature of over 5°C persists during the first 72 hours after execution.

Older, hardened soil mix material may also be damaged if the water freezes in the pores and cracks, expands and creates stresses. The frost sensitivity of the hardened walls is largely determined by the pore structure and dimensions of the cracks. The risk of frost damage is generally greater on horizontal surfaces than vertical planes, since the degree of saturation of the pores is usually greater in these surfaces.

Thaw salts

If thaw salts make ice melt, they extract heat from the environment. If this (endothermic) reaction occurs on a soil mix surface, the temperature of the outer layer will decrease significantly (thermal shock) and the top layer of the soil mix material may peel off. In case of long-term use of thaw salts, the outer layer can additionally become saturated with water, making the risk of frost damage increase further.

Moreover, the nature of the thaw salts can affect the damage process and chlorine ions from the salts may increase the risk of corrosion of steel components in the soil mix walls.

Erosion and wear

The surface of the soil mix can be affected by various erosion mechanisms, such as wear and cavitation. Wear is usually caused by a repeated mechanical movement across the surface, but also by grating effects caused by heavier elements (e.g. sand) that are present in the water or wind. The aggressiveness of these elements increases as their speed, roughness, hardness and size increase. Cavitation occurs when high speed water flows along/impacts on a wall. In case of uneven walls, the parallel flow of the water will be disturbed and move away from the soil mix surface, locally creating a very low pressure. If this pressure is lower than the vapour pressure of the water, air bubbles are formed that implode when they move to a higher pressure zone, causing the wall surface to become damaged.

Fire

For information about fire, please refer to §6.13.

9.4   Damage from steel corrosion

9.4.1   Steel and soil mix material

In general, one can assume that the cement-bound soil mix material is alkaline. In a strong alkaline environment, a passivation layer is formed on the steel surface, consisting of a dense iron oxide. This protects the underlying iron molecules against corrosion, by making them unreachable for moisture and oxygen.

217

This passivation layer can be affected by carbonation of the cement-bound material and/ or by the presence of chlorides. When cracks cause the steel to be directly exposed to the environment, corrosion will occur at a faster pace.

Corrosion is a complex chemical process in which oxygen and iron atoms in the presence of water react to rust (iron hydroxide ($Fe(OH)_2$). Given that rust takes a volume that is several times larger than that of steel, cracks will occur in the soil mix material and pieces will start to crumble. Moreover, corrosion causes a reduction in the cross-section of the steel beams and/or reinforcement rods, decreasing the bearing capacity/strength of the structure. These different causes of corrosion of metal parts in the soil mix wall are described below.

Governing factors of corrosion rate
In relation to an indication of steel corrosion rates, applied in soil mix walls, part 5 of Eurocode 3 (EN 1993-5) provides additional information. This information can also be found in CUR 166. See I - Tables 9.1 and 9.2.

I - Table 9.1. Corrosion (mm) of piles and sheet piles in soils and backfills with or without ground-water (per exposed side)[*].

Required design working life (years)	5 [***]	25 [***]	50	75	100
Undisturbed natural soils (sand, silt, clay, schist,...)	0.00	0.30	0.60	0.90	1.20
Polluted natural soils and industrial sites	0.15	0.75	1.50	2.25	3.00
Agressive natural soil (swamp, marsh, peat,...)	0.20	1.00	1.75	2.50	3.25
Non-compacted and non-agressive fills (clay, schist, sand, silt,...) [**]	0.18	0.70	1.20	1.70	2.20
Non-compacted and aggressive fills (ashes, slags,...)	0.50	2.00	3.25	4.50	5.75

[*] From EC 3 part 5. The listed values are for orientation. The actual values are dependent on local conditions.

[**] Corrosion rates are lower in compacted embankments than in non-compacted ones. For compacted embankments, the listed values should be divided by 2.

[***] Values for 5 and 25 years are based on measurements. The other values have been extrapolated.

I - Table 9.2. Corrosion (mm) of piles and sheet piles in fresh and salt water (per exposed side)[*,**].

Required design working life (years)	5 [***]	25 [***]	50	75	100
Common, fresh water (around the water line)	0.15	0.55	0.90	1.15	1.40
Very polluted fresh water (around the water line)	0.30	1.30	2.30	3.30	4.30
Sea water in temperate climate (splash zone and low water zone)	0.55	1.90	3.75	5.60	7.50
Sea water in temperate climate (permanent immersion zone)	0.25	0.90	1.75	2.60	3.50

[*] From EC 3 part 5. The listed values are for orientation. The actual values are dependent on local conditions.

[**] In water with ebb and flow, the highest corrosion rates occur at the height of the splash zone and the low water zone, see figure 9.2.

[***] Values for 5 and 25 years are based on measurements. The other values have been extrapolated.

The following values are given in EN 1993-5 for the loss in thickness due to corrosion in the atmospheric zone: normal atmospheric conditions: 0.01 mm/years; maritime conditions (nearby sea): 0.02 mm/year.

For an assessment of the risk of corrosion of steel beams in the soil mix walls, different exposition conditions of (parts of) the wall can be distinguished (see figure 9.2):

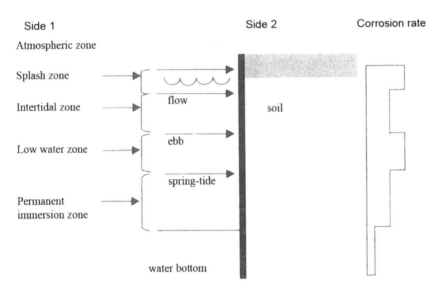

I - Fig. 9.2. Different zones along soil mix walls in soil and water.

The following notes should be given for these tables and figure:
- Steel parts in soil mix walls are generally covered by soil mix material. The degree of protection of the steel by the soil mix is partly dependent on the type of binding agent, dosage, water-cement factor, soil type, groundwater, any contaminations, etc.
- In case of wrongly placed beams and situations in which beams at the location of wide cracks are exposed to environmental conditions, one should take into account the absence of protection of the steel against corrosion.
- In the verification of the lifetime, one can theoretically make a distinction between steel components that are fully enveloped by soil mix material and locations with wide cracks.
- When installing the walls in salt (ground)water, one must take into account the chloride content in relationship to the cement mass in the water pores.
- The risk of corrosion of steel in wall parts (enveloped by soil mix) under the lowest groundwater level is extremely small. This is due to the lack and/or limited supply of oxygen. Nevertheless, in the absence of corrosion protection measures, a minimum corrosion rate is observed, see §6.14.
- When cracks in the soil mix material with a width of > 0.2 mm are located on the steel surface, local corrosion should be expected.

- When the walls are included in the soil (not excavated) and no corrosion protection measures are applied, the values included in I - Tables 9.1 and 9.2 can be applied.

9.4.2 *Carbonation initiating steel corrosion*

Due to the high pH value of young soil mix material, a passivation layer (a layer of iron hydroxide) may form around the steel beams, which – to a certain degree – protects against corrosion (rust formation). As a result of the reaction of the CO_2 from the air with the free calcium in the soil mix material, the pH will decrease from approx. 13 to below 9. This reaction is known as carbonation. If the pH value of the soil mix material drops below the value of 9, the passivation layer will be affected, and the corrosion process may start.

The carbonation front, that is to say the boundary between the carbonated and non-carbonated material, will evenly penetrate the soil mix. For concrete, the associated deposition of calcium carbonate ($CaCO_3$) may result in an improvement in the density of the concrete structure and a slight increase in the compressive strength. One may assume that this also applies to the soil mix material. However, as soon as the carbonation front reaches the steel components, the passivation layer becomes unstable and loses its protective nature against corrosion. It is generally believed that steel corrosion initiated by carbonation is present over large lengths and more or less evenly (generalised corrosion).

The speed at which the carbonation front penetrates the soil mix material (carbonation speed) is dependent on the composition of the soil mix and climatic conditions. The carbonation reaction can only take place in an aqueous environment, whereas the diffusion of CO_2 occurs 10000 times faster in dry material (air in the pores) than in wet material (pores filled with water). The carbonation rate is therefore greatest in soil mix that is alternately dry and wet, but still predominantly dry. A greater carbonation depth is generally present at the height of the cracks and corners. The carbonation rate decreases over time because the CO_2 must penetrate deeper and deeper into the material and pores become constricted by the deposition of calcium carbonate. For soil mix surfaces that are completely and permanently covered by a soil layer and/or groundwater, the risk of carbonation initiating steel corrosion is rather limited.

Carbonation is also described in Fib Model Code 'Service Life Design, report 34'. The input parameter for the diffusion resistance is unknown for the various compositions of the soil mix walls. Because the carbonation rate is related to the porosity of a material, this corrosion mechanism cannot be excluded.

9.4.3   Corrosion promoters

Chlorides

Despite the fact that steel components in the soil mix material may be protected against corrosion because of a high pH value, corrosion may still occur in non-carbonated soil mix. The reason for this is an excessive chloride concentration. These chlorides may be absorbed during execution of the wall due to their presence in the components of the soil (in the sand, water). In addition, in the course of years, chlorides may gradually penetrate into the soil. This phenomenon is mainly found in coast constructions or components exposed to thawing salts.

These chlorides may, even with a pH value greater than 9, penetrate the passivation layer around the reinforcement and, in the presence of sufficient water and oxygen, give rise to local corrosion. This form of corrosion is referred to as pitting corrosion ('pitting') and is dangerous because it can quickly decrease the dimension of the steel components. In addition, during this process, only little corrosion product is created, so that the 'warning' effect of the protruding cover and/or the cracks is postponed. After the corrosion reaction, these chlorides are released into the soil mix, after which they can immediately initiate a new reaction.

For the creation of new concrete, the standard EN 206-1 allows for a specific chloride content and gives a classification for concrete in accordance with the maximum admissible chloride content. Chloride content class Cl 0.40 is applicable to reinforced concrete or concrete with confined metals, while the class Cl 0.20 is used for concrete with prestressed steel reinforcement. For soil mix, no critical chloride levels are currently known. A high cement content in the soil mix has a positive influence, whereas the high porosity has a negative influence.

For a project-specific verification, the chloride content must generally be expressed on the basis of the ratio of the number of chlorine ions ($[Cl^-]$) relative to the number of hydroxide ions ($[OH^-]$). Since the amount of hydroxide ions is difficult to determine, the levels are usually expressed relative to the cement mass. If the cement-bound material carbonates, one can determine a decrease in the pH (and an increase of the ratio $[Cl^-]/[OH^-]$ for the same Cl^--content), whereas the cement mass (and also the chloride content relative to the cement mass) remains unchanged. This example shows that the critical chloride level relative to the cement mass for carbonated soil mix is lower than with non-carbonated soil mix.

For determing the risk (time-dependent) of corrosion initiated by chloride, one may also refer to CUR-Leidraad 1 (RCM value).

When installing soil mix walls in salt (sea) water, one must take into account that the pores are fully saturated with chloride-containing (ground) water.

On the basis of many research results, for reasonably compacted concrete with blast furnace cement and a maximum water-cement factor of 0.50, a percentage of 1% relative to the cement mass as critical concentration limit is applied for the occurrence of progressive reinforcement corrosion. For this reason, during the construction of new buildings a maximum chloride content of 0.4% relative to the cement mass is applied. No related empirical data is available for soil mix material. Based on a higher water-cement factor, a greater porosity and a higher risk of inhomogeneities, maximum chloride levels for soil mix material will likely be lower.

Corrosion caused by recovery
Corrosion generally occurs after execution of concrete surface repair work and is subsequently also expected to affect the soil mix material. The reason for this is that repair work may cause a significant increase in the corrosion potential of the steel in the repaired zone. The locations in the peripheral zones that acted as moderate cathodes before repair, may afterwards become highly anodic, increasing the risk of corrosion occurring in these zones due to the potential difference.

9.4.4 Corrosion caused by stray currents

As described earlier, corrosion is an electro-chemical process. In the vicinity of power stations, transformers and tram and/or railway lines, there are usually lots of stray currents that may disrupt the electro-chemical equilibrium of the steel components. This may result in the creation of additional anode zones, which in turn may give rise to corrosion.

9.5 Design considerations

On the basis of the assessment of the risk factors for durability during the design phase, one should take appropriate measures to safeguard the intended lifetime of the soil mix wall. Guidelines for measures that may be taken during the execution and design phase are provided in §6.14.

9.6 Execution in accordance with the design

When executing soil mix walls, the (variation in) soil composition is often an uncertain factor. Meanwhile, the soil composition is of major influence on the final properties and hence the lifetime of the walls.

A suitable method for gaining insight into the expected properties is the creation of a test wall/element. On the basis of tests carried out on this test wall/element, more information can be obtained on the properties of the wall to be realised in the actual project.

The test wall/element also provides information about, for example, the placement tolerance to be realised and how the intended cover can be managed.

After completion of the wall, the basic principles for the design of the soil mix wall must be validated on the basis of tests (such as the required compressive strength).

9.7   Management

9.7.1   Management - generalities

Measures that could be considered include:
- Including water-retaining walls for permanent polders in a global register (click-register) to prevent further drilling.
- Incorporating wire in the head of the wall, so that one can obtain the position of the walls on the site via click-register.
- Aspects to consider for adjacent buildings: how close can one responsibly drive/ vibrodrive a foundation?

9.7.2   Management – soil mix walls with a temporary function

In this case, the following measures may be considered:
- Periodical monitoring (visual inspection) of the degradation of the soil mix material, on the wall surfaces that are exposed to ambient air.
- Periodical monitoring (visual inspection) of the degradation of steel components in the wall surfaces (cracking, corrosion products).
- Recommended periodicity: during 1st year: 1x per quarter, 2nd year: 1x per half year.

9.7.3   Management – soil mix walls with a permanent function

The following measures are recommended for this:
- See walls with a temporary function. Depending on the measures taken to protect the soil mix wall against environmental influences during its lifetime, visual inspection may be necessary. If necessary, the following periodicity is recommended: during 1st year: 1x per quarter, 2nd year: 1x per half year, subsequently 1x per year. Note that the interval for inspections is clearly shorter than in comparable inspections of concrete constructions.
- A monitoring plan should be created. The parameters to be monitored and measurement frequency are different on a case-to-case basis. The reliability class/risk class of the work, the expected lifetime, the measures taken to protect the soil mix wall, the exposure limit conditions, etc. are all factors that must be taken into account when creating the monitoring plan. It is desirable to monitor the functioning of the walls during the first period of their lifetime in accordance with the requirements. Based on the estimation of possible deviating behaviour, recovery measures may be included in the monitoring plan. For more information about the monitoring of soil mix walls, please refer to Rutherford et al. (2005).
- When applying a cathodic protection system for the protection of steel elements, use specific monitoring techniques as described in BRL 2834.

223

9.8   Maintenance

The following applies for maintenance:
- Design and execute maintenance measures on the basis of results of control measurements and on the basis of a risk assessment of identified defects.
- Measures to be taken in the event of fire damage.

Consequences of damage

An appropriate recovery method must be chosen depending on the cause and condition at the time damage is detected. In case of early detection (small cracks and/or rust spots), one must determine whether the imperfections are merely an aesthetic problem or whether they are a sign of technical defects. Subsequently one can determine whether:
- a simple and (relatively) inexpensive recovery and/or application of an (additional) protective layer may suffice;
- the imperfections can eventually lead to dangers to passers-by (e.g. loosening of wall parts) or lead to the instability of the structure;
- the execution of repair works can be delayed on the basis of safety and economic arguments;
- the cause of the shortcomings may lead to a collapse of the structure in the long term, in the context of which only a (non-complete) replacement of the structure may provide a solution.

Chapter 10 Conclusions and recommendations (aspects that require further research)

This handbook is the result of more than 2 years of hard work by the SBRCURnet/BBRI committee 'Soil mix walls'. It has been the committee's intention to create clear and pragmatic guidelines for the design and execution of soil mix walls, a technology that has become increasingly popular in recent years.

It was a challenging task to clearly define the application area of soil mix walls and integrate guidelines in a coherent fashion in a context of national and European standards that are widely available for other foundation technologies and materials such as steel and concrete, but to a limited extent (or not at all) for soil mix material.

With pride, the committee can therefore say it has successfully completed its job. Progress and agreements have been made in many areas. The committee has always had as its goal to take positions that comply with the current state of knowledge about soil mixing, but is simultaneously open to future developments of this sustainable technology.

Some key issues that were realised in this handbook and which the committee wishes to pay particular attention to are summarised below.

- The different functions (earth-retaining, water-retaining, water cut-off, bearing, temporary and permanent functions) that a soil mix wall can fulfil and the requirements and preconditions to be met are clearly defined. It is important to mention that for permanent soil mix walls, the basic principle is that the wall should be protected against environmental influences during its lifetime (e.g. with a protection barrier/ wall). This basic principle can only be complied with under strict conditions.
- Procedures have been created on how to deal with a design based on pre-estimated values, which can only be verified after completion of the wall.
- Procedures and methods are available for testing the soil mix material, and to derive a characteristic value of the compressive strength from a data set of compression tests. One way of dealing with the (excessively) negative impact of tests on samples with large soil inclusions has also been incorporated.
- An agreement has been reached on a method that can be used for the verification of the pressure arch that occurs in the soil mix material between the steel beams when soil and water pressures press against the wall.
- A method of including the composite bending stiffness of soil mix and steel in the wall calculations has been proposed. Additionally, a design method is available (based on the EN 1994-1-1), that – to a limited extent and under strict preconditions – calculates the interaction between steel and soil mix material. The above methods are based on an extensive test campaign of the BBRI on real scale soil mix elements.

- Guidelines related to tolerances and quality control of the execution process are included in the handbook. Moreover, guidelines are provided on the number and type of control tests that should be carried out on the produced soil mix material. In addition, the number of control tests depends on the function and lifetime of the soil mix walls, the risk class of the work and the empirical data available for the deep mixing contractor.

Despite the fact that a lot of knowledge is already available (see the extensive part 2), the creators of this handbook regularly discovered insufficient insights that require further research. It should not come as a surprise that a number of these insights are related to the long-term performance and durability of soil mix walls. In summary, one may say that the following aspects require additional research:
- the execution methods to achieve a qualitative and homogeneous mix in certain soil conditions and with a specific mixing tool (execution method, minimum mixing energy, definition of an equivalent blade rotation number per type of mixing system, determining criteria per soil type, ...)
- durability of soil mix applications in general and under influence of all degradation mechanisms that may occur and that are summarised in §9 (part 1 of the handbook) and in §2.4 (part 2), including:
  - the alkaline nature of soil mix material: influencing factors and the extent to which soil mix can be considered as corrosion protection;
  - the execution of soil mixing in contaminated soils: quantifying the influence of different chemical substances on the setting, hardening time, long-term strength and stiffness;
  - (new) identification methods and protocols for chemical substances or constituents in the soil and the groundwater that pose a potential risk for the setting reaction, the strength development and/or the long-term strength;
  - protection methods against exposure of soil mix to the environment (drying, frost-thawing, ...);
  - effects of long-term exposure of soil mix to ambient air, frost-thaw cycles in relationship to the execution method, binder type and dosage, soil type, etc.
- behaviour of soil mix in case of fire;
- the axial geotechnical bearing capacity of soil mix elements: factors for base resistance, shaft friction, etc.;
- soil mix walls with reinforcement cages instead of beams;
- other applications of soil mix technology, e.g.
  - as soil improvement or for underpinning;
  - as rigid inclusions in a pile mattress system (LTP) concept;
  - as tensile elements;
  - as block stabilisation,
  - ...

In addition, it is recommended to create a database with empirical data of previously realised walls. Such a database could contain information such as location, execution data, strength and stiffness of the produced soil mix, environmental impact, potential execution problems, ...

Finally, it should be noted that the wall calculation in Belgium and the Netherlands are often performed with a spring model (e.g. D-sheet). These calculation models have not yet been adapted to the specific calculation methodology for soil mix walls elaborated in this handbook. In the long-term, it is therefore advisable to integrate a number of specific soil mix-related aspects from the design methodology in chapter 6 into such calculation models. The following aspects are included:

- the method of verifying the arching effect in soil mix material;
- the method of determining the composite bending stiffness;
- the method for determining the internal forces (M, N, V), which under certain circumstances takes into account the interaction between the soil mix and the steel beams;
- integration of applications of soil mix walls in the permanent (= use) phase, in which, relative to the temporary function, adjusted factors should be applied (long-term reduction of stiffness and strength of soil mix, coefficient of neutral soil pressure, ...);
- opportunity to divide the applied pressures between the soil mix wall (soil pressures) and the protection barrier/wall (water pressure) when creating a protection barrier/ wall;
- ...

Appendix 1 - Determination of the compressive strength of the soil mix material

Characteristic value of the compressive strength - problem definition
According to Denies et al. (2013), the determination of the characteristic value can be divided into two categories. The first category describes the characteristic value as a lower limit, e.g. 5 % quantile of a theoretical statistical distribution. The second uses the average value of the population in combination with a safety factor.

Based on the lower limit x %
As discussed in Denies et al. (2013), the first method focuses on the calculation of the characteristic compressive strength as the X % lower limit on the basis of a statistical distribution function. Nevertheless, in practice, it is often mistakenly assumed that the datasets of the UCS values of soil mix material are distributed normally (see I - Fig. T1.1a). The characteristic UCS value is then incorrectly calculated as the X % lower quantile of the normal distribution with parameters that correspond to the data set. Moreover, this often results in negative and therefore unusable characteristic UCS values. The mathematically correct solution would be applying the most suitable standard distribution function, e.g. a lognormal distribution if the distribution is skewed and/or does not contain subpopulations. The X % lower limit can subsequently be calculated on the basis of this theoretical distribution function, as illustrated in Denies et al. (2012) for a lognormal distribution (see I - Fig. T1.1b). A parameter β may have to be added to the UCS values in order to obtain an optimum consistency with normal distribution after transformation. However, this approach may be too complex for application in specific situations and it is not always possible to define the most suitable standard distribution function for a particular set of data (e.g. with a limited number of cored samples).

The second method for determining the X % lower limit is based on the cumulative frequency curve of the original experimental data set and is therefore independent from a theoretical distribution function. Please note that in order to apply this method, sufficient data points must be available (at least 20 samples are required for an accurate determination of the 5% lower limit without extrapolation). This approach may seem quite simple, but any other method may result in great uncertainty. Figure T1.2 shows the cumulative frequency curve for the UCS values of the data set shown in I - Fig. T1.1.

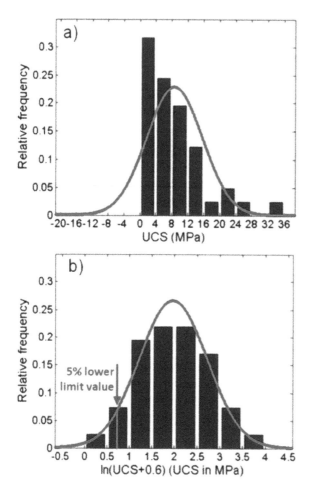

1 - Fig. T1.1. a) Distribution of the UCS values of 41 cores of soil mix material from a site in Gent (Belgium) and the associated theoretical Gauss curve.

b) Distribution of the logarithm of the UCS values increased with β = 0.6 from the same site and associated Gauss curve. The vertical line indicates the 5 % lower limit value, in accordance with Denies et al. (2012).

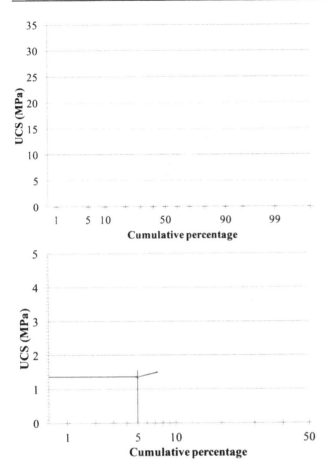

1 - Fig. T1.2. Cumulative frequency curve of all UCS values of the data set of the site in Gent:
a) Complete curve.
b) Enlargement of the part below 50 %: representation of the construction for the evaluation of the
5 % lower limit value. Denies et al. (2013).

On the basis of an average value with safety factor

A second method to determine the characteristic UCS value is the use of the average value of the data set in combination with a safety factor:

$$f_{c,k} = \alpha \overline{q_{uf}}$$ (T1.1)

in which $\overline{q_{uf}}$ is the average UCS value and α a factor that represents a certain reliability and safety level ($\alpha < 1$). In the formalised design method (DIN 4093, August 2012) that is used in Germany, the characteristic UCS value is defined as a minimum value of three parameters:

$$f_{c;k} = min\ (f_{m,min}\ ;\ \alpha f_{m,mittel}\ ;\ 12\ MPa)$$ (T1.2)

in which $f_{m,min}$ is the minimum UCS value and $f_{m,mittel}$ the arithmetical average of UCS value from a series of at least 4 samples. α is determined in function of $f_{c,k}$: α is equal to 0.6 with $f_{c,k} \leq 4$ MPa and 0.75 with $f_{c,k} = 12$ MPa (linear interpolation is required for intermediate values). This method is described in detail by Topolnicki and Pandrea (2012). If the characteristic value $f_{c,k}$ is lower than 4 MPa, additional creep tests must be conducted with a load of $f_{c,k}$ /2 such as described in annex B of the DIN 4093 standard. The design strength for calculations with the concept of partial safety factors is subsequently calculated as follows:

$$f_{c,d} = 0.85\frac{f_{c,k}}{\gamma_m}$$ (T1.3)

in which 0.85 is a factor that takes into account the permanent situations and γ_m the material safety factor as described in part 1 of Eurocode 7 (1.5 for permanent and temporary load situations and 1.3 for accidental load situations). For temporary situations, the design strength is calculated without the 0.85 coefficient.

As described in Topolnicki and Pandrea (2012), the maximum allowed compressive stress is 0.7 x $f_{c,d}$ and the maximum allowed shear stress is 0.2 x $f_{c,d}$ if the compressive and shear stresses are calculated independently of each other (so without 3D stress analysis).

For this second method – on the basis of the average UCS value with safety factor – Denies et al. (2012) noticed that the definition of the most suitable average (arithmetical average, median, etc.) must be dependent on the type of data set distribution. In addition, problems may occur in case of a limited number of samples, skewed populations and the existence of sub-populations. Figure T1.3 compares the characteristic UCS value, calculated with the aid of the cumulative frequency curve (CC method) and with respect to the DIN method.

The ratio of the two characteristic values is assumed as a function of the number of tested samples for each considered data set. At least 20 samples are required to conduct the statistical analysis on the cumulative frequency curve. As shown in I - Fig. T1.3, the characteristic UCS value is always greater when they are calculated with the aid of the cumulative frequency curve (all ratios of the two characteristic values are greater than 1). I - Fig. T1.3 shows the results for two different X % lower quantiles: X = 5 % and 10 %.

After all, for the first category of approaches (on the basis of the lower limit value) a value for X % must be defined. A more detailed analysis is required to determine whether a 5% lower limit, as often mentioned in part 1 of Eurocode 7, is a representative characteristic value for the strength of the soil mix material. A major problem is namely the representativeness of the cored samples with respect to the in situ executed soil mix material as discussed earlier in §2.3.10, §2.3.11 and §2.3.12 in part 2 of the handbook.

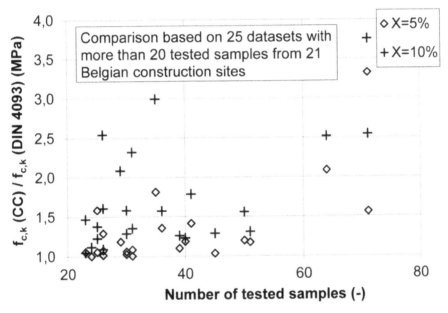

I - Fig. T1.3. Ratio of the characteristic values ($f_{c,k}$ (CC) and $f_{c,k}$ (DIN 4093)) as function of the number of tested samples, Denies et al. (2013).

Calculation procedure for determining the characteristic compressive strength of the soil mix material

In practice, after execution of the soil mix, the actual compressive strength must be verified on the basis of UCS tests on cored material.

The procedure used for this is illustrated in I - Fig. T1.4 and is described in more detail in the text.

On the basis of a specified design value of the compressive strength of the soil mix material, the same procedure can also be applied to determine which average and/or characteristic compressive strength the deep mixing contractor must reach during the production process.

In this calculation procedure, the characteristic value of the compressive strength of the soil mix material can be determined in two ways, depending on the number of samples that are available for the mechanical characterisation of the material. The minimum number of samples is indicated in Chapter 8 of part 1.

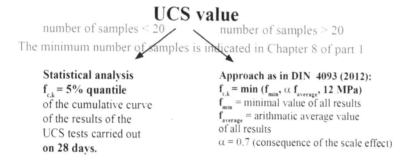

UCS value

number of samples < 20   number of samples > 20

The minimum number of samples is indicated in Chapter 8 of part 1

Statistical analysis	Approach as in DIN 4093 (2012):
$f_{c,k}$ = 5% quantile	$f_{c,k}$ = min (f_{min}, α $f_{average}$, 12 MPa)
of the cumulative curve	f_{min} = minimal value of all results
of the results of the	$f_{average}$ = arithmatic average value
UCS tests carried out	of all results
on 28 days.	α = 0.7 (consequence of the scale effect)

Rule of Ganne et al (2010) for the selection of the samples:
All samples are primarily tested and after the test the rule is applied
on the basis of the observation of the operator of the test.

I - Fig. T1.4. Design procedure for determining the UCS characteristic value of the soil mix material.

The first approach concerns the calculation of the characteristic value on basis of a statistical analysis. As discussed above, the definition of a statistical theoretical distribution is not evident for all practical cases. A more pragmatic approach is to determine the characteristic value of the compressive strength as the 5% quantile of the cumulative curve of the results of the UCS tests carried out on in situ samples after 28 days of hardening. This approach is only permitted if at least 20 test results are available to establish the cumulative curve. If fewer than 20 test results are available, one should execute a extrapolation to derive the 5% quantile value, which could potentially lead to inaccurate values. This approach is illustrated in I - Fig. T1.2.

The second approach is based on the content of DIN 4093 and can be applied if one has fewer than 20 test results of the UCS of cored material. In this case, the characteristic value is determined as the minimum of three values:
- the minimum value of all results of the UCS tests
- the arithmetic average value of all results of the UCS tests multiplied by a constant reduction factor α
- a maximum value of 12 MPa

The constant reduction factor α is equal to 0.7. This factor is determined on the basis of the results of the tests on large soil mix blocks and, in principle, takes into account the scale effect (see §2.3.12 in part 2 of the handbook).

After conducting all UCS tests, it becomes permissible to carry out a limited selection based on the observations of the operator of the UCS tests and as described in Ganne et al. (2010) – see §2.3.11 in part 2 of the handbook. The results of the cored samples with unmixed soft soil inclusions larger than one-sixth of the diameter of the core may be rejected. In addition, no more than 15% of test specimens from one specific site may be rejected. This possibility to reject results of test samples is based on the observation that a soil inclusion of 20 mm or less will not affect the behaviour of a soil mix structure. However, a soil inclusion of 20 mm in a test sample with a diameter of 100 mm can significantly influence the test result on the core, as well as the deduction of a characteristic compressive strength from a data set. Naturally, this precondition is only applicable if it is assumed that the soil mix construction will not contain soil inclusions larger than one-sixth of the width of the soil mix construction.

The influence of unmixed soft soil inclusions in the soil mix material on its mechanical characteristics is discussed in §2.3.11 in part 2 of the handbook, amongst other things on the basis of the results of the numeric analyses. These analyses have demonstrated the relevance of the above selection rule.

In function of the soil type, figure T1.5 illustrates the dispersion of the characteristic values calculated according to the above-described approach. These characteristic values come from 42 sites tested within the framework of the BBRI "Soil Mix" project (2009-2013). I - Fig. T1.5 shows the gross values without correction for the age of the test samples at the moment of the UCS test. Figure T1.6 shows the same information as I - Fig. T1.5, the only difference being that the characteristic values were corrected in order to obtain the strength after 28 days using equations (2.3) and (2.4) in part 2 of the handbook. For the sand, a value of 0.71 was taken for the empirical factor s in these equations. For loam and clay soils this value was 0.98. These are the s-values that were obtained from the lab test campaign of the BBRI (see paragraph 2.3.2d in part 2 of the handbook).

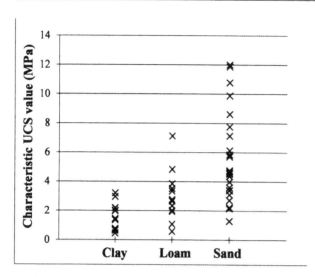

1 - Fig. T1.5. Dispersion of the characteristic values in function of the soil type for the data sets from the BBRI "Soil Mix" project (2009-2013). The values have not been corrected for age.

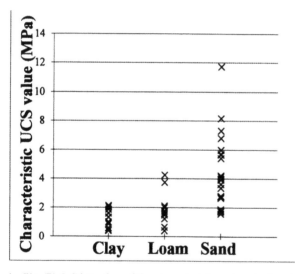

1 - Fig. T1.6. Dispersion of the characteristic values in function of the soil type for the data sets from the BBRI "Soil Mix" project (2009-2013). The values are corrected to obtain the strength after 28 days.

Appendix 2 - Horizontal bending stiffness and moment capacity of soil mix walls - Background information and calculation examples

T2.1 Horizontal bending stiffness - generalities

The large-scale bending tests of the BBRI (see §2.3.13 in part 2 of the handbook) have invariably demonstrated:
- that the bending stiffness of a reinforced soil mix element is significantly larger than the bending stiffness of only the steel beam;
- that – on the other hand – once the cracking moment is exceeded, the global bending stiffness quickly drops to a value that is significantly smaller than the bending stiffness of the uncracked soil mix section.

The rapid return to a completely cracked stiffness is contrary to the behaviour of concrete beams, in which we see a more gradual transition from the uncracked stiffness to the cracked stiffness (as a result of tension - stiffening). Given the low cracking moment, for soil mix walls it is recommended to take into account the crack formation (and in the long term also the creep) for determining the bending stiffness that is applied in the elastoplastic wall calculation (calculation internal forces and deformations).

In §6.6 of the handbook, two calculation methods are described for determining the effective bending stiffness EI-eff that can be entered into the wall calculations:
- In approach 1, based on the methodology from EN 1992-1-1, it is recommended to determine the effective reduced bending stiffness EI-eff to be determined as the average of the uncracked stiffness EI-1 and the cracked stiffness EI-2 of the composite section, or therefore:

$$EI\text{-}eff \ = \ \frac{EI\text{-}1 + EI\text{-}2}{2} \qquad\qquad\qquad (T2.1)$$

in which:
  ◦ both in EI-1 and EI-2, the bending stiffness of the reinforcement is taken into account
  ◦ the uncracked stiffness EI-1 is calculated as described in §6.6.4.1
  ◦ EI-2 is calculated as described in §6.6.4.2

- In the (simplified) approach 2, the effective bending stiffness is calculated as the sum of the stiffness of the steel beam and the stiffness of the compressed soil mix zone, in which it is assumed that the neutral line runs through the middle of the beam; this results in the following simplified formula.

$$EI\text{-}eff = E_a I_a + E_{sm} \left[\frac{b_{cl} \cdot \left(\frac{h_{sm}}{2}\right)^3}{3} \right] \tag{T2.2}$$

Here one should make a distinction between the long term and short term bending stiffness:

- ◦ in the short term, the E-modulus can be applied for the soil mix material as deduced from compression tests with measurement of the deformations on the sample or determined via the empirical correlations with the compressive strength, as described in §6.3.7 in part 1 of the handbook and §2.3.3 in part 2 of the handbook;
- ◦ in the long term, due to the creep behaviour, the average E-modulus of the soil mix material can be reduced by a factor $(1+\Psi)$, or the bending stiffness of the soil mix material can be completely ignored and one can only take into account the bending stiffness of the reinforcement.

T2.2 Influence of the bending stiffness on the calculated internal forces and deformations

Approach 1 above proposes to take the average of the bending stiffness of the uncracked and cracked cross-section for the bending stiffness in the wall calculations. In order to obtain an indication of the impact of the use of this average, reduced bending stiffness in the wall calculation on the calculated internal forces, anchor forces and deformations, please see the following two (for a non-anchored and single anchored wall) wall calculations (in SLS) for different bending stiffnesses:

- the bending stiffness of the beam+soil mix uncracked: EI-1
- the EI-eff, in accordance with §6.6.4 determined as the average of the cracked and uncracked composite cross-section
- the obtained value EI-eff +/- 25%

Example 1 – non-anchored wall
Basic information:

- Wall: panel wall, thickness h_{sm} = 0.55 m
- Beam: IPE270 all 1.1 m; E_a = 210000 MPa
- Soil mix characteristics E_{sm} = 4000 MPa;
  ($f_{sm,m}$ ≈ 3.5 MPa; $f_{sm,k}$ ≈ 2.42 MPa), no creep reduction
- Wall calculations conducted in SLS

Basic calculation result with uncracked bending stiffness EI-1 per m' wall:

EI-1 = (5790E+4 x 210000) Nmm² + 1.52E+10 x 4000) Nmm² = 7.30E+13 Nmm²

EI-1/m = 7.30E+13 Nmm² /1.10 m = 6.64E+13 Nmm² /m

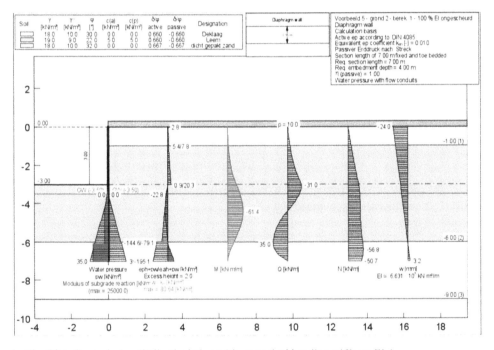

I - Fig. T2.1. Example 1a - Wall calculation with uncracked bending stiffness EI-1.

Calculation with the effective cracked bending stiffness EI-eff per m' wall:

EI-1/m -uncracked: 7.30E+13 Nmm² /1.10 m = 6.64E+13 Nmm² /m

EI-2/m - cracked: 2.91E+13 Nmm² /1.10 m = 2.65E+13 Nmm² /m

EI-eff = (EI-1 + EI-2)/2: 5.10E+13 Nmm² /1.10 m = 4.64E+13 Nmm² /m

The wall calculations for the effective cracked bending stiffness are listed below (see I. Fig T.2.2).

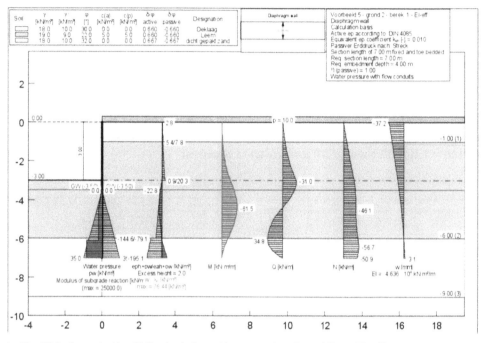

I - Fig. T2.2. Example 1b - Wall calculation with average bending stiffness EI-eff.

Finally, the calculation is repeated with a bending stiffness of 125% of EI-eff and of 75% of EI-eff. The results are summarised in I - Table T2.1.

I - Table T2.1. Summary of wall calculations for different examined wall bending stiffnesses.

Case (-)	EI (Nmm²/m')(x10¹³)	Head displacement (mm)	Bending moment M (kNm/m)	Shear force Q (kN/m)
EI-1 uncracked	6.64	-24.0	-61.4	-31.0/+35.0
100% EI-eff	4.63	-37.2	-61.5	-31.0/+34.8
125% EI-eff	5.78	-34.4	-61.4	-31.0/+34.8
75% EI-eff	3.71	-40.7	-61.5	-31.0/+34.8

For the non-anchored wall, the bending stiffness seems to have almost no effect on the internal forces, which could be expected as soon as the displacements are of such a magnitude that the active soil pressures are completely mobilised and therefore no longer change in function of the deformation. The application of the effective cracked bending stiffness logically results in significantly larger head displacements than in cases where the uncracked bending stiffness

240

is applied. Whether or not the deformations are considered excessive should be taken into account in the verification of the wall deformations.

Example 2 – Single anchored wall
Basic information:

- Wall: panel wall, thickness h_{sm} = 0.55 m
- Beam: IPE330 all 1.1 m; E_a = 210000 MPa
- Soil mix characteristics: E_{sm} = 6000 MPa;
  ($f_{sm,m} \approx 5.25$ MPa; $f_{sm,k} \approx 3.63$ MPa); no creep reduction
- Wall calculations conducted in SLS

Basic calculation result with uncracked bending stiffness:

EI-1 = (11770E+4 x 210000) Nmm² + (1.52E+10 x 6000) Nmm² = 11.55E+13 Nmm²

EI-1/m = 11.55E+13 Nmm² /1.10 m = 10.50E+13 Nmm² /m

The wall calculations with uncracked bending stiffness are indicated below (see I. Fig T2.3.).

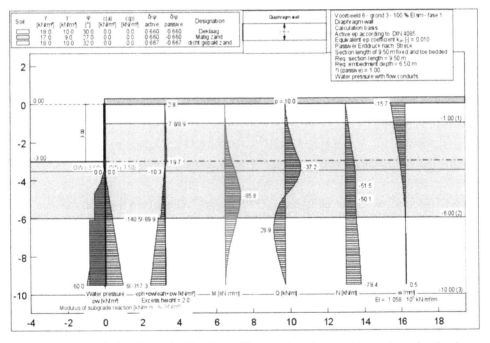

I - Fig. T2.3. Example 2a – Uncracked bending stiffness - phase 1: excavation up to anchor level.

241

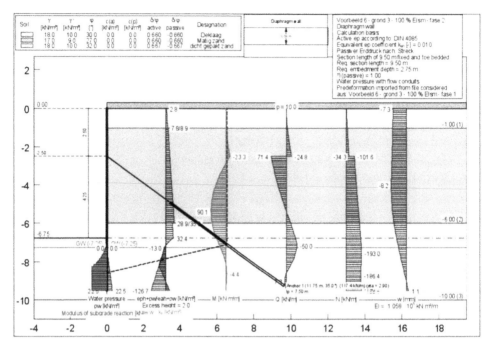

I - Fig. T2.4. Example 2a – Uncracked bending stiffness - phase 2: boring and pretensioning of the anchors and excavation up to final level.

Calculation with the effective cracked bending stiffness EI-eff per m' wall:

EI-1/m - uncracked: $11.55E+13$ mm^4 /1.10 m = $10.50E+13$ mm^4 /m

EI-2 - cracked: $4.87E+13$ mm^4 /1.10 m = $4.42E+13$ mm^4 /m

EI-eff = (EI-1 + EI-2)/2: $8.21E+13$ mm^4 /1.10 m = $7.46E+13$ mm^4 /m

The wall calculations for the effective cracked bending stiffness are listed on the next page (see I. Fig T2.5 and I. Fig. T2.6).

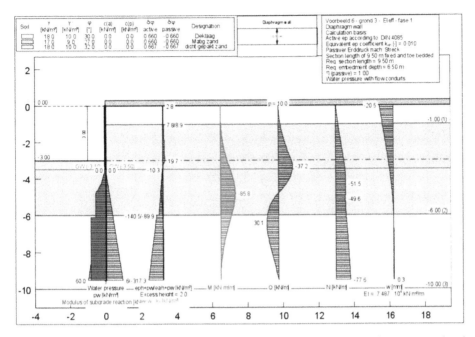

I - Fig. T2.5. Example 2b – Effective bending stiffness - phase 1: excavation up to anchor level.

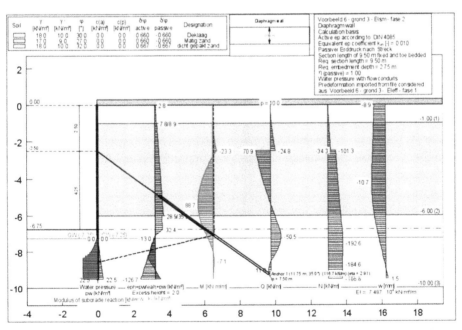

I - Fig. T2.6. Example 2b – Effective bending stiffness - phase 2: boring and pretensioning of the anchors and excavation up to final level.

Case	EI	Head displacement s1 Field displacement s2	Bending moment M	Anchor force Q
(-)	(mm⁴/m') (x10¹³) $(mm^4/m')(x10^{13})$	(mm)	(kNm/m)	(kN/m)
EI-1 uncracked	10.50	-15.7/-8.2	-85.8/+90.1	117.4
100% EI-eff	7.46	-20.5/-10.7	-85.8/+88.7	117.5
125% EI-eff	9.32	-17.2/-9.2	-85.8/+89.6	117.1
75% EI-eff	5.60	-26.1/-13.7	-85.9/+87.5	116.2

The calculation has also been repeated with a bending stiffness of 125% of EI-eff and of 75% of EI-eff. The results are summarised in I - Table T2.2.

The influence of the bending stiffness on the internal forces is also very limited in the example of a single anchored wall. The same applies for the anchor forces. Similarly, here the bending stiffness mainly affects the calculated wall displacements, especially the displacements in the first construction phase (excavation up to anchor level).

T2.3 Comparison of the calculated bending stiffnesses EI-eff with the BBRI test data

The EI-eff bending stiffnesses have also been calculated with method 1 for the wall geometries that have been tested in the large-scale bending tests of the BBRI (see §2.3.13 in part 2 of the handbook). The results thereof are summarised in I - Table T2.3.

The calculation method of EI-eff constantly gives higher values than those found in the bending tests. Therefore, for the calculation of the internal forces this approach is on the safe side. On the other hand, more optimistic values are therefore deduced for the calculation of the wall deformations.

1 - Table 12.3. Comparison of the calculated bending stiffnesses with the values obtained from BBRI large-scale bending tests.

	site	type	EI_a	Calculated without creep			Calculated with creep: Esm/2.0			Deduced from bending tests	
				EI-1	EI-2	EI-eff	EI-1,kr	EI-2,kr	EI-ef	EI-test 100 MPa	EI-test 180 MPa
			kNm^2 $\times 10^3$	kNm^2 $\times 10^3$	kNm^2 $\times 10^3$	kNm^2 $\times 10^3$	kNm^2 $\times 10^3$	kNm^2 $\times 10^3$	kNm^2 $\times 10^3$	kNm^2 $\times 10^3$	kNm^2 $\times 10^3$
1	Heverlee	CSM1	16.3	107	41.6	74.3	62.4	32.7	41.6	40	36
2	Heverlee	CSM2	16.3	107	41.6	74.3	62.4	32.7	41.6	39	36
3	Aalst	CSM1	16.3	114.9	43.4	79.1	66.1	33.6	43.4	30	27
4	Aalst	CSM2	16.3	114.9	43.4	79.1	66.1	33.6	43.4	39	32
5	Aalst	CSM3	16.3	114.9	43.4	79.1	66.1	33.6	43.4	47	42
6	Aalst	CSM4	16.3	114.9	43.4	79.1	66.1	33.6	43.4	59	30
7	Leuven	TSM1	8.17	35.6	16.5	26.1	21.7	13	16.5	15	14.5
8	Leuven	TSM2	8.17	35.6	16.5	26.1	21.7	13	16.5	13.5	13
9	Leuven	TSM3	-	-	-	-	-	-	-	18.5	13.5
10	Leuven	TSM4	-	-	-	-	-	-	-	13	8
11	Blankenberge	C-mix 1	2.76	13.4	5.15	5.15	8.05	4.32	5.15		
12	Antwerp	C-mix 1	12.16	73.3	30	9.3	42.4	22.6	30		
13	Antwerp	C-mix 2	-	-	-	-	-	-	-		
14	Antwerp	C-mix 3	-	-	-	-	-	-	-		
15	Knokke	C-mix 1	12.16	44.1	23.5	33.8	27.8	18.2	23.5		
16	Limelette	C-mix 1	5.82	51.3	16.7	34.0	28.4	12.6	16.7	10.5	10
17	Limelette	C-mix 2	5.82	29.6	13.5	82.3	17.6	10.2	13.5	14.5	15

EI_a: the stiffness of the steel beam alone

EI-1:   the uncracked stiffness of the composite cross-section without creep

EI-2:   the cracked stiffness of the composite cross-section without creep

EI-1,kr   is the uncracked stiffness of the composite cross-section with creep
          (E-modulus of the soil mix divided by a factor of 2.0)

EI-2,kr   is the cracked stiffness of the composite cross-section with creep
          (E-modulus of the soil mix divided by a factor of 2.0)

EI-test   is the stiffness obtained from the tests at a working stress of
          +/- 100 N/mm² and 180 N/mm²

T2.4   Resistance moment and moment capacity – general observations

The large-scale bending tests conducted by the BBRI (see §2.3.13 in part 2 of the handbook) have consistently demonstrated that the moment capacity (= moment of resistance) is significantly greater than the moment capacity of the steel beam alone: both the maximum elastic moment of resistance and the maximum plastic moment of resistance. Results are discussed in more detail in §T2.5.

If the conditions allow to do so, a consideration of the moment capacity for the composite section 'steel soil mix' is completely sensible. The calculation method can be based on the requirements in EN 1994-1. In case the cooperating effect of the soil mix material is ignored, the moment capacity of the soil mix wall can be calculated on the basis of the requirements in EN 1993. A number of values must be determined for the calculation of the moment capacity:

1. the effective width b_{c2} for the structural verification
2. the characteristic compressive strength f_{ck} of the soil mix material
3. the safety factors (material factors) on the steel and soil mix material
4. the potential uncertainty factors on geometry and compressive strength
5. the potential factors for long-term effects.

T2.5   Effective width for the verification of the internal forces

For determining the effective width b_{c2} for the verification of the internal forces, please refer to the criteria described in §6.7.2. Some important points to remember:

1. For panels walls, criterion 1 (boundary in relationship to the axis-to-axis distance between the zero moment points and the axis-to-axis distance between the steel beams) is usually subordinate to criterion 3 (boundary due to the adhesion) and/or criterion 4 (maximum 2x the flange width), and rarely or never governing.
   The axis-to-axis distance between the zero moment points is included in the assessment of the adhesion. Here one must make a distinction between the distance between the zero moment points from the field moments, and the (usually shorter) distance from the zero moment points above and below the support points.
2. With respect to column walls, it is best only to consider the reinforced columns (and ignore the non-reinforced ones) for the calculation of the moment capacity. This can be explained as follows:

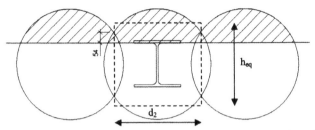

I - Fig. T2.7. Illustrative boundary of the effective width b_c for column walls.

I - Fig. T2.7 shows the plastic neutral line for a column wall with diameter 550 mm and a HEA240 beam. The PNL will often run through the upper flange of the beam. Everything under the PNL can be ignored. Everything above the PNL should be included. However, it is doubtful that the full plastic pressure force can develop itself in the non-reinforced columns. After all, shear forces have to be transferred from the reinforced column onto the non-reinforced column. These shear forces must be transferred at the height of the intersection between both columns and above the neutral line. Due to the round shape of the columns and the fact that the PNL is usually located very high, the height of the surface along which the shear forces must be transferred is very small. For the calculation of the moment capacity of column walls, it is therefore best to completely ignore the non-reinforced columns.

T2.6 Evaluation of the large-scale bending tests performed at the BBRI

For the 13 test panels with steel beams on which the BBRI conducted a large-scale bending test (see §2.3.13 in part 2 of the handbook), moment capacities were calculated with the methods described in§6.7. The overview of the maximum measured bending moments and of the calculated moment capacities are included in I - Table T2.4. In this case, NO boundary is applied for the adhesion, nor on the effective width which is considered equal to the width of the test sample.

1 - Table T2.4. Comparison of the calculated moment capacities and values obtained from BBRI large-scale bending tests.

					Test Results		Calculated values			
	site	type	profile	steel	$f_{sm,k}$ MPa	M_{max} kNm	M(Rk,a,el) kNm	M(Rk,a,pl) kNm	M(Rk) kNm	M(Rd) kNm
1	Heverlee	CSM1	HEA 240	S235	4.5	326	158.65	174.98	261.35	226.87
2	Heverlee	CSM2	HEA 240	S235	4.5	376	158.65	174.98	261.35	226.87
3	Aalst	CSM1	HEA 240	S235	4.65	286	158.65	174.98	263.89	228.43
4	Aalst	CSM2	HEA 240	S235	4.65	303	158.65	174.98	263.89	228.43
5	Aalst	CSM3	HEA 240	S235	4.65	306	158.65	174.98	263.89	228.43
6	Aalst	CSM4	HEA 240	S235	4.65	290	158.65	174.98	263.89	228.43
7	Leuven	TSM1	IPE 240	S235	5.7	168	76.21	86.15	123	112
8	Leuven	TSM2	IPE 240	S235	5.7	162	76.21	86.15	123	112
9	Leuven	TSM3	6 Ø14		5.7	133	-	-	-	-
10	Leuven	TSM4	6 Ø14		5.7	95	-	-	-	-
11	Blankenberge	C-mix 1	IPE180	S235	23	65	34.38	39.10	59.17	54.25
12	Antwerp	C-mix 1	IPE 270	S235	8.03	103	100.79	113.74	193.1	166.38
13	Antwerp	C-mix 2	Cage		15.63	147	-	-	-	-
14	Antwerp	C-mix 3	Cage		10.37	90	-	-	-	-
15	Knokke	C-mix 1	IPE 270	S235	4.62	143	100.79	113.74	166.9	150.7
16	Limelette	C-mix 1	IPE 220	S235	10.7	139	59.22	67.07	126.55	106.7
17	Limelette	C-mix 2	IPE 220	S235	9.7	178	59.22	67.07	131.24	109.3

$f_{sm,k}$:   the characteristic UCS of the soil mix material, obtained on large test samples

M_{max} :   the maximum moment measured in the large-scale bending test

M(Rk,a,el) :   the elastic moment capacity of the steel beam alone

M(Rk,a,pl) :   the plastic moment capacity of the steel beam alone

M(Rk) :   the characteristic moment capacity of the composite section, with the safety coefficient on the soil mix equal to 1.0

M(Rd) :   the design value of the moment of resistance of the composite section, with a safety of 1.5 on the soil mix material and 1.0 on the steel, and a factor of 0.85 for long-term effects

Some important aspects to take note of for the calculations and for the comparison of the measured and calculated moment capacity:

- For the CSM walls, calculations have been made with a section of 1200×550 mm^2.
- For the TSM columns and C-mix columns, calculations have been made respectively with a square section of: $0.89 * 550$ mm = 489 mm and $0.89 * 650 = 578$ mm (which slightly deviates from the regulations listed in §T2 above).
- In most cases, the ratio M_{max} /M(Rk) is between 1.08 en 1.44. This means that the plastic calculation of the composite cross-section can be justified.
- For test 12, the measured value M_{max} is not greater than the plastic moment of the steel beam itself. This because in this test the beam is located on the edge of the column, the beam was not even fully enveloped by soil mix. This cannot possibly lead to a good interaction between the beam and soil mix.
- For the tests 11, 16 and 17, the extremely high compressive strengths of the soil mix material (up to 23 N/mm^2) – in combination with a small beam - are remarkable. Theoretically, this causes the plastic neutral line to be located above the beam, something which is physically impossible. For these tests, we have theoretically reduced the thickness of the wall so that the PNL can run through the beam:
  - for test 11: h_{eq} = 260 mm; this results in a ratio of M_{max} /M(Rk) = 1.1
  - for test 16: h_{eq} = 460 mm, this results in a ratio of M_{max} /M(Rk) = 1.1
  - for test 17: h_{eq} = 485 mm, this results in a ratio of M_{max} /M(Rk) = 1.35

Appendix 3 - Example calculations of the arching effect in Plaxis 2D

A 2D analysis has been made in a finite elements calculation to determine the arching effect and to verify the expected arching effect pattern.

Soil mix wall
- Thickness 0.55 m
- Beams IPE 360 axis-to-axis 1.22 m
- Mohr Coulomb material criterion
- E = 2000 MPa
- Compressive strength = 2.0 MPa
- c = 1000 kPa
- $\varphi = 0.1°$
- Tensile strength = 2000, 100, 0 kPa

Steel IPE 360
- Web: h = 360 mm, t = 8.0 mm
- Flange: b = 170 mm, t = 12.7 mm

1 - Table T3.1. Plaxis parameters "beam" elements.

	Unit	Flange	Body
Thickness	mm	12.7	8.0
Modulus of elasticity	kN/m²	2.1×10^8	2.1×10^8
Surface A	m²/m	12.7×10^{-3}	8.0×10^{-3}
EA	kN/m	2.67×10^6	1.68×10^6
Moment of inertia I	m⁴/m	171×10^{-9}	42.7×10^{-9}
EI	kNm²/m	35.9	8.97

Model
- Evenly distributed load on the wall of 50 kPa;
- Support of IPE beams by means of 'Fixed End Anchors' (not visible in execution); the wall itself is not supported;
- The material model with a 'Tension Cut Off' of 0.1 MPa is based on the assumption that the tensile strength remains constant in the event of increased lengthen. In actuality, crack formation occurs when exceeding the tensile strength and no residual strength remains.

Results

The calculation results are graphically depicted in figures T3.1 to T3.13 and summarised in I - Table T3.2. In Soil mix 01, compressive and tensile strength are nowhere near close to being reached. In Soil mix 02, the tensile strength is achieved in the zone above the beams. Because this zone is relatively small, the stress distribution in the wall is almost equal to Soil mix 01. In Soil mix 03, the tensile strength is equal to 0, as a result of which a compression arch must developed itself. The corresponding maximum compression stress is almost twice as great as in the two other models.

The analytical model by Mathieu et al. (2012) has been used to determine the maximum compression and shear stress:

$$\sigma_{max} = 2p \left[1 + \frac{D^2}{4H^2} \right] \tag{T3.1}$$

$$D = 1.22 - 0.17 = 1.05 \ m$$

$$H = \frac{0.55 - 0.36}{2} + 0.36 = 0.455 \ m$$

$$p = 50 \ kPa$$

$$\sigma_{max} = 233 \ kPa$$

$$\tau_{max} = \frac{pD}{2H} \tag{T3.2}$$

$$\tau_{max} = 58 \ kPa$$

It is concluded that the maximum compressive stress corresponds with the Plaxis model Soil mix 03. The maximum shear stress is greatly underestimated by Mathieu et al. (2012). In Plaxis model Soil mix 03, this is approx. 100 kPa. It is clear that the simple model developed in Mathieu et al. (2012) cannot determine the maximum value.

I - Table T3.2. Plaxis 2D calculation results of the arching effect in the soil mix wall

Model	Soil mix 01	Soil mix 02	Soil mix 03
Tensile strength [kPa]	2000	100	0
Parameter			
Max. stress mid cross-section, compression [kPa]	79	79	228
Max. stress mid cross-section, tensile [kPa]	70	70	0
Stress at the base of the pressure arch [kPa]	75	75	135
Max. principal stress in the pressure arch [kPa]	130	130	220
Max. deviatoric stress in the pressure arch [kPa]	120	115	200
Max. shear stress in the pressure arch [kPa]	60	58	100

Soil mix 01 no tension cut off

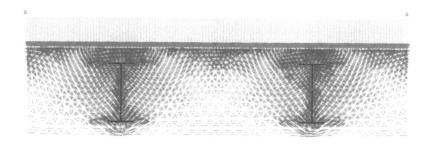

Total principal stresses (scaled up $0.500*10^{-3}$ times)
Maximum value = 175.9 kN/m^2 (Element 2405 at Stress point 28851)
Minimum value = -184.4 kN/m^2 (Element 296 at Stress point 3543)

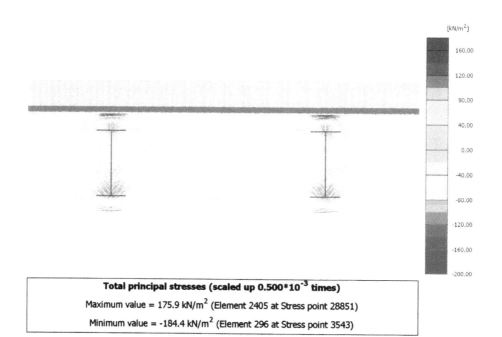

Total principal stresses (scaled up $0.500*10^{-3}$ times)
Maximum value = 175.9 kN/m^2 (Element 2405 at Stress point 28851)
Minimum value = -184.4 kN/m^2 (Element 296 at Stress point 3543)

I - Fig. T3.1. Plaxis 2D calculation results: soil mix 01 – total principal stresses.

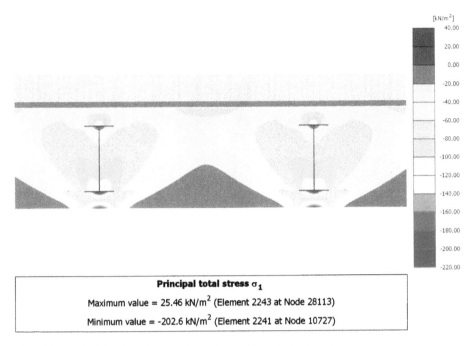

Principal total stress σ₁

Maximum value = 25.46 kN/m² (Element 2243 at Node 28113)

Minimum value = -202.6 kN/m² (Element 2241 at Node 10727)

1 - Fig. T3.2. Plaxis 2D calculation results: soil mix 01 – principal total stress σ₁.

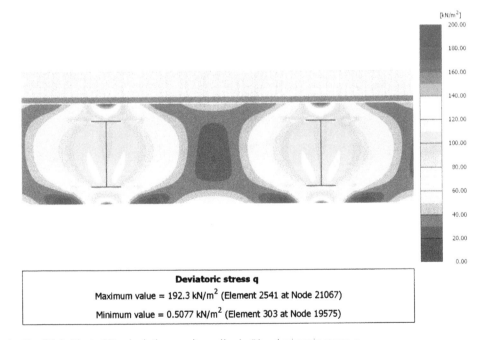

Deviatoric stress q

Maximum value = 192.3 kN/m² (Element 2541 at Node 21067)

Minimum value = 0.5077 kN/m² (Element 303 at Node 19575)

1 - Fig. T3.3. Plaxis 2D calculation results: soil mix 01 – deviatoric stress q.

Total normal stresses (cross section) σ_N (scaled up $1.00*10^{-3}$ times)
Maximum value = 69.90 kN/m^2
Minimum value = -78.65 kN/m^2
Equivalent force is -2.471 kN/m at position (0.000, 1.592) m

Total normal stresses (cross section) σ_N (scaled up $2.00*10^{-3}$ times)
Maximum value = 12.29 kN/m^2
Minimum value = -75.34 kN/m^2
Equivalent force is -21.40 kN/m at position (-0.467, 0.214) m

1 - Fig. T3.4. Plaxis 2D calculation results: soil mix 01 – total normal stresses.

Soil mix 02 tension cut off 100 kPa

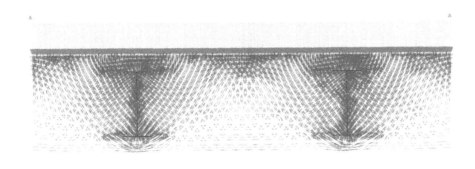

Total principal stresses (scaled up 0.500*10⁻³ times)

Maximum value = 100.0 kN/m² (Element 1946 at Stress point 23348)

Minimum value = -185.5 kN/m² (Element 296 at Stress point 3543)

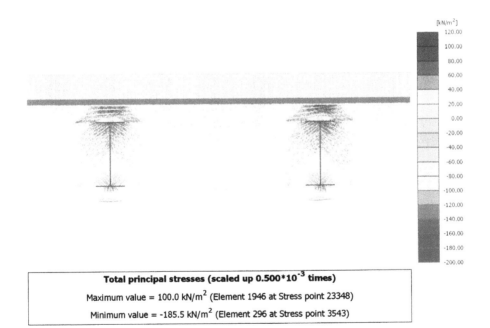

Total principal stresses (scaled up 0.500*10⁻³ times)

Maximum value = 100.0 kN/m² (Element 1946 at Stress point 23348)

Minimum value = -185.5 kN/m² (Element 296 at Stress point 3543)

I - Fig. T3.5. Plaxis 2D calculation results: soil mix 02 – total principal stresses.

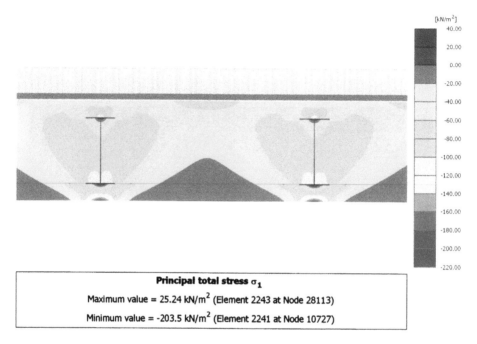

| [kN/m²] |
| 40.00 |
| 20.00 |
| 0.00 |
| -20.00 |
| -40.00 |
| -60.00 |
| -80.00 |
| -100.00 |
| -120.00 |
| -140.00 |
| -160.00 |
| -180.00 |
| -200.00 |
| -220.00 |

Principal total stress σ_1

Maximum value = 25.24 kN/m² (Element 2243 at Node 28113)

Minimum value = -203.5 kN/m² (Element 2241 at Node 10727)

I - Fig. T3.6. Plaxis 2D calculation results: soil mix 02 – principal total stress σ.

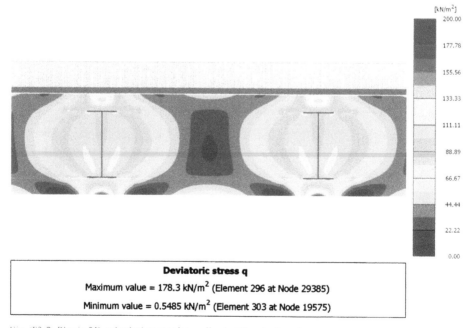

| [kN/m²] |
| 200.00 |
| 177.78 |
| 155.56 |
| 133.33 |
| 111.11 |
| 88.89 |
| 66.67 |
| 44.44 |
| 22.22 |
| 0.00 |

Deviatoric stress q

Maximum value = 178.3 kN/m² (Element 296 at Node 29385)

Minimum value = 0.5485 kN/m² (Element 303 at Node 19575)

I - Fig. T3.7. Plaxis 2D calculation results: soil mix 02 – deviatoric stress q.

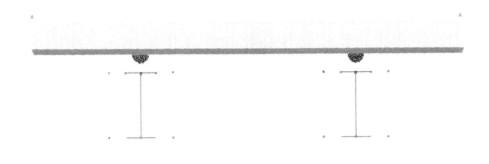

Plastic points

■ Failure point □ Tension cut-off point

▼ Cap point ◆ Cap + hardening point

▲ Hardening point

1 - Fig. T3.8. Plaxis 2D calculation results: soil mix 02 – plastic points.

Total normal stresses (cross section) σ_N (scaled up $1.00*10^{-3}$ times)

Maximum value = 70.32 kN/m^2

Minimum value = -79.46 kN/m^2

Equivalent force is -2.600 kN/m at position (0.000, 1.539) m

Total normal stresses (cross section) σ_N (scaled up $2.00*10^{-3}$ times)

Maximum value = 11.90 kN/m^2

Minimum value = -74.97 kN/m^2

Equivalent force is -21.75 kN/m at position (-0.467, 0.217) m

I - Fig. T3.9. Plaxis 2D calculation results: soil mix 02 – total normal stresses.

Soil mix 03 tension cut off 0 kPa

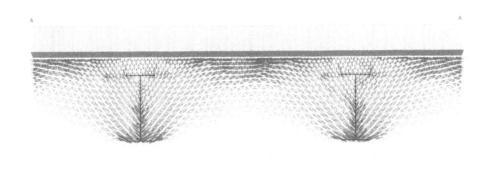

Total principal stresses (scaled up 0.200*10⁻³ times)

Maximum value = $1.000*10^{-3}$ kN/m^2 (Element 2355 at Stress point 28250)

Minimum value = -730.2 kN/m^2 (Element 2347 at Stress point 28155)

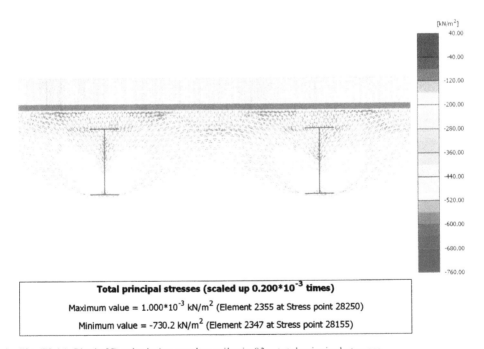

Total principal stresses (scaled up 0.200*10⁻³ times)

Maximum value = $1.000*10^{-3}$ kN/m^2 (Element 2355 at Stress point 28250)

Minimum value = -730.2 kN/m^2 (Element 2347 at Stress point 28155)

I - Fig. T3.10. Plaxis 2D calculation results: soil mix 03 – total principal stresses.

261

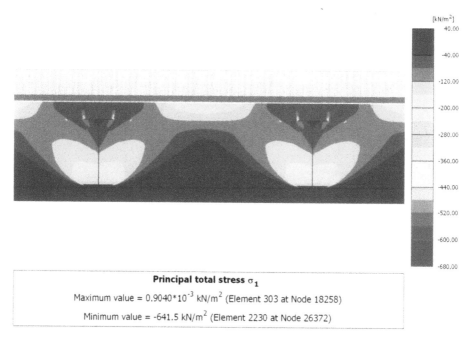

Principal total stress σ₁

Maximum value = $0.9040*10^{-3}$ kN/m² (Element 303 at Node 18258)

Minimum value = -641.5 kN/m² (Element 2230 at Node 26372)

1 - Fig. T3.11. Plaxis 2D calculation results: soil mix 03 – principal total stress σ₁.

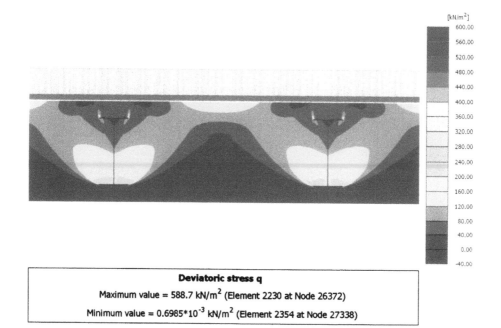

Deviatoric stress q

Maximum value = 588.7 kN/m² (Element 2230 at Node 26372)

Minimum value = $0.6985*10^{-3}$ kN/m² (Element 2354 at Node 27338)

1 - Fig. T3.12. Plaxis 2D calculation results: soil mix 03 – deviatoric stress q.

Total normal stresses (cross section) σ_N (scaled up $0.500*10^{-3}$ times)
Maximum value = -1.126 kN/m^2
Minimum value = -227.5 kN/m^2
Equivalent force is -27.99 kN/m at position (0.000, 0.449) m

Total normal stresses (cross section) σ_N (scaled up $1.00*10^{-3}$ times)
Maximum value = -0.1709 kN/m^2
Minimum value = -134.9 kN/m^2
Equivalent force is -31.98 kN/m at position (-0.491, 0.223) m

I - Fig. T3.13. Plaxis 2D calculation results: soil mix 03 – total normal stresses.

Appendix 4 - Example calculations based on 2 case studies

T4.1 Introduction

During the final phase of editing this handbook, the editors thought it would be useful to apply the proposed design rules, described in Chapter 6, in two case studies. The inclusion of the two case studies below in this appendix 4 has, thereby, the end goal of supplying the reader, on the basis of these examples, with a quantified illustration of how to definitively apply the various rules, and to supply guidance for the use of the proposed methodologies and formulas, the relevant correlations and factors, and the various steps that must be taken into consideration in the design. Moreover, the focus on these case studies by a number of enthusiastic committee members has also proven to be of great help in further clarifying the topics discussed in Chapter 6.

The 1st case study concerns a **cantilever soil mix panel wall** for the support of a single subsoil construction layer. The soil type is mainly sandy. The groundwater was temporarily lowered from -2.0 m-mv to -3.5 m-mv (mv: surface of the soil). The wall has a temporary water-retaining function, and in the use phase an earth and water-retaining function and is also vertically loaded. A protection barrier/wall is realised for the finishing, with an intermediate open (drained) cavity.

The 2nd case study focuses on an **anchored soil mix column wall** for a building with 2 subsoil construction layers. The excavation up to -6.5 m-mv mainly occurs in the quaternary loam cover layer. The wall is temporarily anchored by a row of tensile anchors, and in the use phase is supported on the bottom plate and 2 upper floor plates. Similarly, here a protection barrier/wall is foreseen with a cavity and intermediate drainage as well.

In both case studies, attention is paid to (both in the temporary phase and use phase):
- determination of the deformation and strength parameters of the soil mix material and soil mix wall;
- calculation of the wall stability and internal forces by applying a D-sheet calculation, in which the appropriate bending stiffnesses EI-eff of the wall are applied;
- verification of the pressure arch stability for the transfer of earth and water pressures onto the steel beams;
- determination of the minimum required reinforcement length;
- verification of the structural strength of the soil mix wall and the required steel beams, both in the temporary and use phase (including the effective vertical load);
- the geotechnical verification of the vertical bearing capacity.

T4.2 Case study 1: Cantilever retaining wall made of soil mix panels

The following case is a case study concerning the various design aspects related to the soil mix wall. A number of simplifications have been implemented for calculating the vertical bearing capacity. The calculations are based on NEN 9997-1.

General assumptions
- Construction is classified into reliability class/risk class RC2 (NL) / RK2 (B)
- Construction phase < 6 months
- Lifetime 50 years

Soil composition, soil parameters and geohydrological assumptions

1 - Table T4.1. Overview assumptions.

Layer no.	Top level [m-mv.]	Soil type	γ'/γ_{sat} [kN/m³]	Φ' [o]	c' [kPa]	k_{h1} [kN/m³]	k_{h2} [kN/m³]	k_{h3} [kN/m³]
1	0	Cover layer	18 / 20	30	0	5000	2500	1250
2	-1	Moderatly dense	17 / 19	27	0	16000	8000	4000
3	-6	Dense sand	18 / 20	32	0	60000	30000	15000

The groundwater level is uniform and at the beginning is -2 m-mv (-2m with regard to the soil surface).

Cross-section construction phase

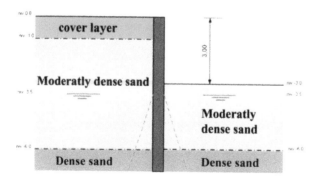

1 - Fig. T4.1 Cross-section of the soil mix wall of case 2 during the governing construction phase including water pressures.

The groundwater level is lowered here both in and outside the construction pit. On the outside, the same hydraulic head is applied in the moderately dense and dense sand layers as on the inside of the construction pit. A uniform upper load of 10 kN/m² is calculated.

266

Cross-section use phase
- 2 levels of floor are applied at -0.25 and -2.75 m-mv.
- Here the wall is equipped with a protection barrier/wall with a cavity and intermediate drainage.
- The drainage is suspended and the groundwater level returns to its original level of -2 m-mv. An upward water pressure is applied to the cellar floor.
- A head beam is applied, with bottom at -0.5 m-mv, which – theoretically – only presses onto the soil mix wall. From this head beam, a vertical load of 100 kN/m' is transferred onto the soil mix wall.

Applicable wall characteristics
A panel wall system with a thickness of 550 mm is assumed. The wall is executed between level 0.0 and -8.0 m-mv. Level -8.0 m-mv is required due to the horizontal stability.

As a starting point for the reinforcement, steel beams of steel quality S235 axis-to-axis 1100 mm, centrally placed (*) are assumed.
(*) in the design phase, the designer is free to decide whether or not to include a placement tolerance (e.g.of 50 mm or more) – see §6.5.2.

The compressive strength is determined on the basis of the following empirical values, which are solely applicable for the studied location.

1 - Table T4.2. Proposed characteristic soil mix compressive strengths in the various soil layers.

Soil type	$f_{sm,k}$ [N/mm^2]
Cover layer	4.0
Moderatly dense sand	4.0
Dense sand	6.0

Here it is a priori assumed that the different soil layers have different compressive strengths in the soil mix material. Depending on the execution method, in actuality these differences will occur to a greater or lesser extent, as a result of which smearing or even complete levelling of the compressive strengths may take place.

For the calculations it is assumed that the levelling/mixing of the different soil types occurs over the height of the wall. Furthermore, considering the geometry and soil stratification, it can be shown that the deflection and the determining field moment will mainly be determined by the properties of the material in the moderatly dense sand layer. On this basis, in the calculation of the earth-retaining functions of the entire wall, $f_{sm;k}$ = 4.0 MPa is assumed, which is equal to $f_{sm;m}$ = 4.0 MPa / 0.7 = 5.71 MPa. It should also be noted that for the detailed calculations (e.g. determination of anchoring length of the reinforcement), the correctness of the above approach should be verified per location. Given the depth of the wall in the sand layer, one should possibly assume the to be expected strength in sand, $f_{sm;k}$ = 6,0 MPa for determining the anchoring length at the bottom of the wall.

Wall calculations

Step 1

Initially, the calculation of the retaining function, in which the stiffness of the soil mix wall is determined on the basis of only the soil mix material, will be considered uncracked(*):

(*) as an alternative, an estimated stiffness of the cracked composite cross-section can also be applied here.

$$E_{sm}=1482*f_{sm;m}^{0,8} \qquad \text{(T4.1)}$$

$$E_{sm} = 1482 * 5.71^{0,8} = 5976 \ MPa$$

$$EI_{sm}/m'=\frac{1}{12}*b_{cl}*h_{sm}^{3}*E_{sm}/b_{cl}=82.8x10^{3} \ kNm^{2}/m' \qquad \text{(T4.2)}$$

On the basis of this calculation, a governing moment of M_d = 177 kNm/m is calculated in the construction phase. With this values, an estimation of the required reinforcement can be made on the basis of the tables from §6.7.6 in part 1. On the basis of these tables , a reinforcement is selected: IPE330. An equivalent HE beam is also possible.

Step 2

On the basis of the reinforcement determined in step 1, the composite stiffness of the wall can be determined, see §6.6 from part 1. The composite stiffness is determined on the basis of method 1:

$$EI\text{-}eff \ =\frac{EI\text{-}1+EI\text{-}2}{2} \qquad \text{(T4.3)}$$

$$EI\text{-}1=\ E_a I_a+E_{sm}(I_{sm}-I_a)=E_{sm}\left[(n-1)I_a+I_{sm}\right]=115.1x10^{3} \ kNm^{2}\ /\ beam\ distance \qquad \text{(T4.4)}$$

$$EI\text{-}2 = E_{sm} * I_2 = 48.8 \ x \ 10^{3} \ kNm^{2}\ /\ beam\ distance \qquad \text{(T4.5)}$$

On the basis of the data, the following applies per m' wall length:

$$EI\text{-}eff \ /m'=\frac{(115.1+48.8)x10^{3}}{(2*1.1)}=74.5x10^{3} kNm^{2}/m'' $$

268

Step 3

On the basis of the calculated composite stiffness determined in step 2, a new calculation of the earth-retaining function is made. The corresponding calculation of the retaining wall, executed on the basis of the Dutch approach in accordance with the CUR 166 step plan is shown below.

I - Table T4.3. Summary of the wall calculations in the construction phase.

Phase no.	Verification type	Displacement [mm]	Moment [kNm]	Shear force [kN]	Mob. perc. moment [%]	Mob. perc. resistance [%]	Vertical balance
1	EC7(NL)-Step 6.1		-0.4	0.4	0.0	14.2	---
1	EC7(NL)-Step 6.2		-0.2	0.3	0.0	14.2	---
1	EC7(NL)-Step 6.3		0.8	-0.8	0.0	14.4	---
1	EC7(NL)-Step 6.4		0.4	-0.6	0.0	14.4	---
1	EC7(NL)-Step 6.5	0.0	0.0	0.0	0.0	9.4	---
1	EC7(NL)-Step 6.5 * 1.20		0.0	0.0			
2	EC7(NL)-Step 6.1		-177.1	187.1	0.0	80.7	---
2	EC7(NL)-Step 6.2		-177.1	187.1	0.0	80.8	---
2	EC7(NL)-Step 6.3		-153.7	127.2	0.0	60.9	---
2	EC7(NL)-Step 6.4		-153.4	125.4	0.0	60.4	---
2	EC7(NL)-Step 6.5	24.4	-82.9	49.4	0.0	25.2	---
2	EC7(NL)-Step 6.5 * 1.20		-99.5	59.3			

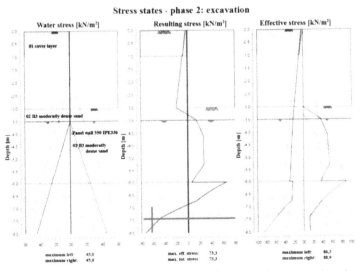

I - Fig. T4.2. Graphic overview normative soil tensions in construction phase (SLS).

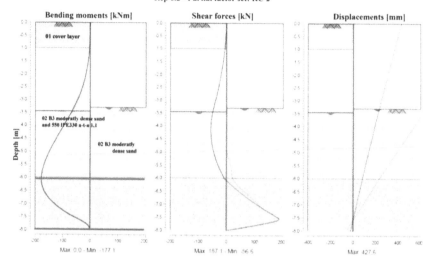

Moments/Forces/Displacements - phase 2: excavation

step 6.2 - Partial factor set: RC 2

1 - Fig. T4.3. Graphic overview of the governing forces in the wall construction phase (ULS).

On the basis of these results:
- $M_{E;d}$ = 177.1 kNm/m';
- $V_{E;d}$ = 187.1 kN/m';
- $\sigma_{hor;rep} \approx$ 62 kN/m²
  (determined on the basis of average over the thickness of the wall – see I - Fig. T4.2)
- Minimal transfer length L_s = (level -6, 1m - level -8,0 m) = 1.9 metre (see I - Fig. T4.3)

Step 4: Constructive verification of the soil mix wall, construction phase
On the basis of the assumptions and calculated internal forces, a constructive verification should be conducted for the soil mix wall.

The following applies for the characteristics of the soil mix material:
- Design value of the compressive strength of the soil mix, $f_{sm;d}$ = 2.42 MPa
  (on the basis of §6.3.3) in which:
  - γ_{sm} = 1.5;
  - α_{sm} = 1.0 (short term);
  - k_f = 1.1 (compressive strength on the basis of empirical data);
  - β = 1.0.
- Design value of the adhesion between soil mix and reinforcement f_{bd} = 0.24 MPa;
- Design value axial of the tensile strength of the soil mix, $f_{sm;td}$ = 0.32 MPa.

270

The following applies for the reinforcement:
- Construction phase, no corrosion;
- $f_{yk} = 235$ N/mm² ;
- IPE 330 (on the basis of class, plastic calculation).

Step 4a: Analysis of the arching effect (§6.4)

1 - Table T4.4. Verification applicability pressure arch (construction with limited a/h$_{sm}$ ratio).

Axis-to-axis steel beams	$a = 1100$ mm
Maximum available height for pressure arch	$h_{bg} = h_a - t_f + c_2 = 330 - 11.5 + 110 = 428$ mm
One should comply with	$a < 3h_{bg}$ $1100 < 1285$ mm; Complies, pressure arch applicable.

1 - Table T4.5. Geometry of the pressure arch.

Height of the pressure arch	$z_{bg} = 0.2\,a + 0.4\,h_{bg} \leq 0.6$ a and $z_{bg} \leq h_{bg} - d_{bg;mid}/3$ $z_{bg} = 0.2 \times 1100 + 0.4 \times 428 \leq 0.6 \times 1100$ and $z_{bg} \leq 428 - 209/3$ $z_{bg} = 358$ mm ($d_{bg;mid} = 209$ mm should be iteratively determined here)
Angle α at the base of the arch	$\alpha = \arctan(4z_{bg}/a) = \arctan(4 \times 358/1100) = 52°$
Thickness of the pressure arch at level of the beam	$d_{bg} = b_f \sin\alpha = 160 \times \sin 52.5° = 127$ mm
Thickness of the pressure arch at mid cross-section	$d_{bg;mid} = b_f \tan\alpha = 160 \times \tan 52.5° = 209$ mm → estimation $d_{bg;mid}$ in determining z_{bg} is sufficient

1 - Table T4.6. Determination of the stresses in the pressure arch.

Max. stresses from earth and water pressure	$\sigma_{hor} = 1.35\,\sigma_{hor;rep} = 1.35 \times 62 = 83.7$ kPa
Max. normal compressive stress at the base of the arch	$\sigma'_{arch;base} = 0.5\,\sigma_{hor}\,(a - b_f - 2_{c1})/(b_f \sin^2\alpha)$ $\sigma'_{arch;base} = 0.5 \times 0.0837 \times (1100-160-2 \times 110)/(160 \sin^2 52.5°)$ $\sigma'_{arch;base} = 0.30$ N/mm²
Max. normal compressive stress in the middle of the arch	$\sigma'_{arch;mid;max} = 2\,\sigma'_{arch;mid;avg} = 2 \times N'_{middle}/d_{bg;mid}$ (with $N'_{middle} = \sqrt{(N'^2_{base} - V^2_{mid})}$ $N'_{middle} = \sqrt{((\sigma'_{arch;base}\,d_{bg})^2 - (0.25\,a\,\sigma_{hor})^2)} = 30.2$ N/mm $\sigma'_{arch;mid;max} = 2 \times 30.2/209 = 0.29$ N/mm²
Max. shear stress in the pressure arch cross-section	$\tau_{Ed} = 0.375\,a\,\sigma_{hor}\cos\alpha/d_{bg}$ $\tau_{Ed} = 0.375 \times 1100 \times 0.0837 \times \cos(52.5°)/127 = 0.17$ N/mm²

Admissible compressive stresses at the base	$\sigma_{RD,max} = k_2 \, v' \, f_{sm,d} = 0.85 \times 0.98 \times 2.42 = 2.03$ N/mm² → greater than maximum stresses $\sigma'_{arch;base} = 0.30$ N/mm², thus complies.
Admissible compression stress middle cross-section	$\sigma_{RD,max} = 0.6 \, v' \, f_{sm,d} = 0.6 \times 0.98 \times 2.42 = 1.43$ N/mm² → greater than maximum stresses $\sigma'_{arch;mid;max} = 0.29$ N/mm², therefore complies
Admissible shear stress in the arch	$\tau_{Rd} = \tau_{Rsm,d} + 0.15 \, \sigma'_{arch;mid;avg}$ $\tau_{Rd} = 0.32 + 0.15 \times 0.14 = 0.34$ N/mm² → greater than maximum shear stress $\tau_{Ed} = 0.17$ N/mm², therefore complies.

→ The pressure arch has sufficient strength to withstand the stress in the arch.

Step 4b: Determination of constructive strength of the soil mix wall (§6.7)

On the basis of the conditions listed in §6.7.2, the criterion related to the adhesion with a transfer length $L_s = 1.9$ m determines the effective width of the soil mix material b_{c2}, which is here set to 0.28 m.

Paragraph 6.7.5 provides the formulas required to determine the M-N diagram of the composite section (soil mix reinforcement).

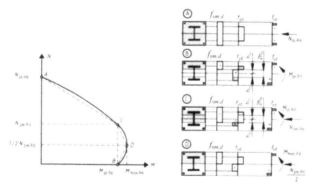

1 - Fig. T4.4. Simplified method for interaction diagram.

Point A:
The maximum absorbable normal force: $N_{pl;Rd} = 1825$ kN.

Point B:
One should check where the plastic neutral line runs.
The calculations show:
Condition 1: $F_{sm;1} = 74$ kN $< F_a = 1471$ kN, so PNL runs through the beam, condition met.
Condition 2: $F_{sm;2} = 81$ kN $< F_{a;w} = 541$ kN, so PNL through the web.
After completing the formulas: $M_{pl;Rd} = 201$ kNm.

272

Point C:
Here it is determined which maximum additional normal force is acceptable for the moment determined in Point B. The calculation shows: $N_{pm;Rd}$ = 368 kN.

Point D:
The maximum acceptable moment: $M_{max;Rd}$ = 214 kNm.
With the data above and the force effect determined in step 3, one can create an M-N diagram that demonstrates that the moment capacity of the wall is sufficient.

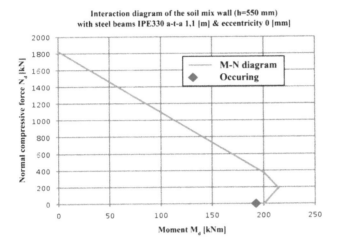

Interaction diagram of the soil mix wall (h=550 mm)
with steel beams IPE330 a-t-a 1,1 [m] & eccentricity 0 [mm]

1 - Fig. T4.5. Simplified M-N interaction diagram of the soil mix wall.

The occurring shear force $V_{E;d}$ = 187.1 x 1.1 = 206 kN/ beam should be fully taken by the reinforcement. The shear force capacity of the steel beam is: $V_{pl;Rd}$ = A_v x f_y / $\sqrt{3}$ = 418 kN. This is more than the occurring shear force.

Step 5 Determination of the minimum required length of the steel beam
At the end of the steel beam, the soil mix wall is located in dense sand. Therefore one may assume that the strength of the soil mix material here is based on the mixing with the dense sand (and not with the moderatly dense sand), where $f_{sm;k}$ = 6 MPa.
When applying a steel beam IPE 330, the cracking moment is:

$$M_{r;sm} = \frac{1}{6} * b_{c2} * h_{sm}^2 * f_{sm;td} = \frac{1}{6} * 0.28 * 0.55^2 * 0.32x10^3 = 4.5 \ kNm \tag{T4.6}$$

273

The level at which this moment is calculated is -7.90 m-mv (-mv, in relation to the soil surface). Taking out an anchoring length of 0.5 m and a minimum extra depth of the soil mix wall of 0.2 m, this results in:

- Underside of beam IPE 330 = -8.40 m-mv;
- Underside of wall maximum of -8.60 m-mv.

The predefined beam IPE330 complies in the construction phase. As a result, no recalculation of the wall is required.

Step 6: Constructive verification of the use phase

The following applies in the use phase:

- When determining the characteristics of the soil mix material and the beam, one should take into account the time effects:
  ○ For the strength of the soil mix material this comes down to α_{sm} = 0.85 (§6.3.3);
  ○ For the stiffness of the soil mix material Ψ = 1.0 (§6.3.7);
  ○ For the beam, one should take into account corrosion for a lifetime of 50 years, 0.6 mm in the situation in question.
- Because corrosion of the beam may take place, when determining the moment capacity, the adhesion between soil mix and steel should be taken at f_{bd} = 0 MPa (§6.3.6). This shows that the occurring moments must be completely absorbed by the steel beam.
- For determining the vertical load transfer, one may assume a composite cross-section of soil mix and steel beam, given that the characteristic compressive strength of the soil mix material is greater than 3 MPa and the requirements are met (see §6.3.6). The vertical load aspects on soil mix and steel can here be determined on the basis of the ratio of the stiffnesses.
- Only the steel beam is taken into account for the verification of the moment capacity, during which a corrosion supplement as mitigating measure is used in calculations for the benefit of durability (see §6.14.2).
- For the verification of the transfer of the moment force, the corrosion supplement is also taken into account for the steel beam. This steel beam should therefore be tested on the combination of moment and normal force (portion on the steel beam).

The calculation of the earth-retaining aspect has been expanded by 1 phase for determining the force effect in the wall in the use phase:

- Phase of upward construction, wall supported by 2 floors, groundwater level to initial situation and uniform load on excavation side to compensate water pressure.
- Normal load of 200 kN/m' on the wall. For this calculation, it has been assumed that the vertical load on the inside of the pit is transferred from the excavation level and that 50% of the load is transferred via this point.

The stiffness of the wall has been adjusted in the calculation, in which the creep factor ψ and a corrosion of 0.6 mm have been incorporated. This results in an adjusted bending stiffness:

$$EI\text{-}eff \ /m' = \frac{(67.0+36.3)x10^3}{(2*1.1)} = 47.0x10^3 kNm^2/m''$$

The related execution of the calculation of the retaining wall, carried out on the basis of the Dutch approach in accordance with the CUR 166 step plan, is shown in I - Table T4.8.

Assessment method B is applied here, in which safety factors and supplements are only applied during the phases to be assessed. In this process, previous phases are representatively calculated.

I - Table T4.8. Summary of the wall calculations in the use phase.

Phase no.	Verification type	Displacement [mm]	Moment [kNm]	Shear force [f' [°]]	Mob. perc. moment [%]	Mob. perc. resistance [%]	Vertical balance
1	Not verified						
2	Not verified						
3	EC7(NL)-Step 6.3		-65.6	-43.5	0.0	37.5	---
3	EC7(NL)-Step 6.4		-68.8	46.1	0.0	38.7	---
3	EC7(NL)-Step 6.5	32.4	-80.4	45.3	0.0	25.2	---
3	EC7(NL)-Step 6.5 * 1.20		-96.5	54.4			

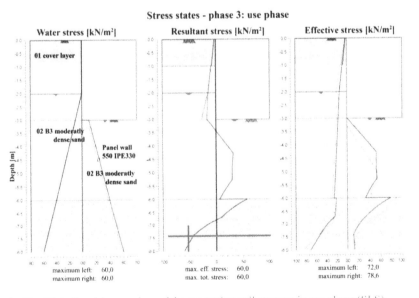

Stress states - phase 3: use phase

I - Fig. T4.6. Graphic overview of the governing soil stresses in use phase (SLS).

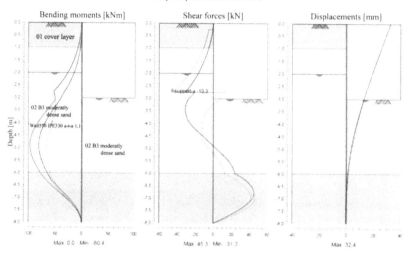

Moments/Forces/Displacements - phase 3: use phase
step 6.5 partial factor set: RC 2

I - Fig. T4.7. Graphic overview of the governing forces in the wall in the use phase (SLS).

On the basis of these calculations, the following governing results are obtained:

- $M_{E;d} = 96.5$ kN/m'
- $V_{E;d} = 54.4$ kN/m'
- $N_{E;d} = 100.0$ kN/m'
- $\sigma_{hor;rep} \approx 52$ kN/m² (determined on the basis of the average across the thickness of the wall);
- Minimum transfer length L_s = (level -4.9 m-mv - level -8.0 m-mv) = 3.1 metres (see I - Fig. T4.6)

Step 7: Constructive verification of the soil mix wall, use phase
For the properties of the soil mix material, the following applies in the use phase:
- Design value of the compressive strength of the soil mix material,
  $f_{sm;d} = 2.06$ MPa (on the basis of § 6.3.3), in which:
  - $\gamma_{sm} = 1.5$;
  - $\alpha_{sm} = 0.85$ (long term);
  - $k_f = 1.1$ (compressive strength on the basis of empirical data);
  - $\beta = 1.0$.
- Design value of the adhesion between soil mix and reinforcement
  $f_{bd} = 0.00$ MPa for the verification of the moment capacity (no interaction);
- Design value of the adhesion between soil mix and reinforcement
  $f_{bd} = 0.1 \times 2.06 = 0.21$ MPa for the verification of the vertical load transfer (considered below the excavation level);
- Design value axial tensile strength soil mix, $f_{sm;td} = 0.27$ MPa.

Step 7a: Analysis of the pressure arch for the use phase (§ 6.4)

This verification has not been expanded upon for this example. The approach is similar to that of step 4a.

Step 7b: Determination of the structural strength of the soil mix wall
for the use phase (§ 6.7)

On the basis of stiffness, the normal force is distributed across the soil mix material and steel beams. The stiffness EA of the soil mix material is 57.6% of the composite stiffness of soil mix-steel.

Of the total normal force $N_{E;d}$ = 100 kN/m', 0.576*100 = 58 kN/m' will therefore be absorbed by the soil mix material.

This results in a normal compression stress of: $\sigma'_{sm;d}$ = 0.11 N/mm². This stress is much smaller than the design value of the compressive strength $f_{sm;d}$ = 2.06 MPa. The soil mix material can absorb this normal force.

The steel beams should be able to absorb the other loads.

- $M_{E;d}$ = 96.5 kN/m' → x 1.1 = 106 kNm/beam
- $V_{E;d}$ = 54.4 kN/m' → x 1.1 = 60 kN/beam
- $N_{E;d}$ = 42.0 kN/m' → x 1.1 = 46 kN/beam

This verification has not been further elaborated upon. Please refer to the formulas in EN 1993-1-1.Here the corrosion should be adjusted by 0.6 mm in the applied cross-section characteristics.

Step 7c: Verification of the adhesion in the zone beneath the excavation

The normal force, which is transferred onto the steel beams at the top, will at the bottom have to be transferred onto the surrounding ground via the soil mix material.

Therefore, one must verify whether the adhesion between the steel and soil mix, in the zone beneath the excavation, is sufficient for the transfer of the vertical load between the steel beam and soil mix material.

The perimeter of the steel beam is

O_a = 4 b_f + 2 h_w + 2 t_f - 2 t_w = 4 x 158.8 + 2 x 308.2 + 2 x 10.3 – 2 x 6.3
= 1260 mm (= 1.26 m).

The minimum required transfer length is:

$$L_{min} = N_{Ed}/O_a \times f_{bd} = 46 /(1.26 \times 0.21 \times 10^3) = 0.17 \, \dot{m} \tag{T4.7}$$

The length of the beam below the excavation level is 5.4 m and is therefore sufficient.

Step 8: Verification of the vertical load capacity of the wall

No further elaboration for this case. For an example, please refer to the elaboration provided for case 2.

T.4.3 Case study 2:
Anchored retaining wall made of soil mix columns

The following is a case study of the various design aspects related to a soil mix wall. Detailed explanation of the anchoring is not included in the geotechnical verification. A number of simplifications have been implemented for calculating the vertical bearing capacity. The calculations are based on NEN 9997-1.

General assumptions:
- Construction is classified into reliability class/risk class RC2 (NL) / RK2 (B)
- Construction phase < 6 months
- Lifetime of 50 years

Soil composition, soil parameters and geohydrological assumptions:

1 - Table T4.7. Overview of the assumptions.

Layer no.	Top level [m-mv]	Soil type	γ' / γ_{sat} [kN/m³]	Φ' [°]	c' [kPa]	k_{h1} [kN/m³]	k_{h2} [kN/m³]	k_{h3} [kN/m³]
1	+0.0	Cover layer	18 / 20	30	0	5000	2500	1250
2	-1.0	Quaternary loam	19 / 19	25	4	6000	3000	1500
3	-6.0	Dense sand	18 / 20	32	0	60000	30000	15000

The groundwater level is uniform and at the start is -2.0 m-mv (in relation to the soil surface).

Cross-section of the construction phase

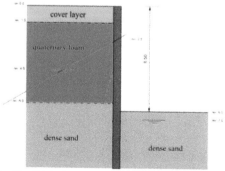

1 - Fig. T4.8. Cross-section of the soil mix wall of the case study 2 during the governing construction phase including water pressures.

278

The groundwater level is lowered here both in and outside the construction pit. On the outside, the same hydraulic head is applied in the dense sand layer as on the inside of the construction pit.

A uniform upper load of 10 kN/m² is taken into account. The wall is temporarily anchored at level -2.0 m-mv and in the use phase rests on 3 floor plates.

Cross-section in the use phase
- In this phase, the anchor is replaced by 3 floor levels at -0.25, -3.25 and -6.25 m-mv.
- Here the wall is equipped with a protection barrier with cavity and intermediate drainage.
- The drainage is suspended and the groundwater level returns to its original level of -2.0 m-mv. An upward water pressure is applied to the cellar floor.
- A head beam is applied, with underside at -0.5 m-mv, which – theoretically – only presses onto the soil mix wall. From this head beam, a vertical load of 200 kN/m' is transferred onto the soil mix wall.

Applicable wall characteristics
A column wall system with columns with a diameter of Ø 530 mm and axis-to-axis 450 mm is assumed. The wall is applied between level 0.0 and -9.6 m-mv.

As a starting point for the reinforcement, steel beams of steel quality S235 axis-to-axis 900 mm, centrally placed (*) are assumed.

(*) in the design phase, the designer is free to decide whether or not to include a placement tolerance (e.g.of 50 mm or more) – see §6.5.2

The compressive strength is determined on the basis of the following empirical values, which are solely applicable for the present case study.

Soil type	$f_{sm.k}$ [N/mm²]
Cover layer	2.0
Quaternay loam	2.0
Dense sand	6.0

Here it is a priori assumed that the different soil layers have different compressive strengths in the soil mix material. Depending on the execution method, in actuality these differences will occur to a greater or lesser extent, as a result of which smearing or even complete levelling of the compressive strengths may take place.

For the calculations it is assumed that the levelling/mixing of the different soil types occurs over the height of the wall. Furthermore, the geometry and soil stratification shows that the

279

deflection and the determining field moment will mainly be determined by the properties of the material in the loam layer. On this basis, in the calculation of the earth-retaining functions of the entire wall, $f_{sm;k} = 2.0$MPa is assumed, which is equal to $f_{sm;m} = 2.0$ MPa $/ 0.7 = 2.86$MPa.

It should also be noted that for the detailed calculations (e.g. determination anchoring length of the steel beam or punching of an anchor plate), the correctness of the above approach should be verified per location. Given the depth of the wall in the sand layer, for determining the anchor length on the bottom, one can assume the to be expected strength in sand, $f_{sm;k} = 6.0$ MPa.

Wall calculations

Step 1
Initially, the calculation of the retaining function, in which the stiffness of the soil mix wall is determined on the basis of only the soil mix material, is considered uncracked(*):

(*) as an alternative, an estimated stiffness of the cracked composite cross-section can also be applied here.

$$E_{sm} = 1482 * f_{sm;m}^{0.8} \tag{T4.8}$$

$$E_{sm} = 1482 * 2.86^{0.8} = 3433 \text{ MPa}$$

$$EI_{sm}/m' = 1/12 * b_{cl} * h_{sm}^3 * E_{sm}/b_{cl} = 2.66 \times 10^4 \text{ kNm}^2/m' \tag{T4.9}$$

On the basis of this calculation, in the construction phase, a governing moment is calculated of $M_d = 159$ kNm/m. With this value, an estimation of the required steel beam can be made on the basis of the tables from §6.7.6 in part 1. On the basis of these tables, a steel beam is selected: HEA240. An equivalent IPE beam is also possible.

Step 2
On the basis of the steel beam determined in step 1, the composite stiffness of the wall can be determined, see §6.6 in part 1. The composite stiffness is determined on the basis of method 1:

$$EI\text{-eff} = \frac{EI\text{-}1 + EI\text{-}2}{2} \tag{T4.10}$$

$$EI\text{-}1 = E_a I_a + E_{sm}(I_{sm} - I_a) = E_{sm}\left[(n-1)I_a + I_{sm}\right]$$
$$= 40.5 \times 10^3 \text{ kNm}^2/\text{beam distance} \tag{T4.11}$$

$$EI\text{-}2 = E_{sm} * I_2 = 24.9 \times 10^3 \text{ kNm}^2/\text{beam distance} \tag{T4.12}$$

On the basis of the data, the following applies per m' wall length:

$$EI\text{-eff} /m' = \frac{(40.5 + 24.9) \times 10^3}{(2 * 0.9)} = 36.3 \times 10^3 \text{ kNm}^2/m'$$

Step 3

On the basis of the calculated composite stiffness determined in step 2, a new calculation of the earth-retaining wall is made.

The corresponding calculation of the retaining wall, executed on the basis of the Dutch approach in accordance with the CUR 166 step plan is shown below.

1 - Table T4.10. Summary of the wall calculations in the construction phase.

Stage no.	Verification type	Displacement [mm]	Moment [kNm]	Shear force [kN]	Mob. perc. moment [%]	Mob. perc. resistance [%]	Vertical balance
1	EC7(NL)-Step 6.1		-0.5	0.7	0.0	13.7	---
1	EC7(NL)-Step 6.2		0.3	0.5	0.0	13.7	---
1	EC7(NL)-Step 6.3		0.8	-1.1	0.0	13.8	---
1	EC7(NL)-Step 6.4		0.4	-0.8	0.0	13.8	---
1	EC7(NL)-Step 6.5	0.0	0.0	0.0	0.0	9.0	---
1	EC7(NL)-Step 6.5 * 1.20		0.0	0.0			
2	EC7(NL)-Step 6.1		-78.7	41.5	0.0	26.8	---
2	EC7(NL)-Step 6.2		-65.5	38.6	0.0	26.8	---
2	EC7(NL)-Step 6.3		-69.9	36.8	0.0	26.3	---
2	EC7(NL)-Step 6.4		-56.6	30.7	0.0	26.2	---
2	EC7(NL)-Step 6.5	17.1	-29.2	-16.9	0.0	15.6	---
2	EC7(NL)-Step 6.5 * 1.20		-35.0	-20.2			
3	EC7(NL)-Step 6.1		-68.9	-74.6	18.9	22.5	---
3	EC7(NL)-Step 6.2		-65.1	-74.7	19.1	22.9	---
3	EC7(NL)-Step 6.3		-68.1	-73.6	18.8	22.1	---
3	EC7(NL)-Step 6.4		-64.8	-74.0	19.1	22.5	---
3	EC7(NL)-Step 6.5	11.5	-68.8	-73.3	12.1	14.3	---
3	EC7(NL)-Step 6.5 * 1.20		-82.6	-88.0			
4	EC7(NL)-Step 6.1		169.8	116.5	100.0	100.0	---
4	EC7(NL)-Step 6.2		169.5	116.6	99.8	99.9	---
4	EC7(NL)-Step 6.3		163.0	112.4	90.6	91.5	---
4	EC7(NL)-Step 6.4		158.7	112.9	91.1	92.1	---
4	EC7(NL)-Step 6.5	9.9	-75.5	78.6	43.1	46.9	---
4	EC7(NL)-Step 6.5 * 1.20		-90.6	94.3			

I - Table T4.11. Summary of the anchor forces in the construction phase.

Stage	Verification of type of	Anchor/strut	
		Force [kN]	State
3	Step 6.1	150.00	Elastic
3	Step 6.2	150.00	Elastic
3	Step 6.3	150.00	Elastic
3	Step 6.4	150.00	Elastic
3	Step 6.5 * 1.20	180.00	Elastic
4	Step 6.1	**227.82**	Elastic
4	Step 6.2	220.26	Elastic
4	Step 6.3	215.26	Elastic
4	Step 6.4	213.74	Elastic
4	Step 6.5 * 1.20	208.13	Elastic
Max		**227.82**	

Stress States - Stage 4: excavation -6,5

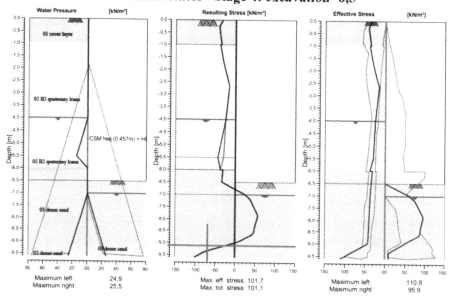

I - Fig. T4.9. Graphic overview of the governing soil stresses in the construction phase (SLS).

Moments/Forces/Displacements - Stage 4:excavation-6,5

Step 6.1 - Partial factor set: RC 2

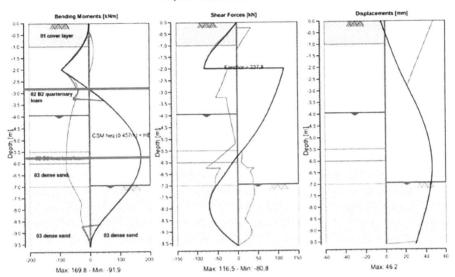

Max: 169.8 - Min: -91.9   Max: 116.5 - Min: -80.8   Max: 46.2

I - Fig. T4.10. Graphic overview of the governing forces in the wall in the construction phase (ULS).

On the basis of these results:
- $M_{E;d}$ = 169.8 kNm/m';
- $V_{E;d}$ = 116.6 kN/m';
- $\sigma_{hor;rep} \approx 70$ kN/m^2
  (determined on the basis of the average over the thickness of the wall – see I - Fig. T4.9)
- Minimum transfer length L_s = (niv. -2.7m - niv. -5.7 m) = 3.0 metre (see I - Fig. T4.3.)
- $N_{E;ank;d}$ = 228 kN/m' (vertical component $N_{E;anch;v;d}$ = sin 30° x 228 kN/m'= 114 kN/m')

Note: the maximum support point moment is $M_{E;d;support\ point}$ = 92 kNm/m', due to the short transfer length this value may be governing/critical in the design if it is greater than the moment capacity of the steel beams.

For the constructive design, it is assumed that:
- The vertical component of the anchor force is evenly distributed across the wall;
- The shear force from the anchor force is only absorbed by the beam, where the anchor force is distributed over two steel beams,
  ($V_{E;d}$ = 1.8*116.5/2 =104.85 kN/beam).

283

Step 4: Constructive verification of the soil mix wall, construction phase

On the basis of these assumptions and calculated internal forces, a constructive verification must be conducted for the soil mix wall.

The following applies for the characteristics of the soil mix material:
- Design value of the compressive strength of the soil mix, $f_{sm;d}$ = 1.21 MPa (on the basis of §6.3.3) in which:
  - γ_{sm} = 1.5;
  - α_{sm} = 1.0 (short term);
  - k_f = 1.1 (compressive strength on the basis of empirical data);
  - β = 1.0.
- Design value of the adhesion between soil mix and reinforcement f_{bd} = 0.12 MPa;
- Design value of the axial tensile strength of the soil mix, $f_{sm;t;d}$ = 0.2 MPa.

The following applies for the steel beam:
- Construction phase, no corrosion;
- f_{yk} = 235 N/mm^2 ;
- HEA 240 (on the basis of class, plastic calculation).

With a column wall, one should assume an equivalent height for the soil mix wall. On the basis of the formulas listed in § 6.3.8 of part 1, one can calculate that $h_{sm;eq}$ = 0.457m.

Step 4a: Analysis of the arching effect (§6.4)

1 - Table T4.12. Verification of the applicability of the pressure arch (construction with a limited a/h ratio).

Axis-to-axis steel beams	a = 900 mm
Maximum available height for pressure arch	$h_{bg} = h_a - t_f + c_2 = 230 - 12 + 113 = 331$ mm
One should comply with	$a < 3h_{bg}$ 900 < 993 mm; Complies, pressure arch applicable.

284

1 - Table T4.13. Geometry of the pressure arch.

Height of the pressure arch	$z_{bg} = 0.2\,a + 0.4\,h_{bg} \leq 0.6\,a$ and $z_{bg} \leq h_{bg} - d_{bg;mid}/3$
	$z_{bg} = 0.2 \times 900 + 0.4 \times 331 \leq 0.6 \times 900$ and $z_{bg} \leq 331 - 260/3$
	$z_{bg} = 244$ mm
	(here, $d_{bg;mid} = 260$ mm should be determined iteratively.)
Angle α at the location of the base of the arch	$\alpha = \arctan(4z_{bg}/a) = \arctan(4 \times 244 / 900) = 47.3°$
Thickness of the pressure arch at the base of the beam	$d_{bg} = b_f \sin \alpha = 240 \times \sin 47.3° = 176$ mm
Thickness of the pressure arch at mid cross-section	$d_{bg;mid} = b_f \tan \alpha = 240 \times \tan 47.3° = 260$ mm
	$\rightarrow$ estimation $d_{bg;mid}$ in determining z_{bg} is well chosen.

1 - Table T4.14. Determination of the stresses in the pressure arch.

Max. stresses from earth and water pressure	$\sigma_{hor} = 1.35\,\sigma_{hor;rep} = 1.35 \times 70 = 94.5$ kPa
Max. normal compressive stress at the base of the arch	$\sigma'_{arch;base} = 0.5\,\sigma_{hor}\,(a - b_f - 2_{cl})/(b_f \sin^2 \alpha)$
	$\sigma'_{arch;base} = 0.5 \times 0.0945 \times (900-240-2 \times 113)/(240 \sin^2 47.3°)$
	$\sigma'_{arch;base} = 0.16$ N/mm^2
Max. normal compressive stress in the middle of the arch	$\sigma'_{arch;mid;max} = 2\,\sigma'_{arch;mid;avg} = 2 \times N'_{middle}/d_{bg;mid}$
	with $N'_{middle} = \surd(N'^2_{base} - V^2_{mid})$
	$N'_{middle} = \surd((\sigma'_{arch;base}\,d_{bg})^2 - (0.25\,a\,\sigma_{hor})^2) = 18.5$ N/mm
	$\sigma'_{arch;mid;max} = 2 \times 18.5 / 260 = 0.14$ N/mm^2
Max. shear stress in the pressure arch cross-section	$\tau_{Ed} = 0.375\,a\,\sigma_{hor} \cos \alpha / d_{bg}$
	$\tau_{Ed} = 0.375 \times 900 \times 0.0945 \times \cos(47.3°) / 176 = 0.12$ N/mm^2

1 - Table T4.15. Verification of the stresses in the pressure arch.

Admissible compressive stresses at the base	$\sigma_{RD.max} = k_2\,v'\,f_{sm.d} = 0.85 \times 0.99 \times 1.21 = 1.02$ N/mm^2
	$\rightarrow$ greater than the maximum stress
	$\sigma'_{arch;base} = 0.16$ N/mm^2, thus complies.
Admissible compression stress middle cross-section	$\sigma_{RD.max} = 0.6\,v'\,f_{sm.d} = 0.6 \times 0.99 \times 1.21 = 0.72$ N/mm^2
	$\rightarrow$ greater than the maximum stress
	$\sigma'_{arch;mid;max} = 0.14$ N/mm^2, thus complies.
Admissible shear stress in the arch	$\tau_{Rd} = \tau_{Rsm.d} + 0.15\,\sigma'_{arch;mid;avg}$
	$\tau_{Rd} = 0.20 + 0.15 \times 0.07 = 0.21$ N/mm^2
	greater than the maximum shear stress
	$\tau_{Ed} = 0.12$ N/mm^2, therefore complies.

$\rightarrow$ The pressure arch has sufficient strength to withstand the stress in the arch.

Step 4b: Determination of the constructive strength of the soil mix wall (§6,7)

On the basis of the preconditions listed in §6.7.2 the effective width of the soil mix material is $b_{c2} = 0.45$ m. Here, the effective width is determined on the basis of the axis-to-axis distance from the columns, being d_2.

Paragraph 6.7.5 provides the required formulas to determine the M-N diagram of the composite section (soil mix reinforcement).

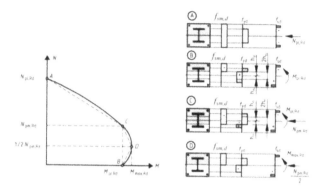

1 - Fig. T4.11. Simplified method for interaction diagram

Point A:

The maximum acceptable normal force: $N_{pl;Rd} = 2046$ kN.

Point B:

One should check where the plastic neutral line runs.
The calculation shows:
Condition 1: $F_{sm;1} = 61.8$ kN $< F_a = 1805.7$ kN, so PNL runs through the beam, condition met.
Condition 2: $F_{sm;2} = 68.3$ kN $< F_{a;w} = 363.1$ kN, thus PNL through the web.

After completing the formulas: $M_{pl;Rd} = 179$ kNm.

Point C:

Here it is determined which additional normal force is acceptable for the moment determined in Point B.
The calculation shows: $N_{pm;Rd} = 249$ kN.

Point D:

The maximum acceptable moment: $M_{max;Rd} = 189$ kNm.

With the above data and the force effect determined in step 3, an M-N diagram can be created, which demonstrates that the moment capacity of the wall is sufficient.

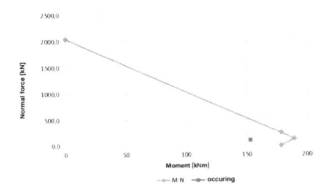

1 - Fig. T4.12. Simplified M-N interaction diagram of the soil mix wall.

The occurring shear force $V_{E;d}$ 116.6 x 0.9 = 104.9 kN/ beam should be fully incorporated into the reinforcement. The shear force capacity of the steel beam is:
$V_{pl;Rd}$ = A_v x f_y / √3 = 341.6 kN. This is more than the occurring shear force.

Step 5: Determination of the minimum required length of the steel beam

At the end of the steel beam, the soil mix wall is situated in dense sand. Therefore one may assume that the strength of the soil mix material here is based on the mixing with sand (and not with loam), where $f_{sm;k}$ = 6 MPa. When using a steel beam HEA 240, the cracking moment is:

$$M_{r;sm} = \frac{1}{6} *b_{c2} *h_{sm}^2 *f_{sm;td}$$

(T4.13)

$$= \frac{1}{6} *0.472*0.457^2 *0.42x10^3 = 6.9 \ kNm$$

The level at which this moment is calculated is -9.15 m-mv. Taking into account an anchoring length of 0.5 m and a minimum extra depth of the soil mix wall of 0.2 m, this results in:
*   Bottom steel beam HEA 240 = -9.65 m-mv;
*   Underside wall maximum of -9.85 m-mv.

The predefined beam HEA240 complies in the construction phase. As a result, no recalculation of the wall is required.

Step 6: Constructive verification in the use phase

The following applies in the use phase:

- When determining the characteristics of the soil mix material and the steel beam, one should take into account the time effects:
  - For the strength of the soil mix material this comes down to $\alpha_{sm} = 0.85$ (§6.3.3);
  - For the stiffness of the soil mix $\Psi = 1.0$ (§6.3.7);
  - For the reinforcement, one should take into account corrosion for a lifetime of 50 years, 0.6 mm in the situation in question.
- Because corrosion of the reinforcement may occur, $f_{bd} = 0$ MPa should be applied (§6.3.6) for the adhesion between the soil mix and reinforcement when determining the moment capacity. This shows that the occurring moments must be completely absorbed by the steel beam.
- For determining the vertical load transfer, no interaction is taken into account above the excavation level, given that the characteristic compressive strength of the soil mix is less than 3 MPa (§6.3.6). Below the excavation level, one may expect interaction. The main point of the above is that over the top part of the wall, the vertical load must be fully absorbed by the reinforcement.

Note: If interaction is to be expected, the vertical load components on soil mix and steel beam can be determined on the basis of the ratio of the stiffnesses.

The calculation of the earth-retaining function has been expanded by 2 phases for determining the force effect in the wall in the use phase. The stiffness of the wall is not adjusted in the calculation, given that a stiff wall results in a conservative calculation of the force effect.

The following phasing is applied:

- Phase of upward construction, wall supported by 3 floors, anchor removed, groundwater level to initial situation and uniform load on excavation side to compensate water pressure.
- Use phase, such as in the phase of upward construction with normal load of 200 kN/m' on the wand. For this calculation, it has been assumed that the vertical load on the inside of the pit is transferred from the excavation level and that 50% of the load is transferred via this point. On the outside of the pit, the top of the sand layer is considered as the top of the positive skin friction zone.

On the basis of the calculations, the following critical results are obtained:

- $M_{E;d} = 79.7$ kN/m'
- $V_{E;d} = 111.3$ kN/m'
- $\sigma_{hor;rep} = 55$ kN/m² (at the location of the excavation);
- $N_{E;d} = 200$ kN/m'

An overview of the force effect (SLS) in the wall for the use phase is provided in I - Fig. T4.13.

288

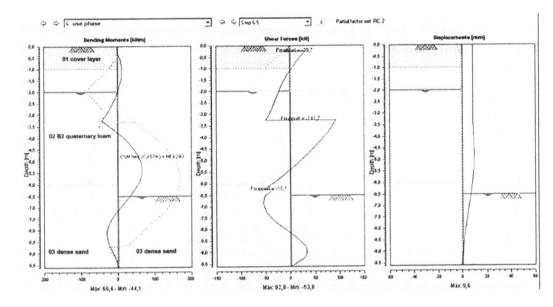

1 - Fig. T4.13. Force effect in the wall, use phase (SLS).

Step 6.1: Constructive verification of the steel beam

Every (corroded) beam is thus affected by the following loads:
- $M_{max;E;d}$ = 79.7*0.9 = 71.7 kNm
- $V_{max;E;d}$ = 111.3*0.9 = 100.2 kN
- N_{Ed} = 200*0.9 = 180 kN

Step 6.1.1: Verification of the shear force
- $V_{E;d} < V_{pl;Rd} = A_v * f_{y;d} / \sqrt{3}$
- 100.2 < 302 → thus complies

Step 6.1.2: Verification of the steel stresses

Above the excavation level, the full normal force should be absorbed by the steel beam:

$$\sigma_{Ed} = \frac{M_{Ed}}{W_{pl}} + \frac{N_{Ed}}{A} = \frac{71.7*10^6}{667.9*10^3} + \frac{180*10^3}{6798.78} \qquad (T4.14)$$

$$= 107.4 + 26.5 = 133.9 \ N/mm^2$$

This shows that $\sigma_{Ed} < f_{yd}$ = 235 N/mm² /1.0, therefore the steel beam complies.

Step 6.2: Constructive verification of the soil mix material

With regards to the soil mix material, the following should be verified:
- Adhesion in zone beneath excavation (sand layer) for the transfer of the vertical load between steel and soil mix;
- Compressive load in comparison with the compressive strength;

- Capacity of the wall for developping the arch effect.

In determining the first two points, one may assume the strength of the soil mix material on the basis of empirical data based on the sand layer, $f_{sm;k} = 6.0$ MPa, this results in:
- $f_{sm;d} = 6.0*0.85/1.1*1.5 = 3.09$ MPa;
- $f_{bd} = 0.3$ MPa.

Step 6.2.1: Verification of the arching effect
For the verification of the arching effect, one should examine the strength at the location of the loam layer. The occurring soil pressure should also be determined here. This verification is similar to the verification conducted in step 4, with adjusted load and strength (long term).

Step 6.2.2: Verification of the normal compressive stress
The theoretical maximum occurring compressive stress in the soil mix material as a result of the normal force is:

$$\sigma_{Ed} = \frac{N_{Ed}}{A} = \frac{200}{0.457*1} = 437.6 \ kN/m^2 \tag{T4.15}$$

This shows that

$\sigma_{Ed} < f_{smd}$ (0.44 MPa < 3.09 MPa) therefore COMPLIES

Note: if one assesses the capacity of the soil mix material at the location of the loam layer, this also complies with $f_{smd;loam} = 1.03$ MPa.

Step 6.2.3: Verification of the adhesion in the zone below the excavation
This is the verification of the transfer length between the steel beam and the soil mix, in which it is assumed that the complete load is transferred through friction. The perimeter of the steel beam is determined on the basis of:

$$O_a = 4*b_f + 2*h_w + 2*t_f - 2*t_w =$$
$$4*238.8 + 2*228.8 + 2*10.8 - 2*6.3 = 1421.8 \ mm \ (= 1.42 \ m) \tag{T4.16}$$

Now that the perimeter and the adhesion are known, the minimum transfer length can be determined:

$$L_{min} = N_{Ed}/O_a*f_{bd} = 200*0.9/(1.4*0.3x10^3) = 0.43 \ m \tag{T4.17}$$

The length of the beam below the excavation level is 3.05 m and is therefore sufficient.

Step 7: Verification of the vertical load capacity of the soil mix wall

The cone resistance in the solid sand layer is $q_c = 15$ MPa. Negative skin friction is ignored. On the inside of the wall, one should take into account the reduction of the cone resistance as a result of the excavation. Since no more vibration occurs after excavation as a result of installing foundation elements, the calculation of the reduction in accordance with NEN 9997-1+C1 can be determined on the basis of the $\sqrt{}$-function.

The wall is classified as stiff, where it is assumed that $\xi_3 = \xi_4 = 1.26$ (1 penetration test). At the location of the base (level − 9.85m-mv, see step 5):

- $\sigma'_{v;excav.} = (9.85\text{-}6.5)*10 = 33.5$ kN/m² ;
- $\sigma'_{v;0} = 1*18 + 1*19 + 4*9 + 3.85*10 = 111.5$ kN/m²

$q_{c;red} = 15*\sqrt{(33.5/111.5)} = 8.2$ MPa.

Determining the base resistance (continuous wall):

$$R_{b;cal;max} = A_{SM}*0.5*\alpha_p*\beta*s*(0.5*(q_{c1;avg} + q_{c2;avg}) + q_{c3;avg})$$

$$R_{b;cal;max} = 0.457*1*0.5*0.5*1.0*0.65*(0.5*(q_{c1;avg} + q_{c2;avg}) + q_{c3;avg})$$

Because the wall forms the boundary of the construction pit, the $R_{b;cal;max}$ is determined on the basis of the average $R_{b;cal;max}$, determined with and without reduction of the cone resistance. This example assumes that at the excavation side the following applies:

$q_{c1;avg} = q_{c2;avg} = 8.2$ MPa and $q_{c3;avg} = 0.5*8.2 = 4.10$ MPa.

On the basis of the outer side, the following applies:
$$R_{b;cal;max} = 0.457*1*0.5*0.5*1*0.65*(0.5*(15000+15000) + 15000) = 2227.9 \ kN/m'$$

On the basis of the inner (excavation) side, the following applies:
$$R_{b;cal;max} = 0.457*1*0.5*0.5*1*0.65*(0.5*(8200+8200) + 4100) = 913.4 \ kN/m'.$$

From this, the average is:
$$R_{b;cal;max} = (2227.9 + 913.4)/2 = 1570.7 \rightarrow R_{b;k} = 1570.7 / 1.26 = 1246.6 \rightarrow R_{b;d} =$$
$$1246.6/1.2 = 1038.8 \ kN/m'.$$

The share of the shaft resistance can be determined on the basis of NEN 9997-C1, where $\alpha_s = 0.006$.

An indicative calculation results in:

$R_{s;cal;max;outside} = 1*(-6.0—9.85)*0.006*15000 = 346.5 \ kN/m'$;

$R_{s;cal;max;inside} = 1*(-6.5—9.85)*0.006*4100 = 82.4 \ kN/m'$;

$R_{s;cal;max} = 429 \ kN/m' \rightarrow R_{s;k} = 340 \ kN/m' \rightarrow R_{s;d} = 284 \ kN/m'$.

The spring constant for the SLS situation can be determined under the assumption that:
- $F_{c;k} = N_d /1.35 = 200 \ kN/m'/1.35 = 148 \ kN/m'$;
- S_{el} is calculated on the basis of a reduction of the reinforcement, where
  - for simplification – the average load is set equal to the maximum load.

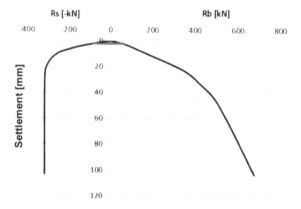

1 - Fig. T4.14. Load settlement diagram GT2.

The calculated settlement of the base of the wall is $s_b \approx 2$ mm.

The elastic reduction is approx. $s_{el} \approx (0.9 \ 148*9.55)/(6799 \times 10^{-6} *2.1 \times 10^8) \approx 0.0009$ m.

This results in $s_1 \approx 2.9$ mm.

$K_{GT2;avg} \approx 148/0.0029 \approx 51000 \ kN/m$.

Part 2
Chapter 1 Introduction

The second part of the handbook gives an oversight of the current state of practice of the deep mixing method which can be found in the international literature. That is among others on the basis of this literature study that the principles for the design and execution of the soil mix walls, presented in the part 1 of the handbook, have been redacted.

Chapter 2 State-of-the-art

2.1   Soil mix walls as retaining structures

In the deep mixing method, a specially made machine is used to mix in place the soil with a binder, often based on cement, in such a way to improve the soil characteristics (e.g. in order to increase its shear strength and/or to reduce its compressibility).

The deep mixing method (DMM) was introduced in the 70's in Japan and in the Scandinavian countries. Since several decades, the deep mixing has been known as a Ground Improvement (GI) technique, as reported in *Porbaha et al. (1998 and 2000)*. Porbaha has notably proposed a terminology for the deep mixing technology, as given in Table 2.1. According to the classification of ground improvement methods adopted by the ISSMGE TC 211 (*Chu et al. 2009*), it can be classified as ground improvement with grouting type admixtures, as illustrated in Table 2.2. A lot of reviews describing various deep mixing systems are available in *Terashi (2003), Topolnicki (2004), Larsson (2005), Essler and Kitazume (2008), Arulrajah et al. (2009), Denies and Van Lysebetten (2012), FHWA (2013) and Denies and Huybrechts (2015)*. In parallel, the results of national and European research programmes have been published in multiple interesting reports (such as *Eurosoilstab, 2002*), while also the European standard for the execution of deep mixing "Execution of special geotechnical works – Deep Mixing" (*EN 14679*) was published in 2005. Most of these research projects focused on the global stabilisation of soft cohesive soils such as clay, silt, peat and gyttja (result of the digestion of the peat by bacteria).

More recently, the deep mixing has increasingly been used for the retaining of soil and water in the case of excavations, as illustrated in Fig. 2.1. As a matter of fact, the construction of soil mix walls represent a more economical alternative to concrete secant pile walls and even in several cases for king post walls (i.e. soldier pile walls). In practice, the soil mix elements (columns or rectangular panels) can be placed next to each other, in a secant way. By overlapping the different soil mix elements, a continuous soil mix wall is realised, as illustrated in Fig. 2.2. Steel H or I-beams are inserted into the fresh soil mix element to resist the shear forces and bending moments. The soil mix material is then designed to span between soldier beams similar to lagging in solider beam/lagging walls (*Taki and Yang, 1991*). The soil mix material ensures the role of the arching effect. The maximum installation depth of the soil mix wall lies – so far – in the order of 20 m. The main structural difference between the soil mix walls and the more traditional secant pile walls is the constitutive material which consists of a mixture of soil and cement, generally called the soil mix material, instead of traditional concrete.

Rutherford et al. (2005) have proposed a historical background of excavation support using soil mix walls. Table 2.3 and 2.4 respectively compare the usual types of excavation support and the ground improvement techniques used for retaining wall construction.

The deep mixing method is also regularly used for the realisation of cut-off walls.

In a more general perspective, *Bruce et al. (1998)* provided a chronology of the evolution of the deep mixing method and proposed a classification of the soil mix systems largely used to date.

A complete list of deep mixing methods and equipment internationally used is available in *Denies and Huybrechts (2015)*.

The purpose of the present chapter is to describe the construction principles and the soil mix material in a first step with a global view but especially with regard to the application of the deep mixing method for earth-water retaining structures and cut-off walls.

Table 2.1. Terminology of the deep mixing family, after Porbaha (1998).

CCP:	chemical churning pile
CDM:	cement deep mixing
CMC:	clay mixing consolidation method
DCCM:	deep cement continuous method
DCM:	deep chemical mixing
DJM:	dry jet mixing
DLM:	deep lime mixing
DMM:	deep mixing method
DSM:	deep soil mixing
DeMIC:	deep mixing improvement by cement stabilizer

In situ soil mixing	
JACSMAN:	jet and churning system management
Lime-cement columns	
Mixed-in-place piles	
RM:	rectangular mixing method
Soil-cement columns	
SMW:	soil mix wall
SWING:	spreadable WING method

Table 2.2. Classification of ground improvement methods adopted by ISSMGE TC211 (after Chu et al., 2009).

D. Ground improvement with grouting type admixtures	
D1. Particulate grouting	Grout granular soil or cavities or fissures in soil or rock by injecting cement or other particulate grouts to either increase the strength or reduce the permeability of soil or ground.
D2. Chemical grouting	Solutions of two or more chemicals react in soil pores to form a gel or a solid precipitate to either increase the strength or reduce the permeability of soil or ground.
D3. Mixing methods (including premixing or deep mixing)	**Treat the weak soil by mixing it with cement, lime, or other binders in-situ using a mixing machine or before placement.**
D4. Jet grouting	High speed jets at depth erode the soil and inject grout to form columns or panels.
D5. Compaction grouting	Very stiff, mortar-like grout is injected into discrete soil zones and remains in a homogeneous mass so as to densify loose soil or lift settled ground.
D6. Compensation grouting	Medium to high viscosity particulate suspension is injected into the ground between a subsurface excavation and a structure in order to negate or reduce settlement of the structure due to ongoing excavation.

Figure 2.1. Soil mix wall with ground and water retaining functions (with the courtesy of CVR nv).

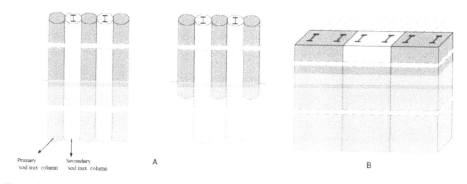

Figure 2.2. Schematic plan view of the secant execution of (A) cylindrical soil mix columns and (B) rectangular soil mix panels.

Table 2.3. Comparison of usual types of excavation support, after Rutherford et al. (2005).

Excavation support system	Advantages	Limitations
Structural diaphragm (slurry) wall	Constructed before excavation and below ground-water, good water seal, can be used as permanent wall and can be used in most soils. Relatively high stiffness. Can become part of the permanent wall.	Large volume of spoils generated and disposal of slurry required. Costly compared to other methods. Must be used with caution or special techniques must be used when adjacent to shallow spread footing.
Sheetpile wall	Constructed before excavation and below ground-water. Can be used only in soft to medium stiff soils. Quickly constructed and easily removed. Low initial cost.	Cannot be driven through complex fills, boulders or other obstructions. Vibration and noise with installation. Possible problems with joints. Limited depth and stiffness. Can undergo relatively large lateral movements.
Soldier pile and lagging wall	Low initial cost. Easy to handle and construct.	Lagging cannot be practically installed below groundwater. Cannot be used in soils that do not have arching or that exhibit base instability. Lagging only to bottom of excavation and pervious.
Secant wall/ Tangent pile wall (similar to DSM walls)	Constructed before excavation and below ground-water. Low vibration and noise. Can use wide flange beams for reinforcement.	Equipment cannot penetrate boulders, requires pre–drilling. Continuity can be a problem if piles drilled one at a time.
Micro–pile wall	Constructed before excavation and below ground-water. Useful when limited right of way.	Large number required. Continuity a problem, low bending resistance.

Table 2.4. Comparison of excavation support using ground improvement techniques, after Rutherford et al. (2005).

Excavation support using GI	Advantages	Limitations
DSM method	Constructed before excavation and below ground-water. Low vibration and noise levels. Fast construction. Reduced excavated spoils compared to slurry (diaphragm) walls. Improved continuity with multi-drill tool.	Difficult with boulders and utilities, spoil generated.

Permeation grouting	Constructed before excavation and below ground-water.	Pre–grouting to control flow of grout through cobbles, does not penetrate soil with more than 15% fines.
Soil nailing	Rapid construction, boulders could be drilled through. Can be used in stiff soils.	Cannot install below groundwater, easements required, cannot be used in soft soils or soils that exhibit base instability. Excavation must have a stable face prior to installation.
Jet grouting	Constructed before excavation and below ground-water.	Difficult with boulders, large volume of spoils generated. Obstructions can obstruct lateral spread of mixing
Soil freezing	Constructed before excavation and below ground-water.	Difficult to install with flowing groundwater and around boulders. Very costly for large area and/or prolonged time. Temporary. Ground heave during freezing and settlement during thaw.

2.2   Construction principles and execution processes

In the deep mixing process, the ground is mechanically (and possibly hydraulically or pneumatically) mixed in place while a binder, based on cement or lime, is injected with the help of a specially made machine. The deep mixing method can be classified according to its execution process. As reported in *Denies and Huybrechts (2015)*, two types of installation methods are generally considered with regard to the way the binder is injected into the ground (with or without water addition): the wet and the dry mixing methods.

According to *Denies and Huybrechts (2015)*, *"in the wet mixing method, which is more frequently applied, a mixture of a binder and water with possibly sand or additives is injected and mixed with the soil. Depending on the type of soil and binder, a mortar-like mixture is created which hardens during the hydration process (Essler and Kitazume, 2008)"*.

In the dry soil mixing process, the binder is directly mixed with the *in-situ* soil. The setting agents directly react with the *in-situ* soil and the present water and form a soil mortar. A State of the Art concerning the dry soil mixing can be found in *Quasthoff (2012)*. He reviewed the construction principles, the equipment and the field of applications of the dry mixing method.

As illustrated in Fig. 2.3, depending on the applications, different improvement patterns can be designed with the wet and dry mixing methods with the help of soil-cement columns, rectangular soil mix panels, continuous barriers or global mass stabilisation.

Other classifications based on construction principles are also available in *Bruce et al. (1998), Porbaha et al. (2001), CDIT (2002), Topolnicki (2004)* and *Essler and Kitazume (2008)*. In parallel, *Larsson (2005)* provides a large description of common systems respectively used in Japan, U.S.A. and Europe for deep mixing.

In Belgium and in The Netherlands, only the wet method is used for the realisation of soil mix walls with the help of columns and panels. The most common systems used in Belgium are the CVR C-mix®, the TSM and the CSM; all three are wet soil mixing systems. They are described in the following paragraphs.

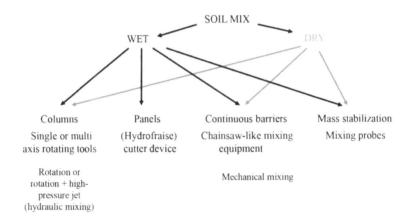

Figure 2.3. Deep mixing systems: classification on the basis of the improvement patterns.

2.2.1 Columns systems

(a) CVR C-mix® system

As described in *Denies et al. (2012a)*, the CVR C-mix® is performed with an adapted bored pile rig and a specific designed shaft and mixing tool. This tool rotates around a vertical axis at about 100 rpm and cuts the soil mechanically. Simultaneously, the water\binder mixture (w\b weight ratio between 0.6 and 0.8), is injected at low pressure (< 5 bar). The injected quantity of binder amounts mostly to 350 and 450 kg binder/m^3, depending on the soil conditions and specifications. The binder partly (between 0% and 30%) returns to the surface. This is called 'spoil return'.

The resulting soil mix elements are cylindrical columns with diameter corresponding to the mixing tool diameter, varying between 0.43 and 1.03 m. For retaining structures, the production rate is about 160 m^2 of soil mix wall per day (single 8 hours shift). In order to increase the production rate, a CVR Twinmix® and a CVR Triple C-MIX® have been designed. A twinmix has two mixing tools, mixing two overlapping cylindrical columns (wall element length of 0.8 to 1.2 m) at the same time. The daily production increases till 210 m^2/day. A CVR Triple C-mix® has three mixing tools in line, with a wall element length of 1.5 to 1.8 m. The production rate increases to 300 m²/day. Figure 2.4 illustrates the CVR C-mix® machine and its Triple version.

(b) SMET Tubular Soil Mix (TSM) system

As reported in Denies et al. (2012a), the TSM technique (see Fig. 2.5) uses a mechanical and a hydraulic way of mixing. Apart from the rotating (around the vertical axis) mixing tool, the soil is cut by the high pressure injection (till 500 bars) of the water\binder mixture with w\b chosen between 0.6 and 1.2. The injected quantity of binder mixture amounts mostly to 200 and 450 kg binder/m³. Part of the binder (between 0% and 30%) returns to the surface as spoil return. The resulting soil mix elements are cylindrical columns with a diameter between 0.38 and 0.73 m. The production rate is about 80 m² of soil mix wall per day. Again, a twin and a triple version exist, leading to wall element lengths respectively varying between 0.8 and 1.4 m or 1.2 and 2.1 m. The production rate is increased till about 180 (twin) and 250 m² (triple) of soil mix wall per day. Figure 2.6 illustrates a retaining structure executed with the help of the triple version of TSM.

(c) Mixed-In-Place (MIP) and SMW methods of Bauer

The soil mix tools of the specialist in foundation engineering machinery Bauer have been recently introduced in Belgium and in The Netherlands. The Mixed-In-Place (MIP) method is illustrated in Fig. 2.7: the drilling and the mixing of the soil with the binder are performed with the help of three independent rotary drives. In the SMW system of Bauer (illustrated in Fig. 2.8) the ground is mixed by three adjacent slightly overlapping augers and mixing paddles.

In addition, other column systems could enter in the near future on the Belgian and the Dutch markets such as the wet soil mixing system of Keller (available in single, double or triple shaft versions), the Single Column Mixing Double Head (SCM-DH) system of Bauer or the Trevi Turbojet system, as recently described in *Topolnicki (2012)*.

2.2.2 Panel systems

As reported in *Denies and Huybrechts (2015)*, the execution of soil mix rectangular panels can be performed with the help of the Cutter Soil Mixing (CSM) system, recently developed by Bauer Maschinen GmbH. According to *Gerressen and Vohs (2012)*, the CSM system is based on the principle of the trench cutter technique. It is mainly used for the construction of cut-off walls, earth retaining structures and for the realisation of ground improvement works. Based on the BAUER Cutter technology, the field of application is extended to hardened strata. While a water/binder (w/b) mixture is injected, the soil is mixed with the w/b mixture, using the cutter wheels as cutting and mixing tool. As illustrated in Fig. 2.9, the two cutting wheels rotate independently about a horizontal axis and cut the soil. At the same time, the water/binder mixture is injected at low pressure (commonly < 5 bar) with w/b ratio chosen in function of the design strength and permeability. The injected quantity of binder amounts mostly to 200 and 400 kg binder/m³. Spoil return usually ranges between 0% and 30%. The CSM system is in operation since 2003 and it is commercially available. The resulting soil mix elements are rectangular panels. In Belgium, these panels have usually a length of 2.4 m and a thickness of 0.55 m, though cutter devices with other dimensions are internationally available. The production rate is about 100 m² to 250 m² of soil mix wall per day.

It can be noted that the Geomix® system of Soletanche-Bachy (illustrated in Fig. 2.10) is based on the combination of the hydromill cutter technology and the soil mixing technique of CSM.

Figure 2.11 illustrates the CSM cutting wheels and the resulting soil mix panel. Figure 2.12 gives an example of a CSM wall.

2.2.3   Trenchmixing systems

As reported in *Denies and Huybrechts (2015)*, the Trenchmixing method produces a soil mix barrier, up to a depth of 10 m, in a single continuous pass which is an advantage particularly in case of a cut-off function (there is no joint or discontinuity in the wall). The Trenchmix uses cutting tools as shown in Fig. 2.13 to mix trenches. It has a dry and a wet method. Figure 2.13 shows the dry method. Trenchmixing tools are commercialized by Soletanche-Bachy (Trenchmix® system) and Siedla-Schönberger (FMI system).

Lareco Nederland BV also install soil mix walls with the help of a specially adapted trench cutting device such as illustrated in Fig. 2.14. In parallel with the addition of a (cement) bentonite slurry, the *in-situ* soil is simultaneously mixed resulting in the forming of a homogeneous soil mix wall. The standard thickness of the wall is 0.35 m but a larger thickness can be reached. This technique is often used for the construction of cut-off walls. Steel beams can possibly installed inside the freshly executed soil mix material in order to ensure a retaining or a structural function.

2.2.4   The mass stabilisation system

The ALLU® mass stabilisation system is developed for mass stabilisation of soil. It has a dry and a wet method. Figure 2.15 illustrates this system, which consists of a power mix mounted on the dipper arm of an excavator with a pair of mixing drums at the end and a pressure feeder mounted on a powered crawler chassis which delivers dry binder into the ground with the aid of compressed air (*ALLU, 2010*).

Figure 2.4. a) The CVR C-mix® machine (single auger system) and b) its Triple version.

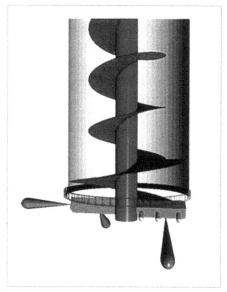

Figure 2.5. Scheme of the TSM system (Smet-F&C).

Figure 2.6. Soil mix wall performed with the help of the triple version of TSM (SMET-F&C).

Figure 2.7. Typical Mixed-In-Place Unit of Bauer (with the courtesy of Bauer).

Figure 2.8. The Bauer SMW method (with the courtesy of Bauer).

Figure 2.9. The cutting wheels of the CSM machine (with the courtesy of Bauer).

Figure 2.10. Geomix² system: the rotating drums carry out the boring and mixing of the soil cement (with the courtesy of Soletanche-Bachy).

Figure 2.11. The cutting wheels of the CSM machine and the resulting CSM panel (Soetaert - Soiltech).

Figure 2.12. Soil mix wall performed with the CSM technique (Lameire Funderingstechnieken n. v.).

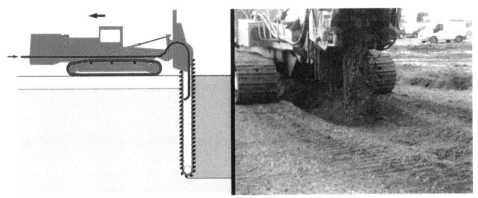

Figure 2.13. Dry Trenchmix ᴿ method (after Borel, 2007)

Figure 2.14. Wet trench cutter device of Lareco Nederland bv.

Figure 2.15. The ALLU ᴿ mass stabilisation system (after Al-Tabbaa, 2012).

2.2.5 Special tools – spreadable soil mixing systems

(a) Keller Foundations FLAPWINGS® system
Within the framework of the European Research project INNOTRACK, SNCF, Keller Foundations and IFSTTAR tested the feasibility of an alternative soil reinforcement technique the FLAPWINGS, based on vertical soil-cement mixed columns with variable diameter (Lambert et al., 2012). Keller Foundations has designed the FLAPWINGS soil mix tool with the purpose of executing soil-cement columns in a railway environment. This is a wet soil mixing system. As reported in Denies and Huybrechts (2015), it consists of a 150 mm core retractable tool able to open up in order to perform the soil mixing phase on a 600 mm diameter column (see Fig. 2.16). The opening and closing of the retrieval blades are ensured with the help of a two way hydraulic jack located in the mixing tool.

(b) Soletanche-Bachy SPRINGSOL® system
Within the framework of the Rufex project (reinforcement and re-use of railway tracks and existing foundations), soil-cement columns were installed with the help of the Soletanche-Bachy 'Springsol' wet soil mixing tool (Guimond-Barrett et al, 2012). As explained in Denies and Huybrechts (2015), this tool is equipped with two mixing blades that spread out under the action of springs (see Fig. 2.17). In its folded configuration, the tool diameter is 160 mm enabling its insertion into a temporary casing. By increasing the length of the mixing/cutting blades, the column diameter can be adapted (e.g. 400 and 600 mm as illustrated in Fig. 2.17). The binder is delivered through outlet holes in the drag bit located at the bottom end of the tool. The main interest of the Springsol tool is the possibility to reinforce the ground under an existing railway track.

The reduced volume of the aforementioned equipment allows good perspective not only in term of reinforcement and re-use of railway tracks but also in a view of underpinning works or for the practice of deep mixing works in tight conditions.

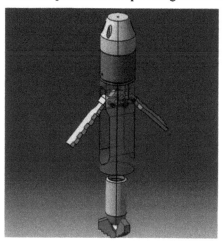

Figure 2.16. FLAPWINGS® system (Keller Foundations), after Lambert et al. (2012).

Figure 2.17. Soletanche Bachy Springsol® systems with 400 mm (left) and 600 mm (right) diameters, after Guimond-Barrett et al. (2012).

In general, the various deep mixing equipment presented in this section allow the execution in a large range of soil types with the following advantages:
- the use of the soil as a construction material,
- a control of the geometry of the soil mix element with depth,
- as the stress relaxation of the soil is limited, soil mix elements can be executed nearby existing constructions,
- contrary to concrete secant pile walls, the execution of the deep mixing method does not depend on the delivery of concrete material and is not affected by delayed supply (e.g. due to traffic jams) of the fresh concrete,
- compared to jet-grouting or slurry walls, the amount of spoil return is more limited and more controllable,
- no important vibrations or noise pollution are caused by the execution,
- it is applicable to on-land and marine projects,
- dewatering is not required,
- execution of soil mix walls is typically faster (also with a relatively easy installation procedure) than other traditional methods resulting in high production capacity,
- the benefit rises with the size of the project.

The drawbacks of the deep mixing method can be summarised as follows:
- the execution is difficult in presence of hardened soils, stones, pack of gravels, overconsolidated clays and peaty soils,
- it is not possible to install oblique soil mix element,

309

- the spoil has to be considered if the execution is performed in cohesive and impervious soils,
- there is a variability of the mechanical properties of the soil mix material.

2.2.6 Installation of geomembranes in soil mix walls

The installation of geomembranes vertically in cement-bentonite slurry walls was developed in the 1980's. The geomembranes were installed vertically in the fresh slurry material just after execution. Adapted steel frames or steel beams were used to sink the geomembranes, coupled to these steel frames, until the installation depth was reached. Since 2006, several projects are conducted wherein these geomembranes are sunk in freshly executed soil mix walls. Several examples of projects involving the installation of geomembranes in soil mix walls are given in the table 2.5 for the sake of illustration (*Gerritsen, 2013*). In parallel of these reference projects, several other applications were realised in other types of slurry wall in The Netherlands and in the rest of the world.

First the installation of geomembranes in soil mix walls was limited to a depth ranging between 6 to 12 m and they were only installed in trenchmixing walls. Within the framework of the project "Haak om Leeuwarden", the large use of geomembranes resulted in the development of a new installation technique involving the use of an adapted steel sinking beam. That technique was further developed within the framework of the reference project of "Aqueduct Westelijke Invalsweg Leeuwaarden" (see table 2.5), wherein soil mix walls were executed for the construction of the entrance installations of a polder construction (see Figures 2.18 to 2.21). In the critical zones of the construction, it was decided to install geomembranes up to a depth of 18 m in the freshly executed soil mix walls. The installation of geomembranes in the soil mix wall is, in the present case, an alternative to cement-bentonite slurry walls equipped with geomembranes as foreseen when a design function of double flood control is imposed (e.g. as for a dam). The main limitation of this technique is the limited depth of installation of both the soil mix wall and the geomembrane (length/stiffness of the sinking beam, manoeuvrability of the crane).

310

Table 2.5. Examples of reference projects: soil mix walls equipped with geomembranes
(Gerritsen, 2013)

Project [-]	Cut-off function [-]	Year [-]	Type wall [-]	Thickness soil mix wall [m]	Type geo-membrane [-]	Range depth geomembrane [m-top level]	Area geo-membrane [m²]
BRM Fascinatio, Cappele along the River IJssel, The Netherlands	Cut-off/ Isolation of contaminants	2006	Trench-mixing	0.3-0.35 m	2 mm HPDE Geolock	7.0 – 9.5	9 500
Dike reinforcement Meerbad, Adcoude, The Netherlands	Cut-off	2007	Trench-mixing	0.3-0.35 m	2 mm HPDE Geolock	6.1 – 7.1	1 650
Landfill Torquay, Torquay, UK	Cut-off/ Isolation of contaminants	2008	Trench-mixing	0.3-0.35 m	2 mm HPDE Geolock	8.5	2 500
Landfill site of Smink, Amersfoort, The Netherlands	Cut-off/ Isolation of contaminants	2009	Trench-mixing	0.3-0.35 m	2 mm HPDE Geolock	11.1 – 12.5	12 000
Aqueduct Westelijke Invalsweg Leeuwarden, The Netherlands	Polder construction, entrance of the aqueduct	2013	Triple mixed-in-place	0.55 m	2 mm HPDE Geolock	18.0	1 830

Figure 2.18. Equipment for the execution of a soil mix wall and installation of the geomembrane in the freshly executed soil mix material (Gerritsen, 2013).

Figure 2.19. Installation of the geomembrane in the freshly executed soil mix material along a previously installed membrane with a "shoehorn" device, a guidance lock device and a swelling cord (Gerritsen, 2013).

Figure 2.20. Installation of geomembranes from a steel working platform (Gerritsen, 2013).

Figure 2.21. Geomembranes just after installation in the soil mix material (Gerritsen, 2013).

The use of geomembranes in a soil mix wall can be considered for the following applications:
- Cut-off walls with severe requirements with regard to the permeability and the leakage flow of the wall.
- Job specifications related to the execution of walls with a design function of double flood control (sandwich construction).
- Permanent applications with guarantees for construction lifetimes of 50 to 100 years.
- A level of expected deformations of the soil mix wall during the works in such way the cracks cannot be closed.
- Aggressive environment, polluted soils.

Special geomembranes are developed for installation at large depths. These geomembranes are made with high-density polyethylene (HDPE) material with a thickness of 2 mm and they have a screen width of 2.5 m. The geomembrane is equipped with a patented male-female connection lock with in the lock a swelling cord pulled along the total length of the element (see figure 2.19).

The use of a geomembrane inside a soil mix wall can certainly improve the impervious character of the wall. Another advantage is the diminution of the risk related to the occurrence of leakage due to the heterogeneity of the wall. *Erkel et al. (2013)* reported results of lab tests supporting that assumption. Within the framework of a test campaign, the leakage flow was measured with several lab test procedures by Kiwa with the help of the Geolocksystem. The measurements were performed under a difference of pressure of 90 kPa (equivalent to 9 metres of water pressure) resulting in the following result: no measurable leakage flow was observed.

In-situ tests can be performed with the help of test membranes integrated into the wall. Two half-waterproof tubes are welded along the full length of the lock of a geomembrane. As a result of the increase of the difference of water height imposed during the test, the leakage flow passing via the lock can be measured with the help of a specialised measurement equipment.

The resistance of the material of the geomembrane against aggressive components (e.g. contaminants) can be very efficient. HDPE-geomembrane of 2 mm of thickness are used in The Netherlands for the sealing of landfills. In the past, the materials constitutive of geomembranes were the subject of a lot of studies regarding their resistance against various types of aggression and regarding their lifetime (*CUR 243 Durability of Geosynthetics*). Within the framework of the project "Aqueduct Westelijke Invalsweg Leeuwarden" (see table 2.5), a study on the durability of this particular application of geomembranes installed in soil mix material was also performed. As a result of these analyses (several material tests and accelerated aging experiments), the expected lifetime for this geomaterial and for the lock devices of the geomembranes is more than 100 years (*Quality Services Testing, 2013*).

Concerning the execution, it is really important to limit the delay between the realisation of the soil mix wall and the installation of the geomembranes in the soil mix material. That involves a good coordination between the different working teams on the field especially regarding the production and the working modalities. Within the framework of the project "Aqueduct Westelijke Invalsweg Leeuwarden", the main concern was related to the hardening of the soil mix material before installation of the geomembrane. Indeed, it was observed that the soil mix material was overly hardened to install the geomembranes to the required depth. In this case, it was then decided to mix one more time the soil mix material to allow the sinking of the geomembrane to the designed depth.

If the soil mix material is only used to allow the installation of the geomembrane in the wall to a required depth, the mix can be designed without cement or with a limited amount of cement. In presence of a mix only made of bentonite, the wall will not harden and the delay between the execution of the wall and the installation of the geomembrane will be less relevant. The drawback of this method is the long-term behaviour of the wall: the medium around the wall will remain plastic during the lifetime of the construction with the following consequence: deformations or setting could be observed as a result of eventual loads due to the traffic or due to the presence of adjacent constructions.

2.3 Hydro-mechanical characterisation of the soil mix material

2.3.1 Governing parameters

Several parameters have an influence on the produced soil mix material. *Terashi (1997)* highlighted the factors affecting the strength of the soil mix material, as illustrated in Table 2.6. The soil mix material quality depends on the cement type and content, on the *in-situ* soil and on the execution process.

The hardening agent is usually a mixture of cement, water, and in several cases bentonite. The water/binder mixture (w/b weight ratio) is also a governing parameter which plays a major role in the mechanical/durability characteristics of the material.

Moreover, the nature of the ground has a huge impact on the strength and uniformity of the material. For example, stiff cohesive soil does not allow an effective mix of the components and can lead to the presence of unmixed material in the soil mix element.

The final product will be the result of a given deep mixing system available on the local market. There are a lot of differences between the various systems – especially with regard to the drilling/mixing tools – and the execution process influences the quality of the soil mix material in terms of strength, uniformity and continuity.

Finally, the exposure conditions of the soil mix elements during their lifetime will have a certain influence on the long term strength gain or on the deterioration of the soil mix material.

Table 2.6. Factors affecting the strength of the soil mix material, after Terashi (1997).

I. Characteristics of hardening agent	1. Type of hardening agent 2. Quality 3. Mixing water and additives
II. Characteristics and conditions of soil	1. Physical chemical and mineralogical properties of soil 2. Organic content 3. PH of pore water 4. Water content
III. Mixing conditions	1. Degree of mixing (Mixing energy) 2. Timing of mixing/re–mixing 3. Quantity of hardening agent
IV. Curing conditions	1. Temperature 2. Curing time 3. Humidity 4. Wetting and drying/freezing and thawing, etc.

2.3.2 Unconfined Compressive Strength (UCS) of the soil mix material

(a) Typical ranges of UCS values

In the context of the European standardization and for the purpose of investigating these questions, in 2009, the Belgian Building Research Institute (BBRI) initiated the "Soil Mix" project in collaboration with KU Leuven and the Belgian Association of Foundation Contractors (ABEF). Within the framework of this research, numerous tests on *in-situ* soil mix material have been performed. A good insight has been acquired with regard to strength and stiffness characteristics that can be obtained with the C-mix®, the TSM® and the CSM systems in several Belgian soils. All the tests were performed on soil mix material from real soil mix walls. All the results and the developments related to the BBRI "Soil Mix" project (2009-2013) are detailed in the following four papers: *Denies et al. (2012a), Denies et al. (2012b), Denies et al. (2012c)* and *Vervoort et al. (2012)*. In a first step, the BBRI "Soil Mix" project (2009-2013) focused on the mechanical characterisation of the soil mix material with regard to the sampling of cores from soil mix walls, its unconfined compressive strength UCS (1073 tests), its modulus of elasticity E (152 tests) and its tensile splitting strength T (95 tests). The purpose of the present paragraph is to present the results of this study. Afterwards, a comparison will be made with other databases available in the international literature.

As reported in *Denies et al. (2012b)*, during the first part of the experimental campaign, cores of soil mix material have been drilled at 38 Belgian construction sites, with different soil conditions and for various soil mix systems. The cylindrical core samples have a diameter ranging between 85 mm and 115 mm. The measurement precision of the core diameter is 0.3 mm. Before testing, they are preserved in an acclimatized chamber with a relative humidity larger than 95% and a temperature equal to 20 ± 2°C, according to the European standard *NBN EN 12390-2*.

UCS tests are performed by a MFL 250 kN loading machine according to *NBN EN 12390-3*. The loading rate amounts to 2.5 kN/s. The height to diameter ratio is 1. This choice was based on the necessity to collect a maximum of cores and was made in order to compare the UCS test results on cylindrical cores with cube strength (*NBN EN 12504-1*). It can be noted that the height to diameter ratio will have an influence on the failure pattern and on the UCS test results.

Figures 2.22 to 2.27 give the histograms of the UCS test results obtained within the framework of the BBRI soil mix project (2009-2013) in function of the soil type and with regard to the execution technique. The age of the samples ranges between 7 and 200 days. No correction regarding to the influence of the sample age has been applied. Table 2.7 gives the minimal and maximal UCS values in function of the soil type and with regard to the execution technique. For the Dutch practice, similar ranges of UCS values are available in *Lantinga (2012)* who provides UCS test results for several case histories.

The density of samples, measured according to *NBN EN 12390-7*, varies between 1372 and 2176 kg/m³. No specific correlation was observed between the density and the UCS. The depth of coring is always larger than 1 m. Indeed, *Ganne et al. (2010)* have observed on different sites that the strength of the soil mix material over the first metre is strongly influenced by the execution process (e.g. infiltration of rinsing water), as illustrated in Fig. 2.28 for a CSM panel in quaternary sand. Hence, the top of the soil mix wall is not representative for the deeper part with regard to its strength. Similar observations were made in the Netherlands for the use of the CSM system in the Dutch soils (*Lantinga, 2012*).

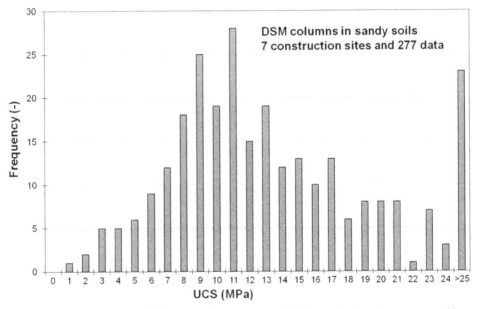

Figure 2.22. Histogram of the UCS test results on core samples from soil mix columns executed in sandy soils. Denies et al. (2012b).

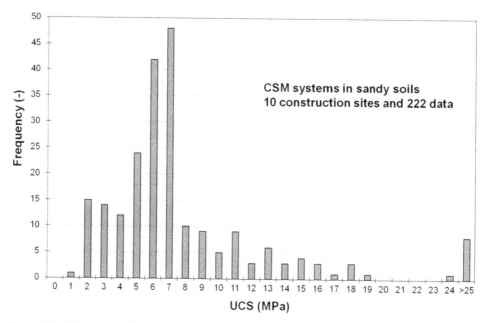

Figure 2.23. Histogram of the UCS test results on core samples from CSM panels executed in sandy soils. Denies et al. (2012b).

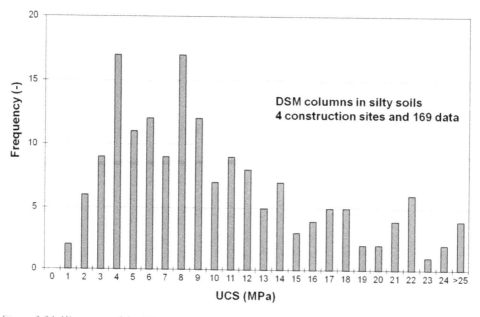

Figure 2.24. Histogram of the UCS test results on core samples from soil mix columns executed in silty soils. Denies et al. (2012b).

318

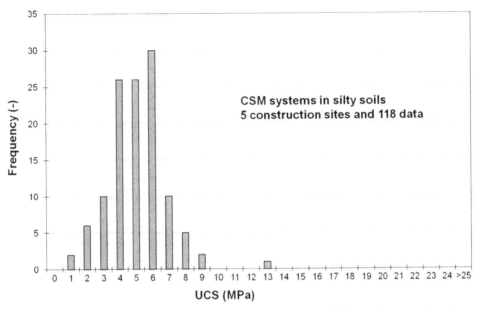

Figure 2.25. Histogram of the UCS test results on core samples from CSM panels executed in silty soils, Denies et al. (2012b).

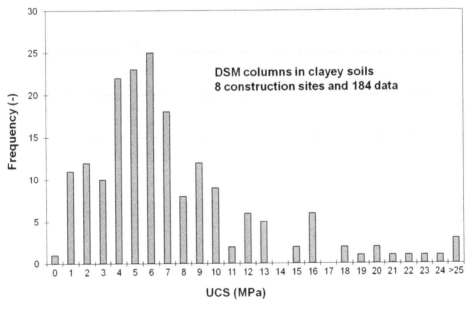

Figure 2.26. Histogram of the UCS test results on core samples from soil mix columns executed in clayey soils, Denies et al. (2012b).

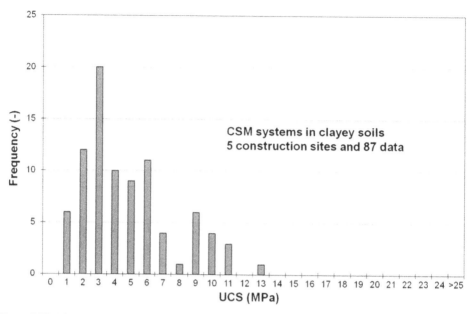

Figure 2.27. Histogram of the UCS test results on core samples from CSM panels executed in clayey soils. Denies et al. (2012b).

Table 2.7. Minimal and maximal UCS values in function of the soil type and with regard to the execution technique, after Denies et al. (2012b).

	Sandy soils		Silty soils		Clayey soils	
	Soil mix columns	CSM systems	Soil mix columns	CSM systems	Soil mix columns	CSM systems
Minimal UCS values	1.32 MPa	1.28 MPa	0.93 MPa	0.66 MPa	0.44 MPa	0.65 MPa
Maximal UCS values	39.90 MPa	32.07 MPa	31.17 MPa	12.63 MPa	33.23 MPa	12.69 MPa

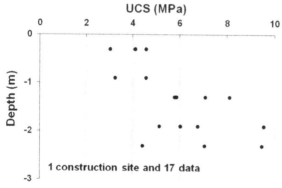

Figure 2.28. UCS test results of samples, cored at different depths (CSM panel in quaternary sand), after Ganne et al. (2010).

At two construction sites (for CSM technique in sandy soils), wet grab sampling was conducted in the first half hour after execution. For wet grab sampling a cylindrical sampler is pushed in the fresh soil mix material. It stays closed until the sampling depth is reached (about 2 m in the present case). At this moment, the sampler opens with a 0.2 m gap. After filling, it is locked and pulled up. The soil mix material is preserved in a cylindrical mould – 113 mm diameter and 220 mm height – in the acclimatized chamber. Two weeks later, soil mix material is *in-situ* cored at the same location and similar depth. The cores and the wet grab samples are tested on the same day (age = 14 days). For all the UCS tests, the height to diameter ratio is 1. Table 2.8 illustrates the UCS test results. The differences between drilled cores and wet grab samples can be explained by the limited number of samples and the lack of uniformity of the samples on the one hand and by the different curing conditions on the other hand. In the following, only tests on core samples are discussed. For several authors and according to the experience acquired within the framework of the BBRI 'Soil Mix' project (2009-2013), cores can be considered more representative than wet grab samples. Nevertheless, coring also presents challenges.

Table 2.8. UCS results of tests on core and wet grab samples (μ is the average UCS value), after Ganne et al. (2010).

	Core samples	Wet grab samples
Site I – CSM element 1	$UCS_1 = 2.33$ $UCS_2 = 2.04$ $UCS_3 = 2.27$ $UCS_4 = 2.85$ $UCS_5 = 2.85$ $\mu = 2.47$ **(MPa)**	$UCS_1 = 2.94$ $UCS_2 = 2.46$ $UCS_3 = 2.44$ $UCS_4 = 2.59$ $\mu = 2.61$ **(MPa)**
Site I – CSM element 2	$UCS_1 = 1.62$ $UCS_2 = 1.63$ $UCS_3 = 1.28$ $UCS_4 = 1.88$ $UCS_5 = 1.90$ $\mu = 1.66$ (MPa)	$UCS_1 = 1.82$ $UCS_2 = 2.00$ $UCS_3 = 1.78$ $UCS_4 = 1.80$ $\mu = 1.85$ (MPa)
Site II – CSM element 3	$UCS_1 = 2.95$ $UCS_2 = 4.53$ $UCS_3 = 4.64$ $UCS_4 = 3.79$ $\mu = 3.98$ (MPa)	$UCS_1 = 3.80$ $UCS_2 = 3.40$ $UCS_3 = 3.66$ $UCS_4 = 3.89$ $\mu = 3.68$ (MPa)
Site II – CSM element 4	$UCS_1 = 5.27$ $UCS_2 = 5.03$ $UCS_3 = 4.12$ $UCS_4 = 5.54$ $\mu = 4.99$ (MPa)	$UCS_1 = 4.18$ $UCS_2 = 3.07$ $UCS_3 = 3.64$ $UCS_4 = 3.69$ $\mu = 3.64$ (MPa)

As reported in *McGinn and O'Rourke (2003)*, a large U.S. database of UCS test results was also created for the Fort Point Channel deep mixing project. At that time, this project consisted in the largest single application of deep mixing in the US, with more than 420 000 cubic metres of Boston blue clay stabilised at the I-90/I-93NB interchange as part of the Fort Point Channel (FPC) crossing. Although the database concerned a unique construction site, a considerable variability in the UCS results was observed. Variations in soil conditions, mixing process and sampling procedures contributed to the variability of data. As discussed in *Rutherford et al. (2005)*, bias in the data reflected in both lower unit weight and higher UCS for the wet grab samples compared with the core samples was related to the small opening of the wet grab sampler. The small opening tended to block unmixed soft soil inclusions outside the sampler. Table 2.9 summarises the data collected for the various samples and tests.

Table 2.9. Summary of UCS test results for the Fort Point Channel deep mixing project (USA), after McGinn and O'Rourke (2003).

Sample description	Number of samples	Arithmetic mean (MPa)	Median (MPa)	Coefficient of variation*
All core samples	823	2.68	1.59	1.30
Core samples with CF = 2.91 MN/m³ and W/C = 0.7	322	2.34	1.17	1.46
Core samples with CF = 2.32 MN/m³ and W/C = 0.9	133	2.84	2.00	0.85

* The coefficient of variation is equal to the quotient of the standard deviation and the mean of the UCS values.

In table 2.9, the computation of the coefficient of variation is based on the assumption that the database is normally distributed. Nevertheless, it is interesting to note that it is not always easy to fit a standard distribution to a given dataset. The most appropriate distribution should be determined, followed by the estimation of its properties. This distribution function can be different for each site and it is even not guaranteed that a suitable standard distribution exists. This question is discussed in the Chapter 6 of the first part of the handbook and in the Appendix 1.

A large database of UCS test results is also available in *CDIT (2002)* for the Japanese practice of deep mixing but the UCS values are often limited to 5 or 6 MPa. It can be explained by the fact that the deep mixing method was for these cases mainly related to soil stabilisation (with increasing of the soil properties) and not to structural applications (as for earth retaining support). The cement contents used for soil stabilisation are then generally lower than for structural applications resulting in weaker strength values. Similar comment can be made about the UCS values from the database detailed in *Bertero et al. (2012)* concerning the mechanical characterisation of the soil mix material used for the reinforcement of the land levees within the framework of the LPV111-project in Louisiana.

Many other references report databases of UCS values. Interested reader can refer to the different reviews as referred in **Section 2.1**. The following paragraphs concentrate on the study of the influence of various governing parameters of the strength of the soil mix material: the mixing energy, the cement factor, the water/cement ratio and the curing time are highlighted.

(b) Influence of the mixing energy
At the present time, a wide variety of different machines and tools are available on the market for *in-situ* deep mixing. As reported in *Topolnicki and Pandrea (2012)*, the mixing technology has to ensure that the soil is mixed sufficiently with the binder to achieve a homogeneous product with a low coefficient of variation for its strength. The quality control of mixing can be performed with regard to the *"Blade rotation number"*, as introduced in the *CDIT (2002)*:

$$BRN = \sum M x \left(\frac{N_d}{V_d} + \frac{N_u}{V_u} \right) \tag{2.1}$$

where BRN is the Blade Rotation Number (1/m), $\sum M$ the total number of mixing blades, N_d the rotation speed of the blades during penetration (rpm), V_d the mixing blade penetration velocity (m/min), N_u the rotational speed of the blades during withdrawal (rpm) and V_u the mixing blade withdrawal velocity (m/min). The BRN evaluates the mixing degree. It gives the total number of mixing blades passes during 1 m of shaft movement (*CDIT, 2002*). In his definition of the BRN, *Topolnicki (2004)* also underlines that a full-diameter blade is counted as two blades. A high value of BRN decreases the coefficient of variation of the strength of the soil mix material. According to *Topolnicki and Pandrea (2012)*, the minimum required BRN depends on the soil type. For cohesive and fine grained soils (loose sands and clays) about 400 (1/m) should be achieved to keep the coefficient of variation for the strength within acceptable limits. In non-cohesive and coarse soils slightly lower values can be sufficient. In practice, the coefficient of variation of the UCS is ranging between 0.2 and 0.6 (*Topolnicki, 2009*). For a set of specific UCS data, *Topolnicki (2009)* has plotted this coefficient of variation in function of its respective blade rotation number, as illustrated in Figure 2.29. Assuming that good mixing quality corresponds to $v = 0.3$ (or less), *Topolnicki (2009)* found that the BRN should exceed the value of 430.

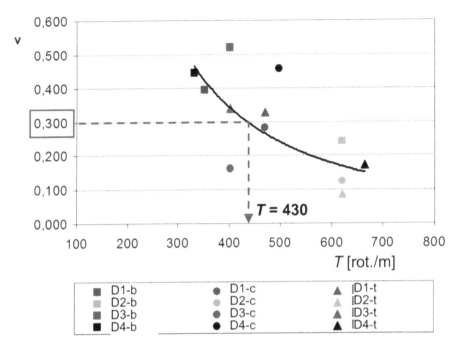

Figure 2.29. The relationship between the coefficient of strength variation v and blade rotation number T, based on field tests, after Topolnicki (2009).

For wet mixing, the European standard *EN 14679* concludes that the mixing energy should be monitored to achieve uniform treated soil. The rotational speed of the rotating unit(s) and the rate of penetration and retrieval of the mixing tool shall be adjusted to produce sufficiently homogeneous treated soil. The European standard *EN 14679* also notes that the current rotation speed of the mixing blades are usually 25 rpm to 50 rpm and the blade rotation numbers are usually greater than 350.

As the BRN was introduced for soil-cement column technique, *Bellato et al. (2012)* propose an equivalent factor for the evaluation of the mixing degree for soil mix material performed with CSM technology: the *"Mixing quality parameter"*, μ, which represents an estimate of the homogeneity achieved in the CSM panel and takes into account the effect of the real soil conditions. According to *Bellato et al. (2012)*, μ (1/m) can be defined as:

$$\mu = \left[\left(\varphi_d R_{d,i} T_{d,i}\right) + \left(R_{u,i} T_{u,i}\right)\right]\frac{N_c M}{100 V_c} \tag{2.2}$$

where $R_{d,i}$ and $R_{u,i}$ are the average rotational speed of the mixing wheels during the penetration and the withdrawal phase, respectively. $T_{d,i}$ and $T_{u,i}$ are the total time taken to blend the soil during the downstroke and the upstroke, respectively. N_c is the number of mixing wheels (2 for the CSM machine). M is the number of mixing element per wheel and V_c

the volume occupied by the cutter. All quantities with the subscript "i" are referred to a single ith layer of predefined thickness. φ_d is called phase factor and takes into account the different role played by the execution process adopted for the realisation of the panel, i.e. one phase ($\varphi_d = 1$) or two phases system ($\varphi_d = 0.5$). Indeed, as explained in *Gerressen and Vohs (2012)*, deep mixing with CSM can be performed in one phase mixing procedure (for applications less than 15m deep in relatively soft ground) or with a two phase system (when mixing deeper panels or penetrating difficult – slow – to mix soils or rocks). In the one phase procedure, the final mixture product consists of cement and water (and possibly bentonite), which is injected on both the down stroke and the upstroke of the machine. In the two phase procedure, just bentonite is used on the down stroke. Once the final depth is achieved, the water/binder mixture is introduced and mixed on the upstroke. This method prevents the mixing tool from being trapped in the panel if the panel construction time exceeds the initial set time of the water/binder mixture.

(c) Influence of the cement content and the water/cement factor
As the degree of mixing is the governing factor of the homogeneity of the soil mix material, its strength is mainly related to the amount of cement (or binder) introduced during the mixing process. According to *Maswoswe (2001)*, the critical factor in the execution of soil mix structures is to maintain the auger withdrawal rate consistent with the grout flow rate. One way to evaluate the success of the procedure and its efficiency is to estimate the cement factor at different locations. The cement factor or cement content is generally noted α. It is the cement mass per cubic metre of soil mix material. On the one hand, the cement factor can be estimated considering the grout flow rate, the auger withdrawal rate and the assumed percentage of grout loss during the process. On the other hand, the cement content is directly related to the water/cement factor, w/c, (or water/binder factor, w/b) of the injected mixture.

As reported in *McGinn and O'Rourke (2003)*, three different w/c factor ratios (0.7, 0.8, and 0.9) and five different cement factors of Portland Type I/II cement (2.2, 2.3, 2.5, 2.6 and 2.9 kN/m³) were used within the framework of the Fort Point Channel DM project. UCS values of soil mix material increased by a factor of 2.5 as the cement content was increased from 1.93 to 2.91 kN/m³ for a constant w/c factor of 0.7. Amazingly, larger UCS values were obtained for an increase of the w/c content. Indeed, the variation of the strength with the w/c factor actually depends on the type of soil, as illustrated in the following study of *Topolnicki (2009)*. In the case of the Fort Point Channel DM project, the increase of the strength observed for growing w/c factor was thus related to the execution of the deep mixing method in the Boston blue clay; in that kinds of soil, a minimal w/c ratio is required with regard to the feasibility/workability in order to obtain a homogeneous and a good quality soil mix produce (see Fig. 2.30).

Topolnicki (2009) has also studied the influence of the cement factor on the UCS of the soil mix material. Based on field test data, he demonstrated that the gain in strength for the soil mix material is almost proportional to the increase of cement content (with a gain of strength of 69% for an increase of 60% of the cement factor). *Topolnicki (2009)* is provided ranges of UCS values in function of the soil type and the cement factor, as illustrated in Table 2.10.

Soil type	Cement content* (kg/m³)	UCS** (MPa)
Sludge	250 – 400	0.1 – 0.4
Peat, organic silts/clays	150 – 350	0.2 – 1.2
Soft clays	150 – 300	0.5 – 1.7
Medium/hard clays	120 – 300	0.7 – 2.5
Silts and silty sands	120 – 300	1.0 – 3.0
Fine-medium sands	120 – 300	1.5 – 5.0
Coarse sands and gravels	120 – 250	3.0 – 7.0

* Cement factor in place: weight of injected cement divided by the total volume of treated ground
and injected slurry

** UCS values are given in term of guaranteed compressive strength at 90% confidence

Topolnicki (2009) concentrated on the influence of the water/cement ratio on the strength
of the soil mix material. His results indicate that the use of cement slurry with 10%
higher density, while keeping mixing work and final cement factor at the same level, has
increased mean strength of stabilised soil of some 38%. This is due to reduced effective
w/c ratio, which accounts for water contained in the untreated soil and in the cement slurry.
Topolnicki (2009) emphasised, however, that this effect is also dependent on the type of soil
and its initial water content, as demonstrated in Figure 2.30, based on laboratory test data.

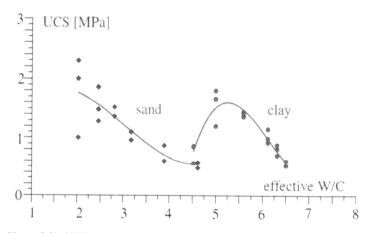

Figure 2.30. UCS test results for laboratory mixed sand and clay-cement samples tested with
different initial water contents, after 28 days of curing (α = 200 kg/m³, density of the cement
slurry ρ = 1.5 g/cm³), after Topolnicki (2009).

Topolnicki (2009) concluded that in sand, UCS steadily decreases with increasing w/c ratio, as could be expected. But clayey soil with too low initial moisture content may be "too dry" for adequate mixing with "too dense" slurry, resulting in poor mixing quality and low UCS. Finally, lower UCS values are also observed for "too wet" conditions, with effective w/c too high because of the too much initial water content of the soil and/or of the slurry. Consequently, in clayey soils, most effective mixing and highest UCS values are obtained when deep mixing is performed with effective water content close to the liquid limit of the clay.

Practically, it can be noted that increasing the w/c ratio delays the set time of the soil mix material allowing for flexibility in the installation of steel reinforcement. But the increase of the w/c factor results in the increase of the volume of spoil return which can require additional transportation and disposal.

Finally, it would be noted that *Bellato et al. (2012)* have recently proposed a procedure for the estimation of the UCS at 28 days of curing with regard to the amount of injected cement. Their approach takes into account the fine content (FC), the mixing quality parameter (μ) and if necessary the effects of the *pH* on the UCS of the soil mix material.

(d) Curing effect: evolution of the strength of the soil mix material with the time
Several relationships have been developed in the past to take into account the effect of the curing time on the strength development of concrete, such as those based on linear and double hyperbolic equations, exponential and logarithmic functions. Within the framework of a large deep mixing project in Bologna (Italy), *Bellato et al. (2012)* have tried to fit their experimental data of CSM treated overconsolidated clays to these empirical relationships, but with unsatisfactory results except the case of the formula provided by *EN 1992-1-1* as already proposed by *Ganne et al. (2010)*:

$$UCS(t) = \beta_{cc}(t) \, UCS_{28days} \qquad (2.3)$$

where UCS(t) expresses the evolution of the strength with the time, t is the time and UCS_{28days} the UCS value at 28 days of curing time. β_{cc} is defined as:

$$\beta_{cc}(t) = \exp\left(s\left(1 - \sqrt{\frac{28}{t}}\right)\right) \qquad (2.4)$$

where s is an empirical factor mainly depending on the type of cement and soil.
According to *Bellato et al. (2012)*, the most satisfactory fit for a curing time larger than 3 days was found with the equation:

$$UCS(t) = \ln(t) - 1 \qquad (2.5)$$

To better represent the increase of the UCS with the curing time observed in the Bologna specimens, they proposed a new empirical equation, based on a double hyperbolic function. This function is composed of two terms. The first one describes the increase of the UCS in the first 28 curing days, whereas the second one defines the development of the long-term strength. This relationship is given by:

$$UCS(t) = \frac{UCS_{28days} \cdot t}{t + K_1 \cdot UCS_{28days}} + \frac{\Delta UCS_\infty \cdot t}{t + K_2 \cdot \Delta UCS_\infty}$$

(2.6)

in which UCS_{28days} can be corrected to take into account the amount of injected cement, the fine content, the mixing quality parameter and the *pH*. ΔUCS_∞ is the strength increment due to long-term reaction products. K_1 and K_2 are two constants dependent on the type of clay and cement used in the treatment (in the case of the CSM treated Bologna overconsolidated clays $K_1 = 1$ and $K_2 = 100$).

Other experimental trends have been observed describing the evolution of the strength of soil mix material with the time. *CDIT (2002)* illustrates the long-term strength development of stabilised clays according to logarithmic relationships. For the Japanese practice of soil stabilisation with the deep mixing technique, *Terashi (2002)* also reports that the strength of the stabilised material in long-term (10 to 20 years) is 2 to 3 times the short-term value but that often concerns soil stabilisation with limited cement or lime content ($\alpha \approx 150$ kg/m³).

Topolnicki (2004) provides various empirical relationships based on experience with real design cases, as illustrated in Table 2.11.

Table 2.11. Empirical relationships describing the evolution of the strength of soil mix material with the time for various soil types, after Topolnicki (2004).

Soil type	Relationships describing the curing time effect on the strength of the soil mix material
	$UCS_{28\ curing\ days} = c.\ 2 \times UCS_{4\ curing\ days}$
Silts and clays	$UCS_{28\ curing\ days} = 1.4 - 1.5 \times UCS_{7\ curing\ days}$
Sands	$UCS_{28\ curing\ days} = 1.5 - 2 \times UCS_{7\ curing\ days}$
Silts and clays	$UCS_{56\ curing\ days} = 1.4 - 1.5 \times UCS_{28\ curing\ days}$

Based on a large review of data, *Filz et al. (2012)* proposes a generalized logarithmic relationship to express the hardening of the soil mix material with time:

UCS (t) = (0.187 ln(t) + 0.375)UCS$_{28curingdays}$

(2.7)

For the Belgian practice, this question was also tackled within the framework of the BBRI soil mix project (2009-2013) with the realisation of a laboratory test campaign. Figure 2.31 illustrates the evolution of the UCS values of laboratory soil mix specimens in function of the curing time. As previously demonstrated by *Ganne et al. (2010)*, the best fit is obtained for equation (2.3) for concrete material as well for sand-cement samples as for loam-cement samples. It can be noted that beyond an initial growing period (126 curing days), there is no more increase of the strength. On the basis of this study, the evolution of the strength of the soil mix material with the time can be described as illustrated in Table 2.12.

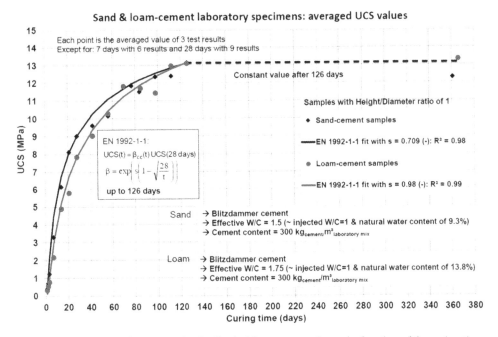

Figure 2.31. Evolution of the strength of soil mix laboratory specimens in function of the curing time, from BBRI soil mix project (2009-2013).

Table 2.12. Empirical relationships describing the evolution of the strength of soil mix material with the time from BBRI soil mix project (2009-2013).

Curing days	$UCS (t) = \beta . UCS_{28curingdays}$	
	Sand-cement laboratory samples	Loam-cement laboratory samples
7 days	$\beta = 0.37$	$\beta = 0.28$
28 days	$\beta = 1$	$\beta = 1$
56 days	$\beta = 1.13$	$\beta = 1.31$
Beyond 126 days	$\beta = 1.46$	$\beta = 1.67$

In a general way, soil mix material show delayed strength development compared to concrete, and show also a long-term increase but both phenomena are dependent on the type of soil and the type of cement. Pozzolanic reaction products should be possibly considered. They can lead to long-term strength development. The prediction of the strength is generally performed before in-situ execution and often based on laboratory experiments. The use of such formulations (equations 2.3 to 2.7) should be then carefully used keeping in mind the limitations of the laboratory specimens to evaluate the real strength of *in-situ* soil mix material.

(e) Influence of the grain size distribution on the strength of the soil mix material
As underlined in *CDIT (2002)*, grain size distribution also contributes to the variation of the UCS value of the soil mix material. *Terashi et al. (1977)* have studied in laboratory this influence on the strength of quicklime treated soil. In the study, Toyoura sand has been added to two different clays resulting in artificial soils with different sand fractions. These soils were then stabilised with quicklime (for cement content of 5 and 10 %). As demonstrated in Fig. 2.32, the UCS values (obtained after 7 curing days) present a peak at a sand fraction of around 40 to 60 %.

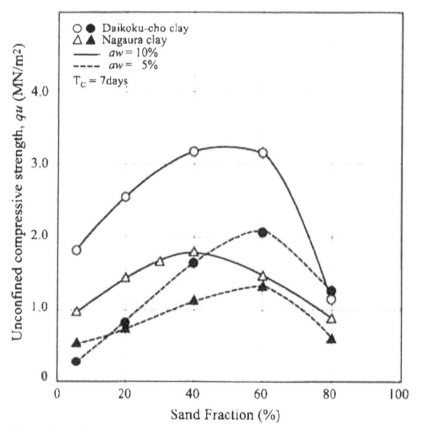

Figure 2.32. Influence of the grain size distribution on the strength of stabilised clays (picture from CDIT, 2002).

Another study concerning the influence of the sand fraction on the strength of stabilised soil is also discussed in *CDIT (2002). Niina et al. (1977)* have performed some laboratory investigation on four stabilised artificial soils with different grain size distributions. Stabilisation was realised with the help of Portland cement with three different cement factors (α). Figure 2.33 presents the results of their laboratory test campaign wherein sand was initially mixed with alluvial clay in order to prepare soils with various grain size distributions. As a result, soils B and C are thus prepared by mixing clay A with sand D. In this case, the UCS tests were performed after 28 curing days. Similarly to the first study, the highest improvement effect is observed for a sand fraction of around 60%.

With regard to these studies concerning the influence of sand fraction, it can be highlighted that they are both performed in laboratories. As a consequence, the effect of the execution is not taken into account and moreover the influence of unmixed (or poorly mixed) soft soil inclusions (mainly present in cohesive soil) is not considered. Indeed, laboratory stabilised soils often present a higher homogeneity than their corresponding one *in-situ*.

Considering the second study, one can question the representativeness and the realism of the results regarding the effect of the cement factor on the measured strengths. A growing strength of the material with a decrease of the cement content is against the reality of the soil mixing practice. It is also possible that the α-values have been inverted.

In light of this analysis, the influence of the sand fraction or the granulometry on the strength of stabilised soils should be reappraised on the basis of tests performed on *in-situ* executed material.

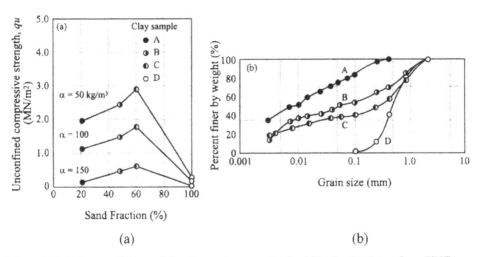

(a)   (b)

Figure 2.33. Influence of the sand fraction on the strength of stabilised soil (picture from CDIT, 2002), (a) UCS test results after 28 curing days in function of the sand fraction of the prepared or natural soil, (b) Grain size distribution of the four tested soils.

2.3.3 Modulus of elasticity of the soil mix material

Within the framework of the BBRI soil mix project (2009-2013), 152 tests were performed in order to determine the modulus of elasticity of soil mix samples cored from real soil mix walls. The tests were performed in an unconfined way with the help of a MFL 250 kN loading machine according to *NBN B 15-203*. The loading rate amounts once again to 2.5 kN/s. The height to diameter ratio is 2, according to *NBN B 15-203*. A selection is made of cores that are visually of a better quality in order to preserve the uniaxial behaviour of the tested samples. The modulus of elasticity was determined in a tangent way varying the applied load between 10% ($\sigma_{10\%UCS}$) and 30% ($\sigma_{30\%UCS}$) of the estimated UCS, as described in detail in *Denies et al. (2012b)*. The loading is then continued to determine the UCS. Figure 2.34 illustrates the correlation between the modulus of elasticity and the UCS of the tested soil mix material, without distinction of the soil type. The samples were tested after a period ranging between 30 and 200 days. Since the aim is to determine the correlation between the modulus of elasticity and the UCS, the test results are not corrected for the age of the samples.

The best fit corresponds to:

$$E = 1482 \ UCS^{0.8} \qquad \text{E and UCS in (MPa)} \tag{2.8}$$

with a coefficient of determination close to 0.80 (-).
Lower and higher 5% quantile estimations of E respectively correspond with:

$$E = 908 \ UCS^{0.8} \qquad \text{E and UCS in (MPa)} \tag{2.9}$$

and

$$E = 2056 \ UCS^{0.8} \qquad \text{E and UCS in (MPa)} \tag{2.10}$$

These estimations are valid for the range 1.5 MPa < UCS < 35 MPa.
In Fig. 2.34, relationships for normal concrete, from *ACI 318-08* and *EN 1992-1-1*, are given for comparison. According to *ACI 318-08*, the modulus of elasticity for normalweight concrete can be defined with regard to the UCS with the help of the following equation:

$$E(psi) = 57000 \sqrt{UCS \ (psi)} \tag{2.11}$$

where E is defined as the secant modulus of elasticity between 0 and 45% ($\sigma_{40\%UCS}$) of the UCS. Based on previous research of *Pauw (1960)*, equation (2.11) is valid for UCS values larger than 2000 psi (or 13.8 MPa).
Eurocode 2 (*EN 1992-1-1*) provides the following relationship for concrete:

$$E(GPa) = 22[UCS(MPa)/10]^{0.3} \tag{2.12}$$

where E is the secant modulus of elasticity between 0 and 40% ($\sigma_{40\%UCS}$) of the UCS. Equation (2.12) is only valid for concrete samples containing quartzite aggregates and for a range of UCS varying between 12 and 90 MPa.

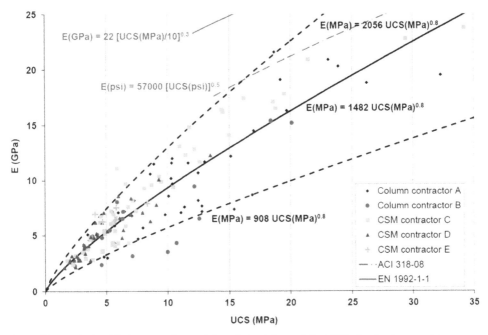

Figure 2.34. Relationship between the modulus of elasticity and the UCS of soil mix material, after Denies et al. (2012b).

It can be noted that in first approximation the modulus of elasticity of soil mix material can be approached with the help of the following relationship:

$$E = 1000\ UCS \tag{2.13}$$

as proposed in Ganne et al. (2010) and valid for the range 2 MPa < UCS < 30 MPa. It can be noted that other correlations between the UCS and the modulus of elasticity of soil mix material are available in the international literature (*CDIT, 2002, Topolnicki, 2004* and *Rutherford et al., 2005*) but they especially concern soil stabilisation applications (with smaller cement content than for earth retaining applications) and often underestimate the modulus of elasticity of the soil mix material with a measurement of the deformations of the sample during the tests based on the relative displacement of the plates of the press.Figure 2.35 presents the results of Fig. 2.34 in function of the soil type.

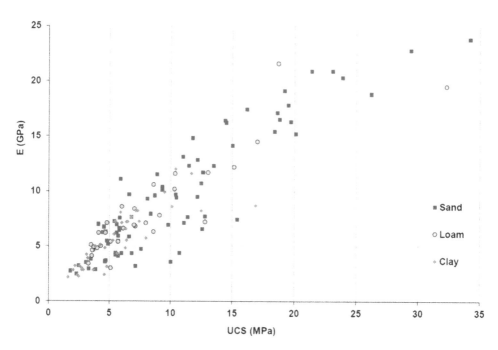

Figure 2.35. Relationship between the modulus of elasticity and the UCS of soil mix material for sandy, loamy and clayey soils, based on the data of Denies et al. (2012b).

2.3.4   Tensile splitting strength of the soil mix material

Figure 2.36 presents the results of the BBRI soil mix project (2009-2013) in term of the tensile splitting strength of the soil mix material. For the determination of the tensile splitting strength (T), sometimes called Brazilian tensile strength, samples with H/D close to 1 have been tested with the help of a MFL 250 kN loading machine (2.5 kN/s of loading rate), according to *NBN EN 12390-6*. Figure 2.36 describes the relationship between the tensile splitting strength (T) and the UCS, without distinction of the soil type. All the samples were cored from real soil mix walls and tested after a period varying between 32 and 200 days. Test results are not corrected for the age of the sample. In Fig. 2.36, experimental results for soil mix cores are compared with well-established empirical relationships for concrete.

According to *Eurocode 2* (EN 1992-1-1), when the tensile strength is determined as the splitting tensile strength, an approximate value of the axial tensile strength, T_a, may be determined as:

$$T_a = 0.9\ T \tag{2.14}$$

Eurocode 2 also provides a correlation with the UCS:

$$T_a = 0.30 \; UCS^{2/3} \tag{2.15}$$

which is only valid for concrete with UCS values less than the UCS of the C50/60 concrete type.

In the engineering practice, the axial tensile strength of concrete is often related to the UCS by the following relationship:

$$T_a = 0.1 \; UCS \tag{2.16}$$

Similar ranges of values were observed in *Bruce et al. (1998)*, *CDIT (2002)* and *Topolnicki (2004)*.

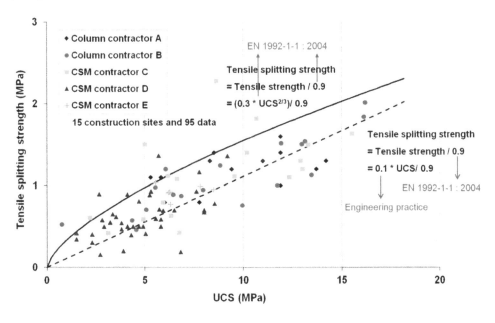

Figure 2.36. Relationship between the tensile splitting strength and the UCS of soil mix material, after Denies et al. (2012b).

2.3.5 *Shear strength of the soil mix material*

Filz et al. (2012) present a design approach based on the design value of the shear strength of the soil mix material. This approach, however, mainly looks at the design of the foundations of dykes and flood barriers and falls outside the scope of this handbook. All issues related to the design of soil mix walls in Belgium and the Netherlands are discussed in detail in the Chapter 6 of the first part of this handbook.

In the international literature, correlations can be found between the UCS and the shear strength. *Topolicki and Trunk (2006)* give the following correlation between the UCS and the shear strength (τ) of soil mix material:

$$\begin{aligned} \tau \ &= 0.40 - 0.50 \text{ UCS where UCS} < 1 \text{ MPa} \\ &= 0.30 - 0.35 \text{ UCS where } 1 < \text{UCS} < 4 \text{ MPa} \\ &= 0.20 \text{ UCS where UCS} > 4 \text{ MPa} \end{aligned} \qquad (2.17)$$

Porbaha et al. (2000) also describe the evolution of the shear strength as a function of the UCS (see Fig. 2.37). Based on direct shear tests performed on soil mix samples, they propose the following formula:

$$\tau = 0.53 + 0.37 \text{ UCS} - 0.0014 \text{ UCS}^2 \text{ where UCS} < 6 \text{ MPa} \qquad (2.18)$$

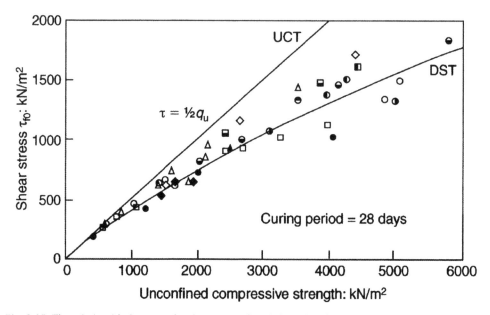

Fig. 2.37. The relationship between the shear strength and the UCS of soil mix material (results of direct shear tests, without vertical load applied to laboratory samples of marine clay mixed with cement after 28 curing days), after Porbaha et al. (2000).

According to *Bruce et al. (1998)*, the shear strength (τ) is limited to 40 to 50% of the unconfined compressive strength (UCS) for UCS values below 1 MPa, but this ratio decreases gradually as the unconfined compressive strength (UCS) increases.

2.3.6 Dynamic effects and the influence of vibrations on soil mix material

Benhamou and Mathieu (2012) describe the use of the deep mix method for the construction of two buildings on a soft alluvial soil in Martinique, France, located in an area with a particularly high risk of earthquakes. To counter soil liquefaction and post-liquefaction damage from landslides along an underground slope, a new type of permanent foundation based on a Geomix caisson (36 × 40 m) was designed. As shown in Fig. 2.61, this concept consists of a grid of Geomix trenches. The strong inertia and the geometry of the caisson device limit displacements of the Geomix panels during earthquakes. Additional shear stresses in the soil and horizontal forces of the structure are concentrated on the Geomix strips so that liquefaction of the surrounding soil is avoided. This treatment is likewise resistant to external post-liquefaction soil movements. Finally, Geomix foundations generally reduce settlement beneath the structure. This approach is also used in Japan, where the risk of earthquakes is very high (*Towhata, 2008*).

In this case, the design is based on the dynamic properties and the shear strength of the soil mix material. In principle, the shear strength of the soil mix material must be higher than the shear stress caused by the earthquake with a certain safety factor.

Information on the dynamic properties of the soil mix material (dynamic shear modulus, damping ratio and cyclic shear strength) can be found in *Kitazume and Terashi (2013)*.

2.3.7 'Steel – soil mix' adhesion

To investigate the adhesion between soil mix material and various steel reinforcements, *in-situ* pull-out tests were conducted within the framework of the BBRI soil mix project (2009-2013) on the basis of *NBN EN 12504-3*. Figure 2.38a presents the test setup. After the execution of the soil mix element, steel reinforcement was suspended from the guidance device and vertically installed into the soil mix. As illustrated in Fig. 2.38b, the top part of the steel reinforcement is made frictionless (over 1 m) using a flexible protection tube in order to eliminate the influence of the first non-representative metre on the results.

Figure 2.39 presents the peak extraction resistance in function of the UCS obtained on soil mix cores, for different types of steel reinforcements.

No information was found in the literature on the effect of corrosion of the reinforcement steel on the adhesion between steel and soil mix (= issue of debonding), especially on elements that are completely underground.

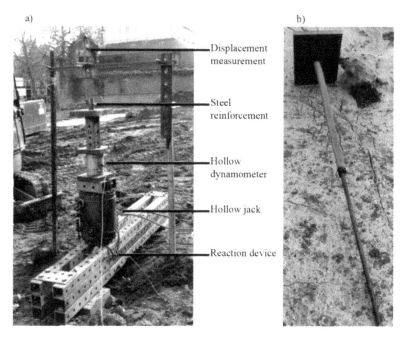

a)

—Displacement
measurement

—Steel
reinforcement

—Hollow
dynamometer

—Hollow jack

—Reaction device

b)

Figure 2.38. a) Pull-out test set-up and b) steel reinforcement with protecting tube,
from Denies et al. (2012b).

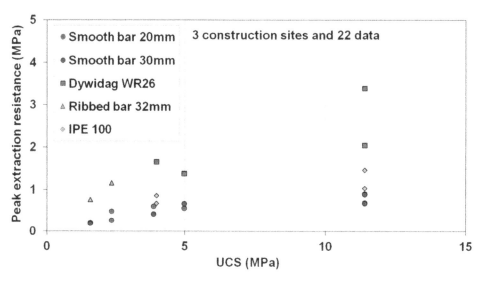

Figure 2.39. Peak pull-out resistance in function of the UCS of cored soil mix material,
from Denies et al. (2012b).

2.3.8 Porosity and petrographic analysis of the soil mix material

Within the framework of the BBRI soil mix project (2009-2013), porosity values were measured according to the Belgian standard *NBN B15-215* and vary between 25 and 65% for all soil types, as illustrated in Fig. 2.40. In order to explain this high range of values, a petrographic analysis is conducted on samples from two construction sites (in silty and sandy soils) with the help of image processing techniques (IPT) and thin section technology. Figure 2.41 gives an example of microscopic analysis of a thin section cut from a soil mix core, where P represents the pores, S the sand grains and C the cement stone. If open cracks, without specific orientation, are observed, they have a limited width varying between 10 and 200 μm.

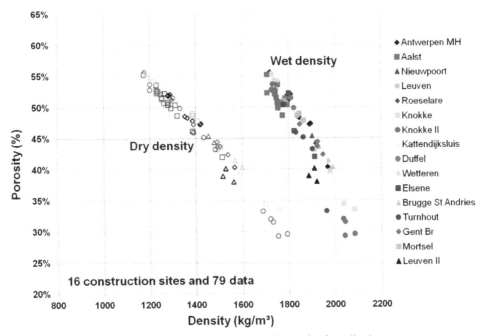

Figure 2.40. Relationship between dry and wet density and porosity for soil mix cores, from Denies et al. (2012b).

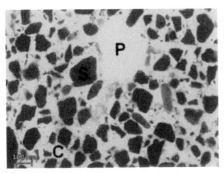

Figure 2.41. Microscopic analysis of a thin section of soil mix material with fluorescent light, from Denies et al. (2012b).

The pores in the soil mix sample are coloured by the resin used in the production of the thin section. All the pores with a surface area higher than 10 μm² are indicated as macropores. They represent around 2.4% of the total surface area. As a result, high porosity values, illustrated in Fig. 2.40, can only be related to the high and homogeneous capillary porosity. The high capillary porosity could result from the high w/c ratio, used for the execution of the soil mix walls. The high hydration level and the presence of portlandite Ca(OH)$_2$ in the soil mix samples consolidate this assumption.

In the study of *Bellato et al. (2012)*, scanning electron microscopy, X-rays powder diffraction and X-rays micro-tomography have been extensively used in order to investigate the degree of homogeneity of CMS treated overconsolidated clays and the spatial distribution of cement throughout the soil matrix. Figure 2.42 illustrates the micro-tomographic analysis of a 8 mm diameter specimen. As reported by *Bellato et al. (2012)*, it is possible to note the relative homogeneity of cemented soil matrix, showing the effectiveness of CSM treatment as mixing method. Because the X-rays generated from the source are absorbed by the specimen according to the element atomic number, some inclusions can be recognized, that basically consist of untreated clay or minerals.

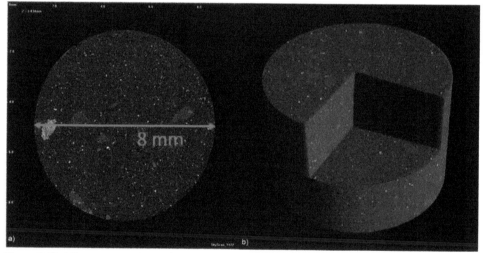

Figure 2.42. Micro-tomography analysis of a CSM treated clay specimen: a) 2D cross section; b) 3D reconstruction, after Bellato et al. (2012).

2.3.9   Permeability of the soil mix material

Within the framework of the BBRI soil mix project (2009-2013), permeability tests were performed on soil mix samples according to the standard *DIN 18130-1*. The samples were cored from real soil mix walls. The coefficient of hydraulic conductivity varied between 10^{-8} and 10^{-12} m/s, regardless of the execution process and the soil conditions, as illustrated in Fig. 2.43 and 2.44. Similar range of values of hydraulic conductivity was reported in

CDIT (2002). In the Netherlands, *Lantinga (2012)* reports measurements of coefficient of permeability varying between 3.5 10^{-11} and 5.0 10^{-12} for a construction site characterised by a sandy soil. No correlation was observed between porosity and permeability for the BBRI data on the contrary of the results of *Taki and Yang (1991)* presenting logarithmic relation-ship between porosity and hydraulic conductivity.

In *Bruce et al. (1998)*, the range of values varied between 10^{-6} and 10^{-9} m/s. *Topolnicki (2004)* presented a range of values for permeability between 10^{-7} and 5 10^{-9} m/s depending on the soil type, as illustrated in Table 2.13.

In overconsolidated clays, *Bellato et al. (2012)* have measured coefficients of permeability around 10^{-11} m/s.

During the construction of the cut-off wall at the Westelijke Invalsweg viaduct in Leeuwarden, a large number of permeability tests were performed (*Gerritsen, 2013*). Amongst other things, the relationship between the compressive strength and the permeability was examined (see Figure 2.45). The test results examined showed that the permeability significantly decreases, from an average of 1.10^{-8} m/s at 7 days, to more than 1.10^{-9} at 56 days. Permeability in several samples fell between 28 days and 56 days. However, the distribution in the permeability also seemed to have increased by 56 days. Based on this, it seems plausible that the permeability of the walls does not decrease significantly after 28 days.

In addition to local measurements on sample material, the permeability of the soil mix walls and of the entire system can also be investigated by way of a full-scale pumping test (see Section 8.8 of the first part of the handbook).

At the Westelijke Invalsweg viaduct, a large-scale pumping test was performed within the polder compartments designed with soil mix walls (*Gerritsen, 2013*). Most of the walls are constructed as simple soil mix walls. The deepest compartment is constructed with a soil mix wall equipped with a geomembrane. Due to the pumping tests, the water level within all compartments was decreased to about 7 metres below the surrounding polder level. During the pumping tests the leakage flow was measured. In addition, the water level within and outside the compartments was also recorded. The dewatering was recorded using extensive measurements with digital divers installed in well pipes. This showed that on setting the polder level for the pumping test, a slight lowering (ca. 0.5 m) occurred in both the water permeable intermediate layer (above the sealing soil layer) and in the deep water permeable layers under the sealing soil layer (impermeable layer). Large-scale leakage or sinking was not observed in the area. However, the measured leakage rate during the pumping test was higher than the flow rate calculated from the wall permeability and the vertical permeability of the bottom seal. On the basis of the measured water levels, it seems plausible that in this case both the impermeable soil layer and the wall are more permeable than expected from the laboratory test results. Evaluation of permeabilities and allocation to the various system components could be done by performing an analysis with a 3-dimensional geohydrological model. Falling head tests performed on lab samples seem to give lower permeabilities than actually observed in whole systems in practice. That phenomenon has been observed in various projects.

The permeability of the walls might be subject to a time-dependent effect related to the possible silting-up of walls, as this is sometimes observed in sheet piling interlocks. To date, however, there are no practical experiences available to quantify this phenomenon in soil mix walls. Quantification could be achieved by checking whether a trend of decreasing leakage rate, based on long-term flow measurement, can be observed in constructions. Whether or not the walls silt-up, the homogeneity of the soil mix wall itself and the soil structure in the environment will also play a major role. If, however, a trend of increasing flow rates is found, this could indicate erosion of the wall surrounding any leaks. These could be detected by using a leak detection method (see Section 8.9 of the first part of the handbook).

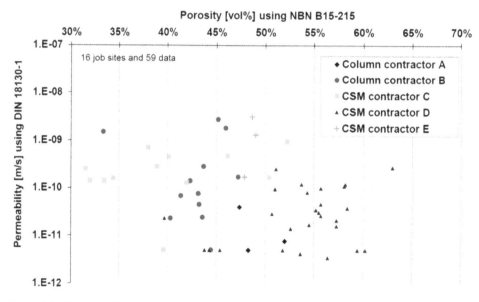

Figure 2.43. Relationship between permeability and porosity for soil mix cores in function of the execution process. Denies et al. (2012b).

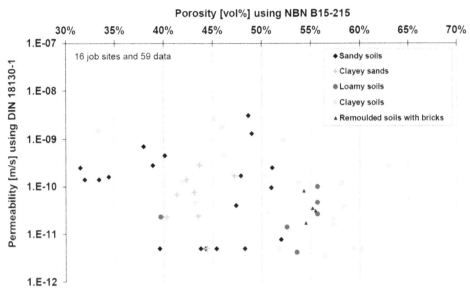

Figure 2.44. Relationship between permeability and porosity for soil mix cores in function of the soil type, based on the data of Denies et al. (2012b).

Table 2.13. Typical field permeability for ranges of cement factor and soil types (data based on soils stabilised with the wet mixing process), Topolnicki (2004).

Soil type	Cement content* (kg/m³)	Permeability (m/s)
Sludge	250 – 400	1×10^{-8}
Peat, organic silts/clays	150 – 350	5×10^{-9}
Soft clays	150 – 300	5×10^{-9}
Medium/hard clays	120 – 300	5×10^{-9}
Silts and silty sands	120 – 300	1×10^{-8}
Fine-medium sands	120 – 300	5×10^{-8}
Coarse sands and gravels	120 – 250	1×10^{-7}

* Cement content in place: weight of injected cement divided by the total volume of treated ground and injected slurry. Note of Topolnicki (2004): Data compiled from Geo-Con., Inc. (1998), FHWA (2001) and Keller Group.

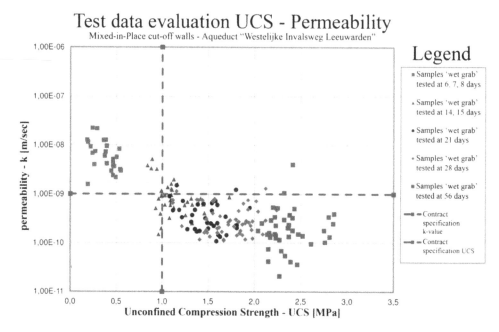

Figure 2.45. Relationship between the unconfined compressive strength and the permeability when executing triple soil mix cut-off walls - Aqueduct Western Invalsweg, Leeuwarden - Witteveen + Bos (Gerritsen, 2013).

2.3.10 Sampling, transportation, handling, storage and preparation of the test specimens

In the deep mixing projects, the design can be based on laboratory test campaign. Laboratory soil-cement samples are then used to study the mechanical properties of the stabilised soil. For that purpose, *EuroSoilStab (2002)* provides detailed laboratory procedures respectively for soil-cement columns and soil stabilisation applications. Materials, equipments and preparation of the laboratory samples are presented. Within the framework of an international collaborative study fully described in Terashi and Kitazume (2011), Grisolia et al. (2012) have analysed the applicability of laboratory molding procedures in function of the initial consistency of the mixture, considering the mechanical properties of the specimens and the repeatability of the test results. Four molding techniques, namely Dynamic Compaction, Tapping, Rodding and No Compaction were used. The results provide a base to select the right molding technique in function of the soil-binder mixture consistency that produces the best specimen's quality for testing.

If the previous works allow the construction of standardized and international test procedures for laboratory soil mix samples, the Quality Control (QC) of the execution process is generally based on laboratory tests performed on core material. Each core sample is characterised by its own history influencing the test result and its interpretation. Beyond the question of the representativeness of the core samples with regard to the *in-situ* executed material, *Denies et al. (2012c)* concentrate on the sampling, the transport, the storage, the handling and the preparation of the soil mix test specimens and propose test procedures in the continuity of the content of the European standard *EN 14679 (2005)* for deep mixing. Table 2.14 illustrates the timeline of the soil mix sample life with regard to the standards or test methodologies supporting its several stages, as used during the BBRI 'Soil Mix' project (2009-2013).

Table 2.14. Timeline of the soil mix samples: procedures followed within the framework of the BBRI 'Soil Mix' project (2009-2013), Denies et al. (2012c).

Sampling	Transportation	Preserving Storage	Handling/Preparation of test specimens/ Test and report	
In-situ	In-situ	In laboratory	In laboratory	
EN 14679: 2005 Execution of special geotechnical works – Deep mixing **EN ISO 22475-1: 2007** Geotechnical investigation and testing - Sampling methods and groundwater measurements - Part 1: Technical principles for execution **ASTM D 420 – 93** Standard Guide to Site characterization for Engineering, Design, and Construction Purposes				
EN 12504-1: 2009 Testing concrete in structures - Part 1: Cored specimens - Taking, examining and testing in compression	**ASTM D 4220 - 89** Standard Practices for Preserving and Transporting Soil Samples	**EN 12390-2: 2009** Testing hardened concrete - Part 2: Making and curing specimens for strength tests	Visual analysis and quantification of soft soil inclusions	**Section 5.2 of Denies et al. (2012c), after Ganne et al. (2011 and 2012)**
			Handling procedure for the preparation of soil mix test specimens	**Section 5.3 of Denies et al. (2012c)**
			Density	**EN 12390-7: 2009** Testing hardened concrete - Part 7: Density of hardened concrete
ASTM D 2113 - 83 Diamond Core Drilling for Site Investigation	**ASTM D 5079 - 90** Standard Practices for Preserving and Transporting Rock Core Samples	**ASTM D 1632 - 87** Standard Practice for Making and Curing Soil-Cement Compression and Flexure Test Specimens in the Laboratory	Unconfined compressive strength, UCS	**EN 12390-3: 2009** Testing hardened concrete - Part 3: Compressive strength of test specimens
			Modulus of elasticity, E	**NBN B 15-203: 1990** Concrete testing - Statical module of elasticity with compression **ISO/FDIS 1920-10: 2010** Testing of concrete - Part 10: Determination of static modulus of elasticity in compression
			Tensile splitting strength, T	**EN 12390-6: 2010** Testing hardened concrete - Part 6: Tensile splitting strength of test specimens
			Ultrasonic pulse velocity, V_p	**ASTM C 597 - 09** Standard Test Method for Pulse Velocity Through Concrete **EN 12504-4: 2004** Testing concrete in structures - Part 4: Determination of ultrasonic pulse velocity
			Porosity	**NBN B 15-215: 1989** Concrete testing - Absorption of water by immersion
			Hydraulic conductivity	**DIN 18130-1: 1998** Laboratory tests for determining the coefficient of permeability of soil

2.3.11 Influence of the unmixed soft soil inclusions in the soil mix material

One major issue is the representativeness of the core samples with regard to the mixed in place soil mix material. There is mainly the question of the influence of unmixed soft soil inclusions on the mechanical behaviour of the soil mix material. Indeed, as a natural material (i.e. soil) is being mixed, it is to be expected that the entire wall is not perfectly mixed and homogeneous: inclusions of unmixed soft soil are present.

Within the framework of the BBRI 'Soil Mix' project (2009-2013), a methodology taking into account these inclusions was developed and illustrated with Belgian case studies (*Ganne et al., 2011 and 2012*). Figure 2.46 gives an overview of the results for 27 Belgian construction sites. The amount of soil inclusions in soil mix material mainly depends on the nature of the soil:
- in quaternary or tertiary sands, it is less than 3.5%,
- in silty (or loamy) soils and alluvial clays, it ranges between 3 and 10%,
- in clayey soils with high organic content (such as peat) or in tertiary (overconsolidated) stiff clays, it can amount up to 35% and higher.

In a view of design consideration, *Ganne et al. (2010)* have proposed to reject all test samples with soil inclusions > 1/6 of the sample diameter, on condition that no more than 15% of the test samples from one particular site would be rejected. This possibility to reject test samples results from the reflexion that a soil inclusion of 20 mm or less does not influence the behaviour of a soil mix structure. On the other hand, a soil inclusion of 20 mm in a test sample of 100 mm diameter significantly influences the test result. Of course, this condition is only suitable if one assumes that there is no soil inclusion larger than 1/6 of the width of the *in-situ* soil mix structure.

For the purpose of studying this question, 2D numerical simulations were performed at KU Leuven with the aim to quantify the effect of soil inclusions on the strength and stiffness of the soil mix material. The following parameters are being considered: size, number, relative position and percentage of soil inclusions. The results of this study are presented in *Vervoort et al. (2012)* and *Van Lysebetten et al. (2013)*. As illustrated in Fig. 2.47, they confirm that soil mix samples with soft soil inclusions larger than 1/6 have a considerable influence on the deduction of the engineering values. Based on this numerical analysis, the "rule of 1/6" as proposed by *Ganne et al. (2010)* seems to be justified.

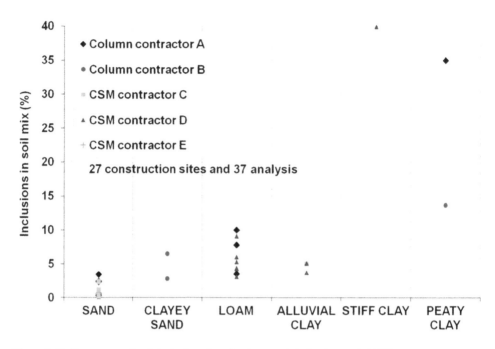

Figure 2.46. Percentage of soil inclusions in soil mix material (Denies et al. 2012b).

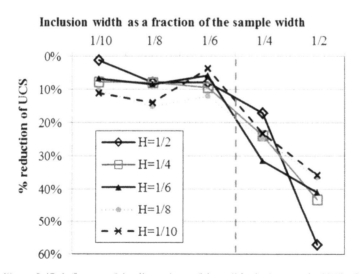

Figure 2.47. Influence of the dimensions of the soil inclusions on the UCS of soil mix material. Results of 2D numerical simulations performed with the help of the Universal Distinct Element Code UDEC of Itasca'. Details of the model are available in Van Lysebetten et al. (2013). H is the ratio between the height of the soil inclusion and the sample diameter, results published in Denies et al. 2013.

As illustrated in Fig. 2.48, the presence of 1% of weak inclusions results in an average reduction of about 3% of the stiffness, while 10% of inclusions results in a 30% reduction (on average) of the stiffness.

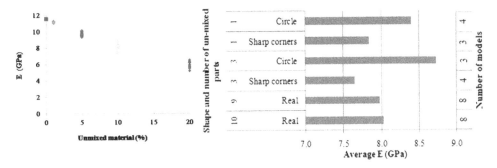

Figure 2.48. Results of linear elastic numerical simulations (Vervoort et al. 2012): a. Variation of Young's modulus as a function of the percentage of unmixed material (surface area); b. Effect of the number of inclusions and their shape on the average Young's modulus for 30 numerical models (10% unmixed material).

Figure 2.49 illustrates the large effect of the unmixed percentage on the UCS of the simulated models. For 1 % of unmixed material the strength is reduced on average by about 20%. For 10% of inclusions the UCS is reduced on average by about 50% and for 20% even by about 70%.

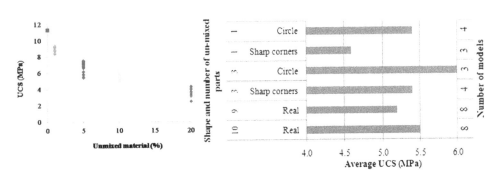

Figure 2.49. Results of elasto-plastic numerical simulations (Vervoort et al. 2012): a. Variation of UCS values as a function of the percentage of unmixed material (surface area); b. Effect of the number of inclusions and their shape on the average UCS value for 30 numerical models (10% unmixed material).

2.3.12 Scale effect – large-scale UCS tests

Apart from traditional core samples (with a diameter around 10 cm), large-scale UCS tests were conducted on rectangular blocks with approximately a square section, with a width corresponding to the width of the in-situ soil mix wall (about half a metre) and with a height approximately twice the width (*Vervoort et al. 2012*). The influence of the scale effect was found to be limited for the Young's modulus while the UCS values of the large samples show a reduction of 30 to 50% in comparison to these of the 10 cm diameter cores. The results of all the tests performed in KU Leuven are presented in Fig. 2.50 for various soil conditions and different execution systems: the CSM and the TSM.

As observed in Fig. 2.50, a linear relationship is observed between the test results obtained from the typical core samples and the large rectangular blocks. Although there is a scatter in the test results, the UCS of the full-scale blocks is about 70% of the average UCS of the typical core samples. It is to note that similar conclusion was observed for soil mix columns in Japan (*CDIT 2002*).

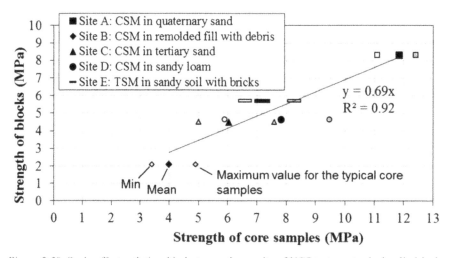

Figure 2.50. Scale effect: relationship between the results of UCS tests on typical cylindrical core samples (10 cm diameter) and on large rectangular blocks tested in KU Leuven (from Denies et al. 2013).

2.3.13 Large-scale bending tests

As part of the BBRI soil mix project (2009-2013), 17 large-scale bending tests were performed on soil mix elements. Eleven soil-cement columns and six CSM panels were first installed *in-situ* (at seven different construction sites with different soil conditions), excavated (after hardening of the soil mix material) and transported to the BBRI laboratory for testing. The purpose of these tests was to determine the bending capacity and bending stiffness of the reinforced soil mix elements. The soil mix elements were subjected to 4-points (see Fig. 2.51) or 4-points (see Fig. 2.52) bending tests.

Figure 2.51. 4-points bending test performed on a half-CSM panel that was reinforced with a steel beam (HEA 240) (BBRI Soil mix project, 2009-2013).

Figure 2.52. 3-points bending test performed on a soil mix column reinforced with a steel beam (IPE 240) (BBRI Soil mix project, 2009-2013)

The following values were recorded during the loading of each soil mix element:

- δ_{max}: (central deflection): deflection of the soil mix element measured in the middle (L/2) by a Linear Variable Displacement Transducer (LVDT),
- $\delta_{supports}$: vertical displacement of the soil mix element at the support beams measured by LVDTs,
- δ_{slip}: slipping of the steel beam relative to the soil mix material. This was done by means of an LVDT,
- the deformations at the bottom (inferior flange) and top (superior flange) of the steel beam were measured by optical fibres
- the applied force was measured by a load cell.

Because one constant loading step only lasted five minutes, creep and durability aspects were not considered in these bending tests.

The test procedure and the results of one bending test are discussed in detail in *Denies et al. (2014)*. An overview of the results of the 17 bending tests performed in the BBRI is given in *Denies et al. (2015)*. The maximum bending moments recorded in the bending tests are given in the overview table A2.3 in the Appendix A2 of the present handbook.

As discussed in *Denies et al (2015)*, the analysis of the results obtained on the 13 tested soil mix elements reinforced with a steel beam reveals a real interaction between the steel beam and the soil mix material. However, the stiffness of the composite section (steel + soil mix) decreases with increasing applied bending moment and the progressive opening of the cracks in the soil mix material during the test. There is a progressive displacement of the neutral axis of the section during the test. This test series shows that, in the range of the flexural bending moments supported by the soil mix wall in practice, the stiffness of the soil mix wall is significantly greater than the stiffness of the steel beam alone. This observation is important because the wall stiffness entered in numerical models has a significant effect on the calculated bending moments in the soil mix wall.

The maximum bending moments reached during the tests are 1.8 to 3 times greater than the bending moments computed only considering the yield strength of the steel beam alone.

The measurements of the steel stresses show a real interaction between the soil mix and the steel: the yield strength in the steel beam ($\sigma = \sigma_{el}$) is really reached (i.e. measured) at bending moments which are 20% to 70% greater than would be the case without the interaction with the soil mix material, i.e. only considering the steel resistance (see detailed conclusions in *Denies et al., 2015*).

2.3.14 *Correlations between mechanical properties of soil mix material*

In the present chapter, hydro-mechanical properties of soil mix material have been discussed. Range of UCS values from field data have been presented and criticized with regard to governing parameters such as the cement content, the w/c factor, the mixing energy and the curing time. Difference was made between core and wet grab samples. Results of tests for the determination of the modulus of elasticity and the tensile splitting strength were also highlighted. As illustrated in Fig. 2.34 and 2.36, both properties can be deduced on the basis of UCS test results respectively with the help of equations (2.8) to (2.10) and (2.14) to (2.16). Indeed, in practice, it is not always possible to conduct tests for the determination of the modulus of elasticity and the tensile splitting strength. UCS tests are quick and less expensive. Table 2.15 present correlations obtained for core samples from real soil mix walls collected within the framework of the BBRI soil mix project (2009-2013). Other correlations available in the literature are also presented.

Parameter	BBRI soil mix project (2009-2013)	Topolnicki and Trunck (2006)	Bruce et al. (1998)
Age of the specimen (days)	Laboratory sand-cement samples $UCS_{7\ days} = 0.37\ UCS_{28\ days}$ $UCS_{56\ days} = 1.13\ UCS_{28\ days}$ $UCS_{126\ days\ and\ more} = 1.46\ UCS_{28\ days}$ Laboratory loam-cement samples $UCS_{7\ days} = 0.28\ UCS_{28\ days}$ $UCS_{56\ days} = 1.31\ UCS_{28\ days}$ $UCS_{126\ days\ and\ more} = 1.67\ UCS_{28\ days}$	$UCS_{28\ days}$ = 2.0 $UCS_{4\ curing\ days}$ = 1.4 – 1.5 $UCS_{7\ days}$ (silt and clay) = 1.5 – 2.0 $UCS_{7\ days}$ (sand) $UCS_{56\ days}$ = 1.4 – 1.5 $UCS_{28\ days}$ (silt and clay)	$UCS_{28\ days}$ = 1.4 to 1.5 $UCS_{7\ days}$ (silts and clays) $UCS_{28\ days}$ = 2 $UCS_{7\ days}$ (sands) $UCS_{60\ days}$ 1.5 times the 28-day UCS, while the ratio of 15 years to 60 days UCS may be as high as 3 to 1. In general, grouts with high w/c ratios have much less long term strength gain beyond 28 days, however.
Shear strength (MPa)		τ = 0.40 – 0.50 UCS for UCS < 1 MPa = 0.30 – 0.35 UCS for 1< UCS < 4 MPa = 0.20 UCS for UCS > 4 MPa	40 to 50% of UCS at UCS values < 1 MPa, but this ratio decreases gradually as UCS increases.
Tensile strength (MPa)	$T_{tensile\ splitting\ strength}$ Eq. (2.14) to (2.16)	$T_{direct\ tensile\ strength}$ = 0.08 – 0.15 UCS with a maximal value of 0.2 MPa	Typically 8 – 14 % UCS
Modulus of elasticity (MPa)	Tangent $E_{10\%\ to\ 30\%\ UCS}$ Eq. (2.8) to (2.10)	Secant $E_{50\%\ UCS}$ = 50 – 300 UCS for UCS < 2 MPa = 300 – 1000 UCS for UCS > 2 MPa	350 to 1000 times UCS for lab samples and 150 to 500 times UCS for field samples
Elongation at maximal force %		ε_u = 0.5 – 1.0 for UCS > 1 MPa = 1.0 – 3.0 for UCS < 1 MPa	
Poisson ratio (-)		ν = 0.25 – 0.45 = 0.30 – 0.40 (usually)	

2.4   Durability of the soil mix material

The durability of soil mix material is an important topic, several aspects of which should be considered. On the one hand, the "durability" concerns all aspects that can be related to the development and/or degradation of the properties (strength, permeability, pH, etc.) of the soil mix material itself. On the other hand, there is the problem of the durability of soil mix that has been executed in contaminated soils or in soils containing components that adversely affect the development of the strength characteristics. In both cases, there may also be an effect on corrosion or an increased corrosion rate of the reinforcing steel.

For the soil mix procedure in particular, the contaminants, or the other components such as chlorides from brackish water, are mixed with the injected binder and the soil and thereby fully integrated into the soil mix matrix. The potential effect of the contaminants or other components is therefore greater than, for example, *in-situ* cast-in-place or prefabricated concrete elements, which are only exposed to the contaminants at the soil interface.

For temporary structural soil mix elements, the problem is primarily one of contaminant effects on the setting process and the development of strength of the soil mix material over time. For permanent structural applications, depending on the project, it is moreover important that the soil mix material can continue to perform its long-term function (arching effect for transferring the forces to the steel reinforcements, long-term permeability, etc.), and that the risk of corrosion of the reinforcing elements due to various factors is taken into account.

It is not the intention in this section to set out a comprehensive study of the durability of soil mix material, but rather to identify the various parameters and factors that may affect its long-term behaviour. This parameter study is mainly based on the PhD study of *Guimond-Barrett (2013)* and on the experience gained from the BBRI Soil mix project (2009-2013).

Figure 2.53 gives an overview of the various significant factors. This figure shows that two antagonistic phenomena are at work. On the one hand, it has been established that the strength of the soil mix material increases with time. This has already been discussed in detail in Section 2.3.2d in Part 2 of the handbook. On the other hand, there are a number of factors that can lead to a progressive degradation of the soil mix material.

The first degradation phenomenon is caused by the diffusion of Ca^{2+} cations from the soil mix elements to the surrounding soil. This phenomenon is the result of a natural tendency of a system towards chemical equilibrium. A decrease in Ca^{2+} cations is observed over time in the zone of the soil mix element that is in direct contact with the soil, while an increase can be observed in the surrounding soil. The progressive decrease of Ca^{2+} cations leads to a reduction in the strength of the soil mix material. However, according to all the literature available, and where referred to in *Guimond-Barrett (2013)*, this only affects a limited zone in the soil mix material. The maximum depth of the affected zone is in all cases less than 8 cm. According to *Topolnicki (2004)*, this effect is also compensated by the long-term increase in the strength of the soil mix material (see Section 2.3.2d in Part 2 of the handbook).

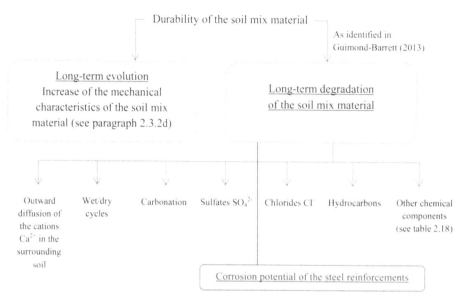

Figure 2.53. Identification of the factors that affect the durability of soil mix material.

A second factor is the effect of the wetting or drying of the soil mix material. This effect is important in practice because one side of the earth-retaining soil mix wall is quickly excavated after execution, thereby exposing them to the surrounding air and weather influences such as rain and insolation. Based on lab observations of these phenomena, it can be seen that the strength (and its development over time) of the soil mix material is especially influenced by the drying cycle. The evaporation of (excess) water adversely affects the strength development of the soil mix material. As a result, even shrinkage cracks cause the strength to reduce rapidly. This phenomenon is irreversible if the sample is continuously kept in dry conditions. *Guimond-Barrett (2013)* has demonstrated in the course of his PhD, that the shear modulus (for minor deformations) decreases almost linearly with the degree of saturation of the soil mix material.

This phenomenon was also found in creep tests performed on soil mix material during the BBRI Soil mix project (2009-2013). Initially these creep tests were performed at a relative humidity of 60%. Even at very low stresses, shrinkage cracks were observed due to the drying of soil mix samples and the strength and stiffness of the soil mix material decreased. However, when the same series of creep tests was performed under saturated conditions, the phenomenon disappeared.

However, in practice, this is not so bad. Soil mix walls are usually in contact with the soil/ groundwater on one side, which usually causes the wall to be capillary saturated. Only on very dry ground and during long-term exposure/drought is there therefore a potential risk of shrinkage cracks in the zone exposed to ambient air.

At present, there is a consensus in Belgium that no special measures are needed for soil mix walls for less than [1 to 2 years] in ambient air to stop them from drying out. For permanent walls, however, a precautionary principle is applied and a protective barrier (concrete facing wall, reinforced shotcrete, etc.) is always installed on the side exposed to ambient air. Moreover, in highly ventilated underground spaces, such as garages, in spite of the capillary nature of the soil mix wall, there may still be progressive degradation of the soil mix material. In Belgium, permanent soil mix walls are not directly exposed to the outside environment, as it is assumed that exposure of soil mix material to freeze-thaw cycles will weaken them.

In the Netherlands, there seems to be a more permissive approach to exposing permanent soil mix constructions to ambient air and even to the outside environment. However, as knowledge of the long-term behaviour is still quite limited, applications of that type still need additional measures, such as a modified binder type, minimum binder dosage and monitoring of the construction, etc.

In the United States, two ASTM standards were available to check the durability of soil mix material against wet-dry cycles and freeze-thaw cycles: the *ASTM D 559-89* standard (Standard Test Methods for Wetting and Drying Compacted Soil -Cement Mixtures) and the *ASTM D 560-89* standard (Standard Test Method for Freezing and Thawing Compacted Soil-Cement Mixtures). However, these two standards were rescinded in 2012, apparently for administrative reasons. *Shihata and Baghdadi (2001)* conducted a series of tests on laboratory samples in accordance with the principles of these standards. Laboratory samples were prepared with three types of sand and with cement contents of 5-9% by weight. These test samples were subjected to freeze-thaw cycles (or wet-dry cycles) and the mass loss measured as a function of time. Their study led to the following conclusions regarding the effect of frost: in general, mass loss rapidly increases during the first three months, after which there is a progressive stabilisation of the samples with a final mass loss of 7-25%. The ASTM standards also refer to experimental research in this area (*Packard and Chapman, 1963 and Packard, 1962*). In general, it can be deduced that the mass loss depends on the binder content (inversely proportional) and the soil type.

Figure 2.54 illustrates the carbonation process in which CO_2 from ambient air reacts with the calcium hydroxides of the soil mix material (as it does in concrete) to form calcium carbonate. In concrete, this process is very well known: carbonation is a slow and continuous process that progressively extends from the outer surface to the inside but whose velocity decreases with increasing diffusion depth. For soil mix material, the effect of this chemical process is less well known. The Guimond-Barrett results (2013) show that carbonation of soil mix material is a slow process, with less than 10 mm penetration depth after 180 days. The penetration depth of the carbonation process appears to be linear against the logarithm of time. In addition, the results show that the rate at which the process develops in the material depends on the age of the soil mix material at the time it is exposed to ambient air and thus to CO_2.

The effect of carbonation on the mechanical characteristics, however, remains unclear. Depending on the source, the strength either increases or decreases (see *Perera et al., 2005*). Carbonation is mainly related to the corrosion of the steel reinforcement, as it causes a reaction that reduces the alkaline content, expressed in pH, of the soil mix material. With regard to steel corrosion, it can be said that the soil mix material does not offer alkaline protection against corrosion if the pH falls below about 9.5.

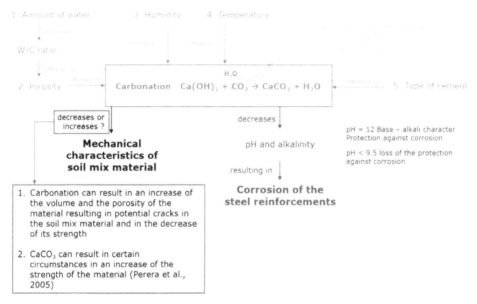

Figure 2.54. Factors determining the carbonation and the consequences for the durability of the soil mix material.

In addition to the effects of wet-dry and carbonation, designers and deep mixing contractors should take into account the presence of contaminants in the soil before starting soil mix operations. Certain contaminants may be harmful to the soil mix material itself (setting, hardening) and/or the steel reinforcements (corrosion). The following section discusses the possible effects of sulfates, chlorides and hydrocarbons.

Figure 2.55 shows that, in the presence of sulfates, a reaction with the calcium components of the binder can occur. The reaction between these calcium components and the sulfates can lead to the formation of ettringite. In turn, ettringite formation can cause swelling reactions to occur in the soil mix material, resulting in cracking. This phenomenon has already been seen in concrete, but in this case the sulfates are not only present around the concrete element, but also effective in the soil mix matrix. To avoid the adverse effects of sulfates, *Guimond-Barrett (2013)* recommended the use of either Portland cement (CEM I) with a reduced amount of tricalcium aluminate (C3A), which is the component specifically affected by sulfates, or the use of blast furnace cement (CEM III).

358

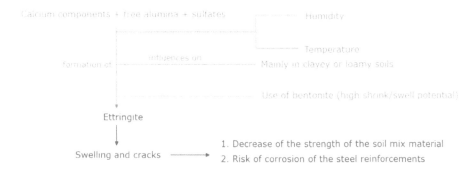

Calcium components + free alumina + sulfates ———— Humidity

—————— Temperature

formation of ———— influences on ———— Mainly in clayey or loamy soils

————————— Use of bentonite (high shrink/swell potential)

Ettringite

Swelling and cracks ————→ 1. Decrease of the strength of the soil mix material
2. Risk of corrosion of the steel reinforcements

Figure. 2.55. Potential risk related to the presence of sufates in the soil mix material or in the surrounding soil.

Drawing on experience from his experiments, *Guimond-Barrett (2013)* emphasised that when bentonite is used, it can flocculate into larger particles. These can then lead to larger pores, which then reduce the strength of the soil mix material.

The effect of chlorides on steel reinforcement elements (corrosion) is relatively well known. However, the influence on the soil mix material itself is less clear. The literature contains contradictory results. In several cases, it is reported that the addition of chlorides to the mix has no effect on the setting or hardening (*Guimond-Barrett, 2013*). In other cases it is shown that chlorides affect both the strength development of the soil mix material (*Horpibulsuk et al., 2012*) and its permeability (*Al-Tabbaa and King, 1998*).

It can be concluded from these tests that the conditions and concentrations used may have had a significant impact on the results. A good insight into this is therefore a prerequisite for reaching correct conclusions.

The same is true for the effect of hydrocarbons on the strength of soil mix. While *Guimond-Barrett (2013)* observed no significant effect from adding hydrocarbons to soil mix, *Cruz et al. (2004)* demonstrated a clear adverse effect of hydrocarbons on the setting process and on the hardening of the soil mix material. However, the concentrations in the two experimental studies are very different. For example, *Guimond-Barrett (2013)* limited the concentration of hydrocarbons in the tested mixtures of soil mix to commonly-occurring concentrations that can be observed on hydrocarbon-contaminated sites in France.

In general, it can thus be concluded that, to date, the scientific research on the effect of contaminants on the setting and strength development of the soil mix material is quite limited and sometimes even leads to contradictory conclusions. So far there is few information available on the effect of contaminants on other important characteristics of the soil mix, such as permeability. For more information, the reader can still refer to the content of the recent articles and researches of Prof. Al-Tabbaa (see Section References).

As part of a recent project in the Netherlands, *Deltares (2005b)* examined the durability, over a period of 100 years, of a dyke reinforced using the MIP method.

The study involved determining the life expectancy of stabilised columns executed in humus rich clay and peat. The tests were based on the principle of 100-year simulation ageing. A monitoring plan was set up for the site, consisting of tests on aged and non-aged test samples. For this, tests were performed to determine the strength parameters, density, permeability and characteristics of the binder matrix. The durability of stabilised soil is naturally directly related to the degradation of the integrity of the binder matrix by dissolution and/or mechanical erosion. This is observed considering the mass loss and/or a decrease in compressive strength and/or changes in permeability.

These tests were performed on specimens subjected to various environmental conditions:
- Permeation of test specimens at 20°C.
- Permeation of a number of test specimens at 12°C (soil temperature).
- Permeation at higher temperature (chemical ageing).
- Permeation at elevated speed (accelerated leaching).

The details of the test procedures are described in *Deltares (2005a)*.Leaching of the material is simulated on specimens placed in a permeation cell. In this case, leaching was measured for a simulated period of about 100 years of permeation through "normal groundwater flow" but at a temperature of 18°C. The permeability of the test specimens was measured regularly during the test.
The density was determined in order to determine any loss of material.
Strength parameters are determined at five different times.

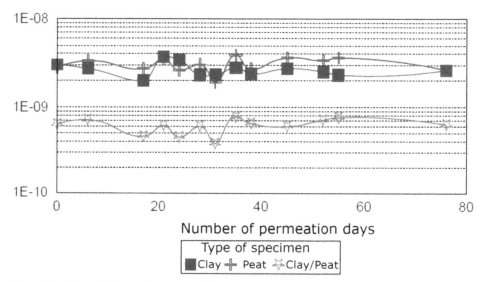

Figure 2.56. The evolution of the permeability under continuous flow through soil mix test specimens. Deltares (2005a).

Figure 2.56 illustrates the evolution of the permeability of the test specimens (built-in for permeation) as a function of time (*Deltares, 2005a*). In addition, the curves show that after 76 days of permeation (about 85 years in practice) the permeability had not changed. The strength of the test specimens was determined as shown in Table 2.16 (*Deltares, 2005a*).

Table 2.16. Results of the tests to determine the strength parameters of the soil mix material (Deltares, 2005a)

Test specimen	Density (kg/m³)	Strength (kPa)	Modulus of elasticity (MPa)	Deformation at failure (%)
Peat				
No permeation	1200	399	107	0.68
17 days	1250	207	11	3.05
55 days	1190	216	35	0.97
76 days	1235	286	45	1.02
Clay				
No permeation	1630	1085	140	1.31
17 days	1600	1255	285	1.06
55 days	1625	1304	269	1.11
76 days	1590	1382	349	0.61

The test specimens corresponding to stabilised peat show great variation. According to *Deltares (2005a)*, this is partly due to the inhomogeneities in the peat such as wood residues, which are not stabilised. As illustrated in Table 2.16, the strength values here are very limited with respect to the compressive strengths of the soil mix walls with a barrier function. These values, however, fall within the range of typical values for soil stabilisation projects.

To assess the effect of ageing, three stabilised soil samples were investigated (see Table 2.17, from *Deltares, 2005a*).

Table 2.17. Overview of the test samples (Deltares, 2005a).

Test specimen ID	TNO thin section	Type	Consistence on delivery
2M	00495	Clay, aged	Soft but shows good cohesion
6M	00496	Peat, aged	Relatively hard and shows good cohesion
6MB	00497	Peat, not aged	Hard, compact and shows good cohesion

Visual inspection of the samples revealed no visible evidence of deterioration, precipitation, cracking, crumbling, etc. To gain insight into the mineralogical composition and internal structure of the mixtures due to ageing, the mixtures were characterised by the following methods:
- Polarisation and Fluorescence Microscopy (PFM),
- Scanning electron microscopy (REM) supplemented with X-ray microanalysis (RMA or EDS),
- X-ray diffraction analysis (RDA or XRD),
- Differential thermal analysis / thermal gravimetric analysis (DTA/TGA).

The PFM analysis shows that the aged stabilised peat sample 6M (as well as in the aged stabilised clay sample 2M) is more porous than the non-aged stabilised peat sample 6MB. This could be due to flocculation of the soil, especially of the clay minerals present, under the influence of the cement. Furthermore, no evidence of deterioration was been found (*Deltares, 2005a*).

REM-RMA analysis also shows that the non-aged sample 6MB has a more closed structure than the two aged samples. However, water permeability does not increase on ageing, indicating that there is probably no erosion by dissolution (*Deltares, 2005a*).

However, REM-RMA analysis also shows that in the aged stabilised peat sample 6M (as well as in the aged stabilised clay sample 2M), there is less ettringite present than in the non-aged stabilised peat sample 6MB and that the crystal morphology had also changed. The extent to which this is an attack that reduces the durability of the ground stabilisers is unclear (*Deltares, 2005a*). However, this is uncertain, and in view of the dispersed way in which ettringite occurs in the matrix of the stabilised samples, its deterioration could lower the stability of these matrices over the long term (*Deltares, 2005a*).

In the research by *Deltares (2005a)* no loss of functionality (permeability and strength) was observed for stabilised clay at macro scale as the test specimens were aged.

For stabilised peat, this applies to water permeability, but the variation in the strength between the test specimens is so great that no conclusion can be drawn from this.

Microscopically, after ageing, it appears that:
- the porosity has increased,
- there is relatively less ettringite present,
- the shape of the ettringite crystals differs from those in the non-aged stabilised peat test specimens.

However, it must be realised that the non-aged sample had been hardened as would occur in an impermeable soil. That is, at 10°C and in the presence of the water that is already present in the sample, while the other sample was aged at 20°C under continuous water permeation. This means that in the non-aged sample the hydration of the binder will only have progressed slowly, while in the aged sample more reaction will have taken place (*Deltares, 2005a*).

Although it is not impossible that the relative reduction and change in the crystal form of ettringite is due to an attack or a deterioration, it can be expected that even with normal continuous hydration, the amount of ettringite will decrease. The crystalline form of ettringite greatly depends on the composition of the solution that is present and a change in crystalline form is therefore not necessarily due to attack (*Deltares, 2005a*).

The increase in porosity is probably due to the soil flocculating under the influence of the cement. As this is not accompanied by an increase in permeability, it does not compromise functionality (*Deltares, 2005a*).

For stabilised clay, the microstructure after ageing is similar to that of aged specimens in another soil stabilisation study (*Nijland et al., 2005*), in which ageing did not lower the functionality (soil stabilisation).

Assuming that the soil stabilisation will degrade primarily as a result of chemical and/or physical erosion due to water permeation, the fact that the permeability does not increase means that the chance of this is small. The variation in strength of the stabilised peat samples clearly shows that in practice the homogeneity of the mixtures deserves attention (*Deltares, 2005a*).

For permanent constructions, the combination of such tests (coupled with tests on potential chemical degradation and wet-dry cycles) could give an idea of the influence of the various environmental conditions on the ageing of the soil mix material and the expected lifetime.

This test campaign took about a year to complete, but after three months it was possible to get a good idea of the life expectancy of the soil mix material for a certain range of execution parameters (cement content, water-cement ratio, etc.).

For the cut-off function of the walls, it is also important that the soil mix material is sufficiently resistant to groundwater flow. Erosion of the hardened material can lead to degradation of the screen, causing the degradation of the cut-off function of the wall. In line with the guidelines of the CUR concerning cement-bentonite walls (*CUR 189*), various forms of erosion can be distinguished for soil mix walls:
1. Internal erosion
2. Contact erosion
3. Chemical erosion

For a further description of erosion types, please refer to *CUR 189*.

In Germany, extensive research was carried out at that time on the hydraulic resistance (internal erosion sometimes referred as leachability) of cement-bentonite walls. An extensive test campaign was performed in laboratory, in which the strength of the materials was varied, as was the hydraulic head across the samples. It can be concluded that the hydraulic resistance to internal erosion is ensured up to hydraulic gradients of $i = 30$ and a minimum unconfined compressive strength larger than 0.3 MPa.

363

At compressive strengths above this value, no internal erosion of the material has been observed in laboratory tests under the influence of either transverse or longitudinal water permeation (*Kleist, 1999*). In addition to the compressive strength and hydraulic gradient, the solid content, curing time and load time all play a role in the erosion resistance. If this is considered to be critical for soil mix walls, specific research must be carried out for this purpose.

Conclusions on the durability/sustainability aspects of the soil mix walls:
- The cement type and cement content affect not only the way the strength of the soil mix material develops over time, but also the resistance of the soil mix material to degradation phenomena as described above.
- According to Guimond-Barrett (2013), the best way to improve the durability of soil mix material is to reduce its porosity as far as possible. The porosity of soil mix material is directly dependent on the water-cement ratio used to make it (see Section 2.3.8 in Part 2 of the handbook). The lowest possible water/cement factor should therefore be pursued, provided that it does not adversely affect the workability and hence the quality and homogeneity of the soil mix material.
- The water-cement ratio currently used is mainly determined by finding the right balance between the costs (which are directly proportional to the amount of cement used), the required workability (directly proportional to the amount of water used), and the durability requirements (depending on the final porosity of the soil mix material).
- Another way to improve soil mix durability is to use high mixing energy during execution, resulting in a more homogeneous mix. Generally, it is assumed that the more homogeneous a soil mix material, the more durable it will be.

Awaiting more research, the following measures should be taken:
- A preliminary study is recommended for temporary soil mix applications on heavily polluted sites. The study consists of identifying and determining the concentrations of the contaminants present in the soil/groundwater that pose a potential risk to the setting and the strength development of the soil mix material. Based on the concentrations present, a laboratory test campaign can be conducted to check the effect of the contaminants on the setting and the strength development in order to quantify the effect of the cement type and cement content. This information can then be used to determine the *in-situ* design mix.
- Based on the currently available knowledge it is not recommended to use permanent soil mix walls with structural functions (vertical bearing capacity or earth retaining function) on highly contaminated sites.
- For the influence on setting and strength development (delaying or accelerating effect), please refer to *CUR 199*.
- In the case of concrete, information is available in the literature regarding components that can either cause setting problems or degrade already hardened concrete (e.g. *EN 1992-1-1*, the table 2.18 from *NEN 8005: 2008* the Dutch supplement to *NEN EN 206*, etc.). This information can momentarily be used to estimate the risks associated with the effects on soil mix material. However, it should be borne in mind that, unlike concrete, the contaminants are integrated into the soil mix matrix.

This section drew attention to the factors that could potentially affect the durability of soil mix material. Although there is relatively little information available in the literature dealing with the effect of these factors on structural soil mix applications, there are many applications where soil mix techniques are used for *in-situ* remediation (*Al-Tabbaa and Evans, 2003; Al-Tabbaa, 2005* and *Al-Tabbaa et al., 2009 and 2012*). In these applications, however, the research is primarily aimed at investigating the efficiency of the soil mix material that is then used for:
 - The construction of containment walls (isolation of the contaminated zone)
 - Stabilisation and Solidification (S/S) applications
 - The construction of Permeable Reactive Barriers (PRBs)

These are three areas of application where the permeability and the resistance of the soil mix material against erosion are essential parameters, with the strength playing only a secondary role. However, interesting information may also be derived from this field of engineering regarding the long-term development of the characteristics of soil mix material used in construction.

Table 2.18. Overview of the degree of aggressiveness of various components for already-hardened concrete (English translation of the content of the table AA.2 from NEN 8005: 2008).

Name	Reaction type[a]	Aggressiveness[b]
Acids		
Acetic acid	O	3-4
Boric acid	O	2
Carbolic acid (phenol)	O/U	2-3
Citric acid	O	4
Phosphoric acid	O	4
Humic acid	O	4
Lactic acid	O	3
Formic acid	O	3
Oxalic acid	O	1
Nitric acid	O	5
Tannin (tanning agent)	O	1-2
Hydrogen fluoride	O	5
Tartaric acid	O	1
Hydrochloric acid	O/C	5
Hydrogen sulphide	O	2
Sulphuric acid	O/E	5

Name	Reaction type[a]	Aggressiveness[b]
Salts and alkalis		
Carbonates of		
Ammonium	U	2
Potassium	E	2
Sodium (soda)	E	2
Chlorides of		
Aluminium	U/C	3
Ammonium	U/C	3
(ammonium chloride)		
Calcium	C	1
Potassium	C/E	1
Copper	C	1
Mercury	C	1
Magnesium	U/C	3
Sodium (brine, salt)	C/E	2
Iron	C	2
Zinc	C	2

Name	Reaction type[a]	Aggressiveness[b]
Salts and alkalis		
Fluorides		
Ammonium	U	4
Hydroxides of		
Ammonium		1
Calcium		1
Potassium (lye)	E	2
Sodium (lye)	E	2
Nitrates of		
Ammonium	U	5
Calcium		1
Potassium (salt peter)	U/E	3
Sodium	U/E	3

Name	Reaction type[a]	Aggressiveness[b]
Sulphates of		
Aluminium	E	4
Ammonium	U/E	5
Calcium	E	4
Potassium	E	4
Copper	E	4
Manganese	E	4
Magnesium	U/E	5
Sodium	E	4
Nickel	E	4
Iron	E	4
Zinc	E	4

Name	Reaction type[a]	Aggressiveness[b]
Petroleum - distillates		
Petroleum spirit		1
Kerosene		1
Naphthalene		1
Petroleum		1
Light oil		1
Heavy oil		1
Diesel oil		1

Name	Reaction type[a]	Aggressiveness[b]
Coal-tar distillates		
Anthracene		1
Benzene		1
Cumene		1
Creosote (oil)	U	2
Creosote	U	2
Paraffin		1
Tar		1
Toluene		1
Xylene		1

Name	Reaction type[a]	Aggressiveness[b]
Vegetable oils		
Almond oil	U	3
Tung oil	U	3
Cotton seed oil	U	3
Coconut oil	U	3
Linseed oil	U	3
Poppy seed oil	U	3
Olive oil	U	3
Groundnut oil	U	3
Rapeseed oil	U	3
Castor oil	U	3
Soya oil	U	3
Turpentine	U	3
Walnut oil	U	3

Name	Reaction type[a]	Aggressiveness[b]
Animal fat and fatty acids		
Bone oil	O	2
Pork fat	O	2
Fish oil	O	2
Slaughterhouse waste	O	3

Name	Reaction type[a]	Aggressiveness[b]
Miscellaneous		
Alcohol		1
Acetone		1
Ammonia (water)		1
Beer	O	2
Bleaching water	C	2
Borax		1
Caustic soda		1
Cider, apple wine	O	4
Ether		1
Essential oil		1
Phenol	U	3
Formaldehyde	U	3
Glucose	U	3
Glycerine	U	2
Honey		1
Wood pulp		1
Potassium -		
Permanganate		1
Chalk		1
Buttermilk	O	3
Carbon dioxide		1
Silage	O	5
Lead		1
Tannins		1

367

Name	Reaction type[a]	Aggressiveness[b]
Miscellaneous		
Molasses, sugar syrup	U	3
Milk		1
Manure	O/U	4
Sugar: dry		1
Sugar solution	U	3
Tetra		1
Toluene		1
Tri (chloroethylene)		1
Urea		1
Urine	O/U	3
Vaseline		1
Fruit juice	O	4
Water glass		1
Whey	O	3
Wine		1
Soft water[c]	O	3
Soap		1
Sulphur		1

[a] O = Oplossing (in Dutch) = Dissolution
   U = Uitwisseling (in Dutch) = Exchange
   E = Expansie (in Dutch) = Dilatation
   C = Corrosie (in Dutch) = Corrosion of the steel reinforcement
[b] 1 = Onschadelijk (in Dutch) = Harmless
   2 = Licht agressief (in Dutch) = Softly aggressive
   3 = Matig agressief (in Dutch) = Moderately aggressive
   4 = Sterk agressief (in Dutch) = Strongly aggressive
   5 = Zeer sterk agressief (in Dutch) = Very strongly aggressive
[c] Water with a total hardness smaller than 0.55 mmol/l, determined according to NEN 6441

REMARK

The "aggressiveness" in real circumstances is determined by the concentration, the pH, the temperature and the degree of refreshing.

2.5   Case histories of soil mix walls

The following paragraphs illustrate several case histories concerning the construction of earth-water retaining walls with the deep mixing method.

CASE STUDY
SOIL MIX NUCLEAR MEDICINE BUNKER MCA in Alkmaar / The Netherlands
Deep mixing contractor: Bauer Funderingstechniek with the Mixed-In-Place technique

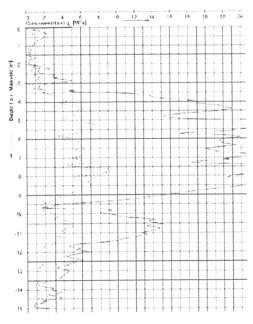

- Location: The Netherlands - Alkmaar.
- Customer: Medisch Centrum Alkmaar.
- Main contractor: Van Wijnen Heerhugowaard BV.
- Execution period: July 2011 - Aug 2011.
- Execution: 950 m² structural Mixed-In-Place wall reinforced with HEB400 and IPE270 steel beams for the construction of the retaining wall adjacent to the Medical Centre in Alkmaar.
- Soil composition: alternating gravelly sand and silty sand followed by clay at a depth of 12m below Amsterdam Ordnance Datum (hereafter NAP).
- Groundwater level: approx. 0.3 m below NAP.
- Ground level: approx. 1.5 m above NAP.
- Excavation level: down to 5.32 m below NAP.
- Level of the base of the wall: approx. 14 m below NAP.
- Soil mix technique: Bauer triple auger technique (MIP).
- Function:

Earth-retaining		Water-retaining		Structural	
Temporary	Permanent	Temporary	Permanent	Temporary	Permanent
V		V			

- Prior suitability test campaign for determining the design mix.
- Monitoring: drilling parameters and tests on samples: density, permeability, compressive strength and modulus of elasticity.

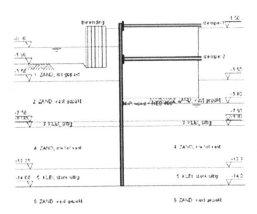

CASE STUDY
SOIL MIX NOORDWESTTANGENT in Leeuwarden / The Netherlands
Deep mixing contractor: Bauer Funderingstechniek with the Mixed-In-Place technique

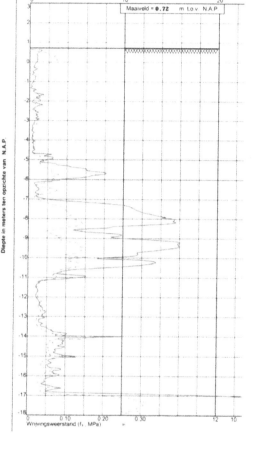

- Location: The Netherlands - Leeuwarden - Westergoawei
- Customer: The Province of Friesland
- Main contractor: Combinatie KWS - VHB
- Execution period:
  - PART 2: 24-1-12 to 12-3-12
  - PART 4: 14-5-12 to 18-6-12
  - PART 5: 26-3-12 to 30-4-12
- Execution: 26 300 m² Mixed-in-place cut-off walls for the construction of 3 underpasses
- Soil composition: from ground level downwards: Clay / Loam, Clay, permeable intermediate sandy layer, Loam, permeable sandy layer
- Groundwater level: approx. 1.00 m below NAP
- Ground level: approx. 0.7 m above NAP
- Excavation level: down to approximately 4.0 m below NAP
- Level of the base of the wall: 13.5 to 17.5 m below NAP.
- Soil mix technique: Bauer triple auger technique (MIP)
- Function:

Earth-retaining		Water-retaining		Structural	
Temporary	Permanent	Temporary	Permanent	Temporary	Permanent
			V		

- Prior suitability test for determining design mix
- Monitoring: Drilling parameters and tests on samples: density, permeability, compressive strength and modulus of elasticity.

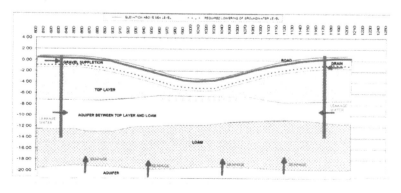

SOIL MIX BALK VAN BEEL Parking Tweewaters in Leuven / Belgium
Deep mixing contractor: CVR with the CVR Triple C-mix® technique

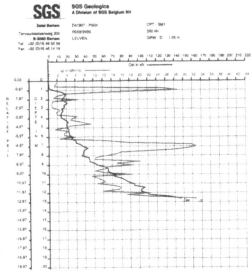

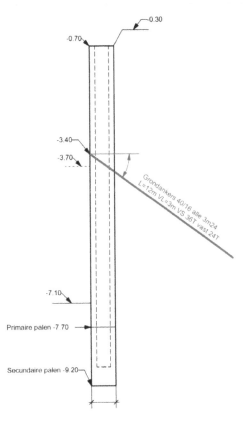

- Location: Belgium. Leuven (Louvain). Vaartkom.

- Customer: Ertzberg.

- Main contractor: Willemen.

- Construction period: April-July 2011.

- Execution: 320 m² retaining wall with one level of ground anchors.

- Excavation level: 7.40 m below ground level.

- Soil composition: 0 to -3.5 m: sandy loam, -3.5 to -6.50 m: loamy sand, -6.50 to -8.00 m dense gravelly sand, below -8.00 m dense sand.

- Water level: 3.50 m below ground level.

- Technique: CVR Triple C mix® diam.: 630 mm

- Function:

Earth-retaining		Water-retaining		Structural	
Temporary	Permanent	Temporary	Permanent	Temporary	Permanent
V		V			V

- Monitoring: drilling parameters and registration of anchoring forces.

- Tests for determining the UCS and the modulus of elasticity of cores. Additional tests (determination of long-term characteristics) on a soil mix column executed in 2002.

SOIL MIX DRENTS MUSEUM in Assen / The Netherlands
Deep mixing contractor: Franki with the Cutter Soil Mix (CSM) technique

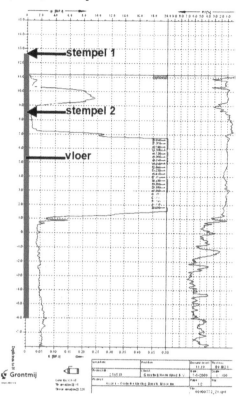

- Location: Drents Museum, Assen, the Netherlands.
- Design: Grontmij Nederland BV (geotechnics) and IB Wassenaar BV (construction).
- Contractor: Franki Foundations Group.
- CSM execution period: 2010.
- Soil composition: dense sand (thickness 10 m) on top of loam.
- Defining levels: ground level 12.0m NAP and ground water level 9.8 m NAP.
- Technique: CSM-hydromill, thickness 550 mm, HEB320 S240GP, centre-to-centre 1.0 m
- Function:

Earth-retaining		Water-retaining		Structural	
Temporary	Permanent	Temporary	Permanent	Temporary	Permanent
V		V		V	

Things to note:
The deformation was monitored during the excavation works. The requirements were
very strict with a permissible settlement of only 5 mm to 10 mm. The monitoring data were used in comparison with 2D and 3D
FEM calculations to check whether the design characteristics were achieved in practice.

CASE STUDY
SOIL MIX ABUTMENT VIADUCT Floriadebrug in Venlo / The Netherlands
Deep mixing contractor: Wedam BV with the Cutter Soil Mix (CSM) technique

- Location: The Netherlands - Venlo - Venrayseweg
- Customer: Floriade - Venlo
- Main contractor: Ballast Nedam Infra Zuid Oost - Eindhoven
- Execution date: July 2011
- Execution: 430 m² by way of bridge deck-anchored CSM earth-retaining wall. Mixing depth 9.5 metres
- Soil composition: Loose to moderately dense (slightly loamy) sand
- Water level: 3.5 metres
- Excavation depth: -3.0 metres
- Soil mix technique: Cutter Soil Mix (CSM)
- Function:

Earth-retaining		Water-retaining		Structural	
Temporary	Permanent	Temporary	Permanent	Temporary	Permanent
V					V

- Permanent walls for two abutments.
- Quality Assurance: strength calculation and consideration on the service life.
- Product used: HS-Blitzdammer 738-S
- Water - Cement factor: 100:100
- Compressive strength obtained: 12 MPa

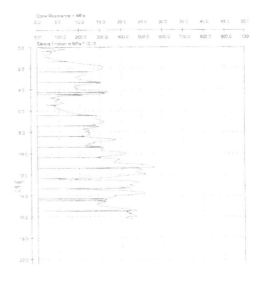

SOIL MIX PARKING GARAGE COREMOLEN Zeestraat in Noordwijkerhout / The Netherlands
Case Study Deep mixing contractor: Wedam BV with the Cutter Soil Mix (CSM) technique

- Location: The Netherlands - Noordwijkerhout - Zeestraat
- Customer: v.d. Poel Vastgoed - Noordwijk
- Main contractor: Heembouw - Roelofsarendsveen
- Execution date: May - November 2011
- Execution: 5 400 m² single-anchored earth-water retaining CSM wall. Mixing depth: 15 metres
- Soil composition: loose to moderately dense sand
- Water level: -3.0 metres
- Excavation depth: -8.0 to -9.0 metres
- Soil mix technique: Cutter Soil mix (CSM)
- Function:

Earth-retaining		Water-retaining		Structural	
Temporary	Permanent	Temporary	Permanent	Temporary	Permanent
	V		V		

- Permanent walls for a two-deck parking garage with temporary grout anchors at -2.6 metres. The function of the grout anchors was subsequently taken over by the parking decks
- Quality assurance: strength calculation and considerations on the service life
- Product used: HS-Blitzdammer 738-S
- Water - Cement factor: 100 :100
- Compressive strength obtained: 10 MPa

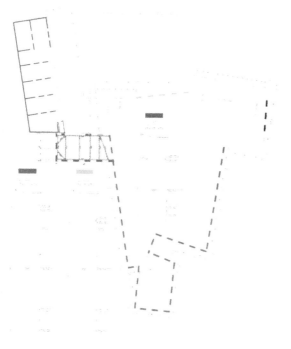

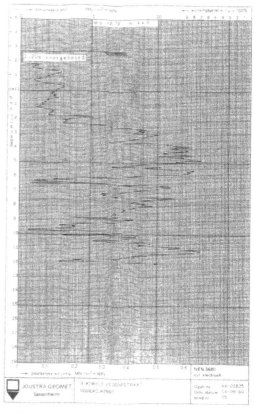

SOIL MIX RETAINING WALL of the switching station Sonsbeeksingel in Arnhem / The Netherlands

Case Study Deep mixing contractor: Wedam BV with the Cutter Soil Mix (CSM) technique

- Location: The Netherlands - Arnhem - Sonsbeeksingel
- Customer: Prorail - Utrecht
- Main contractor: Van Spijker Bouw - Meppel
- Execution date: February - May 2010
- Execution: 1 000 m² by way of double-anchored CSM earth-retaining wall. Mixing depth 16.5 metres
- Soil composition: loose to moderately dense sand
- Water level: -16.0 metres
- Excavation depth: -11.5 metres
- Soil mix technique: Cutter Soil Mix (CSM)
- Function:

Earth-retaining		Water-retaining		Structural	
Temporary	Permanent	Temporary	Permanent	Temporary	Permanent
V		V			

- Temporary retaining wall along railway track for the construction of a new substation
- Quality Assurance: strength calculation, determination of the flexural strength
- Product used: HS-Blitzdammer 738-S mod 1
- Water - Cement factor: 95:105
- Compressive strength obtained: 18 MPa
- Flexural strength: 3 MPa

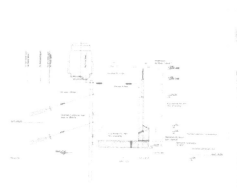

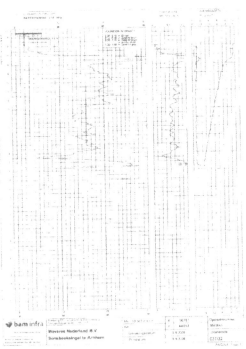

CASE STUDY
SOIL MIX IMEC TOWER in Leuven (Louvain) / Belgium
Deep mixing contractor: Lameire with the Cutter Soil Mix (CSM) technique

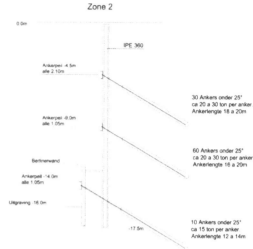

- Location: Belgium - Leuven (Heverlee) - Koning
  Bouwdewijnlaan.
- Customer: IMEC.
- Main Contractor: THV Imec Toren.
- Construction date: July - September 2011.
- Execution: 2245 m² triple anchored CSM retaining wall with start
  at -17.5 metres.
- Soil composition: (lightly loamy) sand to dense sand.
- Water level: -9 metres.
- Excavation depth: -16 metres.
- Soil mix technique: Cutter Soil Mix (CSM) with a width of
  55cm.
- Function:

Earth-retaining		Water-retaining		Structural	
Temporary	Permanent	Temporary	Permanent	Temporary	Permanent
V			V		

- Monitoring: drilling parameters + testing:
  determination of UCS obtained on core samples and on large-
  scale excavated blocks (55/75/120), modulus of elasticity,
  permeability, tensile splitting strength, porosity, creep
  (see Denies et al. 2014).
- Large-scale bending tests on half an excavated CSM panel
  (see Denies et al. 2015).
- Acoustic measurements.
- Determination of the percentage of unmixed soft soil inclusions.

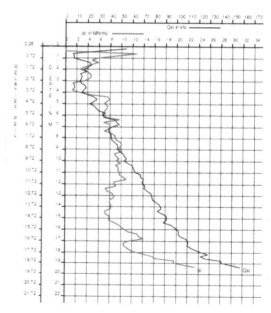

CASE STUDY
SOIL MIX WESTLANDGRACHT in Amsterdam / The Netherlands
Deep mixing contractor: Lareco

- Location: Westlandgracht, Amsterdam
- Customer: Waternet
- Main Contractor: Lareco Nederland BV
- Execution period: April 2010
- Execution: 2 220 m² down to 6.0 metres below ground level
- Soil mix technique: trench mixing tool
- Soil composition: top layer of sand, with clay and peat below the sand layer
- Water level: -2.0 metres
- Function: Dyke improvement
- k-value obtained : 3.4 x 10⁻⁹ m/s
- Compressive strength obtained: 1.1 MPa
- Function:

Earth-retaining		Water-retaining		Structural	
Temporary	Permanent	Temporary	Permanent	Temporary	Permanent
			V		

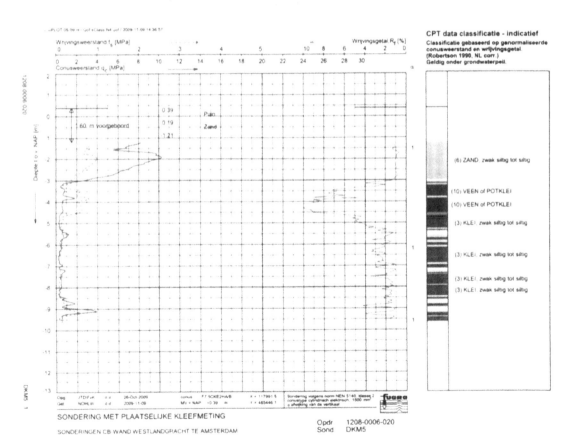

377

CASE STUDY
SOIL MIX DE MUNT in Roeselare / Belgium
Deep mixing contractor: SOILTECH with the Cutter Soil Mix (CSM) technique

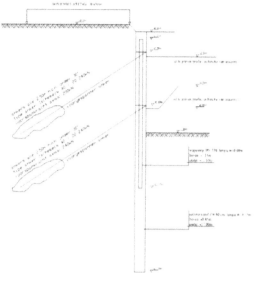

- Location: Belgium - Roeselare, Henri Horriestraat
- Customer: CBS INVEST
- Main contractor: VAN LAERE NV
- Construction date CSM: July-September 2010
- Execution: 7 178 m² double-anchored CSM retaining wall
  starting at 17.0m below ground level
- Soil composition: sandy loam and loamy sand with
  Yperian 15.0 m below ground level
- Water level 3.0 m below ground level
- Excavation level: 7.50 m below ground level
- Soil mix technique: Cutter Soil Mix (CSM), wall width 800 mm
- Function:

Earth-retaining		Water-retaining		Structural	
Temporary	Permanent	Temporary	Permanent	Temporary	Permanent
V			V		V

- Monitoring: Drilling parameters, pumping pits flow.
- Tests: coring during excavation (core samples of 113 mm
  diameter) for determining UCS, permeability and modulus of
  elasticity.
- Tests: coring at 16.00m below ground level by diagonal drilling
  for determining UCS and permeability.
- Results included in the BBRI SOIL MIX project (2009-2013) –
  VLAIO/IWT 080736.

SOIL MIX HOPMARKT in Aalst / Belgium

Deep mixing contractor: SOILTECH with the Cutter Soil Mix (CSM) technique

- Location: Belgium - Aalst - Hopmarkt.
- Customer: INTERPARKING.
- Main contractor: MBG nv.
- Construction date CSM: February - March 2012.
- Execution: 4 500 m² CSM starting at 20.00m below ground level; 225 m sheet piling wall AZ26-AZ36 with a length of 15 m with one level of anchors.
- Soil composition: sandy loam and loamy sand with an aquitard stratum 19 metres below ground level.
- Water level: 3.50 m below ground level.
- Excavation level: 10.80 m below ground level.
- Soil mix technique: Cutter Soil Mix (CSM) with a wall width of 550 mm.
- Function:

Earth-retaining		Water-retaining		Structural	
Temporary	Permanent	Temporary	Permanent	Temporary	Permanent
V		V			

- Vibration-free insertion of the final earth-water retaining sheet pile wall.
- Monitoring: drilling parameters.
- Tests: determination of UCS obtained on core samples and on large-scale blocks (panels 55/75/120), modulus of elasticity, permeability (see *Denies et al. 2014*).
- Large-scale bending tests performed on excavated half CSM panel reinforced with steel beam (typical or with reinforced adhesion) (see *Denies et al. 2015*).
- Results included in the BBRI Soil Mix project (2009-2013) – VLAIO/IWT 080736.
- For more information about the design solution: see *Denies and Huybrechts (2015)*.

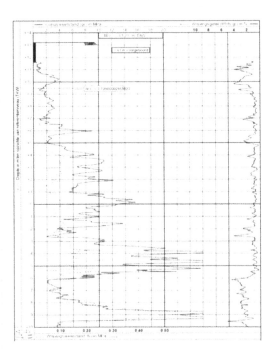

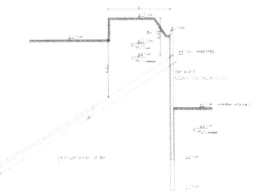

CASE STUDY
SOIL MIX FIETSERSTUNNEL (cycle tunnel) in Mol / Belgium
Deep mixing contractor: Smet F&C with the triple Tubular Soil Mix (TSM) technique

- Location: Belgium - Mol - Zuiderring N71
- Customer: 3V Veilig Verkeer Vlaanderen
- Main contractor: VBG
- Deep mixing contractor: Smet F&C
- Date of execution: 2010
- Execution: soil mix walls for 3 cycle tunnels under the Mol
  by-pass, soil mix wall built of soil mix columns of 0.63 m
  diameter, 55 cm centre-to-centre, 1 of every 2 columns reinforced
  with steel beam, length of columns 5 - 7 m, two levels of ground
  nails in the open part, the section in the tunnel is excavated under
  a roof plate, which serves as a support, a concrete protection wall
  and a concrete top beam are finally applied
- Soil type: quaternary sand
- Groundwater level (m): 1m below ground level
- Excavation level (m): variable 3-5 m
- Soil mix technique: tubular soil mixing (triple - TSM)
- Function:

Earth-retaining		Water-retaining		Structural	
Temporary	Permanent	Temporary	Permanent	Temporary	Permanent
V					V

- Monitoring: average compressive strength obtained on cores:
  19 MPa.
- Averaged modulus of elasticity: 15 GPa.
- Monitoring during execution: registration of drilling parameters
  and checking grout density.
- Monitoring at excavation: not known.

PRINCIPEDOORSNEDE: OPEN GEDEELTE

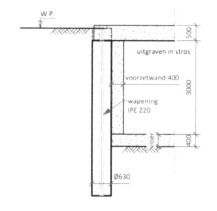

PRINCIPEDOORSNEDE: TUNNEL-GEDEELTE

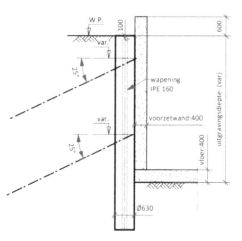

2.6 Other fields of application of the deep mixing method

In the present handbook, the authors mainly focus on the use of the deep mixing method for the construction of soil mix walls for earth-water retaining structures and cut-off walls. Nevertheless, the deep mixing method presents a broad versatility with a large field of geotechnical applications. The table 2.19, from *Denies and Huybrechts (2015)* provides an oversight of the main applications of the deep mixing method as internationally used. The table 2.20 associates the different fields of application with specific literature references.

Table 2.19. Main application of the deep mixing method as used worldwide, table from Denies and Huybrechts (2015).

Soil reinforcement and foundations
    as an alternative to classical foundation,
    for underpinning with the help of the spreadable systems,
    for the realisation of the foundation of linear structures such as railway tracks and
        pipelines with the help of the trenchmixing systems.

Earth/water retaining walls
    for excavation,
    as silo structure,
    as pit for (micro)tunnelling activities.

Cut-off walls, floodwalls and reinforcement of land levee and embankment

Slope stabilisation and landslide mitigation

In-situ remediation
    for PRB walls,
    for containment walls,
    for soil treatment by stabilisation/Solidification S/S.

Global mass stabilisation
    for the total shallow treatment of an area (e.g. for industrial installations)

Liquefaction mitigation
    with the construction of soil mix caissons,
    with specific arrangement of isolated soil mix elements and soil mix walls.

Land reclamation
    particularly in the case of near shore construction (port and harbour facilities and man-made islands).

Table 2.20. Examples of literature references for each field of application of the deep mixing method.

APPLICATIONS	EXAMPLES OF REFERENCES
Soil reinforcement (ground improvement) and foundations (structural soil mix element with a bearing function)	Chapman et al. (2012), Eurosoilstab (2002), Guimond-Barrett et al. (2012), Iwasa et al. (2015), Lambert et al. (2012), Leach (2014), Li et al. (2015), Maghsoudloo and Can (2014), Maswoswe (2001), Peixoto et al. (2012), Pinto et al. (2012), Topolnicki (2009), Topolnicki (2015), Topolnicki and Soltys (2012), Varaksin et al. (2016)
Earth/water retaining walls	Denies et al. (2012a), Denies et al. (2012b), Denies et al. (2014), Denies et al. (2015), Denies and Huybrechts (2015), Ganne et al. (2010), Maswoswe (2001), Rutherford et al. (2005), Pinto et al. (2012), Sondermann (2014), Yang et al. (2015)
Cut-off walls, floodwalls and reinforcement of land levee and embankment	Arnold (2015), Bertero et al. (2012), Bertoni et al. (2015), Borel (2012), Bruce (2012), Carnevale et al. (2014), de Jong et al. (2015), FHWA (2013), Filz et al. (2012), Gerritsen (2013), Kondoh et al. (2015), Kynett et al. (2015), Leoni and Bertero (2012), McGuire et al. (2012), Ortigao et al. (2015), Paul (2014), Pye et al. (2012), Shrestha et al. (2015)
Slope stabilisation and landslide mitigation	Gaib et al. (2012), McGuire et al. (2012), Pinto et al. (2012)
In-situ remediation	Al-Tabbaa and Evans (2003), Al-Tabbaa (2005), Al-Tabbaa et al. (2009), Al-Tabba et al. (2012), Al-Tabbaa et al. (2014), Al-Tabbaa et al. (2015), Barber and Miller (2015), Chapman et al. (2015), Guimond-Barrett (2013), Helson (2017), Perera et al. (2005)
Global mass stabilisation	ALLU (2010), Forsman et al. (2015), Martel and George (2015), Wilk (2015), Zitny and Deklavs (2015)
Liquefaction mitigation	Benhamou and Mathieu (2012), Hall et al. (2015), Matsui et al. (2013), Takahashi et al. (2014), Tokunaga et al. (2015), Yamashita et al. (2013), Yan et al. (2015)
Land reclamation	CDIT (2002), Holm et al. (2015), Kitazume and Terashi (2013), Sasa et al. (2015), Tagushi et al. (2015), Terashi and Kitazume (2015)

The reader can also refer to the following State-of-Practice Reports providing an oversight of the different fields of application of the deep mixing method (bearing function, earth retention, liquefaction mitigation, embankment...): *Terashi (1997), Bruce et al. (1998), FHWA (2001), CDIT (2002), Terashi (2003), Topolnicki (2004), Larsson (2005), Essler and Kitazume (2008), Denies and Van Lysebetten (2012), Kitazume and Terashi (2013), Denies and Huybrechts (2015) and the DFI Guidelines (2016).*

Finally, *Denies and Huybrechts (2015)* provide a list of references detailing various soil-mixing equipment used internationally, as illustrated in Table 2.21.

Table 2.21. Deep mixing methods and equipment used internationally, table from Denies and Huybrechts (2015).

Bruce et al. (1999)	DJM, Lime-cement columns, TREVIMIX DRY
CDIT (2002)	CDM, DJM
Topolnicki (2004)	DJM, Nordic Method, TREVIMIX DRY, Shallow Soil Mixing (SSM) method, CDM, DCM, SCC, Hayward Baker-Keller mixing tools, Bauer mixing tools, Mix-in-Place (MIP), SMW, DSM, Colmix, Spread Wing (SWING), JACSMAN, Hydramech, TRE-VIMIX Wet, TURBOJET, GEOJET, FMI (cut-mix-injection) machine
Larsson (2005)	CDM, Colmix, CSM, TRD, Hayward Baker-Keller mixing tools, SCC, Geo-Solutions tools, Raito tools, Schnabel DMW (Deep Mix Wall), May Gurney tools, Trevi tools, Rectangular 1 (Cutting wheels), Rectangular 2 (Box Columns), FMI (cut-mix-injection) machine, DJM, Nordic Method, Shallow Soil Mixing (SSM) method, ALLU mass stabilisation mixing tools, SWING, JACSMAN, LDis
Kitazume and Terashi (2013)	CDM, DJM
FHWA (2013)	CDM, DJM, DSM, SMW, TREVIMIX Wet, Colmix, Soil Removal technique, SSM, ISS Auger Method, RAS Column Method, Rectangular 1 (Cutting wheels), Rectangular 2 (Box Columns), SAM, Cementation, Single Axis Tooling, Rotomix, CSM Method, Spread Wing (SWING), JACSMAN, LDisGeoJetTM, Hydramech, RAS Jet, TURBOMIX/TURBOJET, TRD, Nordic Method, TREVIMIX DRY, MDM (Modified Deep Mixing), Dry Soil Mixing Mass, Schnabel DMW (Deep Mix Wall)

2.7   International QC/QA procedures

2.7.1   QC/QA activities – definition

As reminded in the recent US guidelines of the Geotech Tools® (2013), control quality of the construction is achieved by meeting pre-established requirements, as detailed in project plans and specifications, including applicable test standards. Quality Control (QC) and Quality Assurance (QA) are terms applied to the procedures, measurements, and observations used to ensure that construction satisfies the requirements in the project plans and specifications. QC and QA are often misunderstood and used interchangeably. Nevertheless, the following definitions distinguish both activities.

QC refers to procedures, measurements, and observations used by the contractor to monitor and control the construction quality such that all applicable requirements are satisfied. The monitoring during execution is then part of the QC activities.

QA refers to measurements and observations performed by the owner or the owner's engineer to provide assurance that the construction has been realised in agreement with the project plans and specifications.

2.7.2 QC/QA activities for deep mixing projects – first steps

In 2001, Maswoswe described the development and the implementation of QC/QA procedures for the installation of 420 000 m³ of soil mix material with the help of triple-auger soil mix rigs, within the framework of the Central Artery/Tunnel (CA/T) project, in Boston. At the beginning of the 90's, it was the largest land-based soil mix installation in the USA, and perhaps in the world. *Maswoswe (2001)* related the **QC activities** to:

- the choice of the **suitable deep mixing equipment**,
- the determination of the **process parameters** ensuring acceptable performances:
  the grout mix composition, the auger rotation (or withdrawal) rate and the grout
  (and drilling water) flow rate,
- and the selection of **procedures** allowing auger penetration and verticality.

If the choice of the deep mixing equipment, the installation parameters and the procedure were left up to the contractors, specifications would require the soil mix material to meet the following **acceptance criteria for QA** related to:

- the minimum and maximum values of **UCS** on wet samples, after a determined
  number of days,
- the control of the **homogeneity and uniformity**,
- the minimum **unit weight** of core samples,
- the **vertical tolerance** with a limitation of the horizontal deviation with depth in
- any direction,
- the control of the **auger penetration** and the **final depth**.

2.7.3 Workflows for deep mixing projects

In 2011, Terashi and Kitazume have proposed an international workflow for deep mixing projects. This workflow is represented in Fig. 2.57, where q_{uf} means UCS of core samples and q_{ul} UCS of laboratory mix samples. These authors discussed the similarities and differences in the QC/QA procedures followed in different countries (Finland, Japan, Sweden, UK, USA and more generally in Europe), and proposed future research needs with this regard.

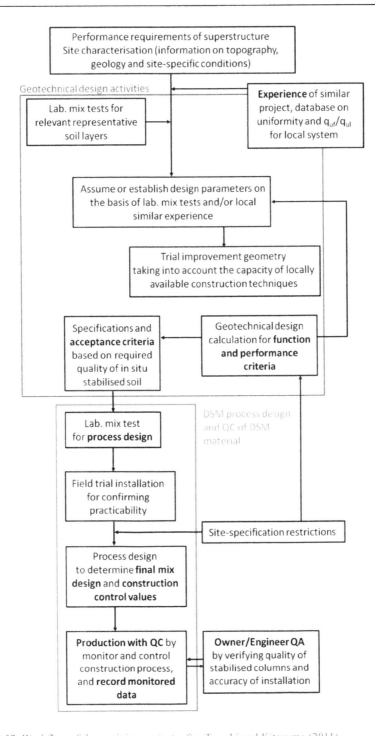

Figure 2.57. Workflow of deep mixing project, after Terashi and Kitazume (2011).

Their study was conducted in collaboration with experts of 45 organisations from seven countries and resulted nowadays in the relation of QC/QA activities to laboratory mix tests, field tests, monitoring and control of the execution parameters and assessment of the final products by measuring the mechanical characteristics of the soil mix material by tests on core samples (usually UCS tests) or by sounding.

The soil mix process design and the QC of soil mix material, as described in Terashi and Kitazume (2011), constitutes a major schedule for the deep mixing contractors with the following steps:
- **laboratory mix** test for process design,
- **field trial installation** for confirming practicability,
- **process design** to determine final mix design and construction control values,
- **production with QC** by monitoring, control and recording the monitored data of the construction process.

Filz et al. (2012) have also presented a flow chart for the design and construction of deep-mixing support systems for embankments and levees, such as illustrated in Fig. 2.58. It includes four main project phases: (1) collection of information, (2) analysis and design, (3) contractor procurement and (4) construction with continuous quality control and quality assurance. Once again the roles of the stakeholders are well specified: the deep mixing contractor is responsible for the continuous QC activities and the quality assurance is the task of the owner of its representative. Moreover, it is really important that the activities related to the quality assurance (lab tests on core or wet samples, *in-situ* testing...) be performed by an organisation (research centre, public service or private company) independent from the contractor.

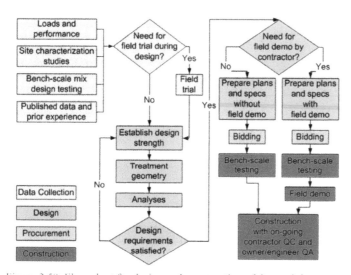

Figure 2.58. Flow chart for design and construction of deep mixing support systems for embankments and levees, after Filz et al. (2012).

If for reduced-scale project this kind of workflow seems to be time-consuming (the design mix is often based on the experience of the deep mixing contractor), for large project the application of this schedule is really fail-safe and cost-effective such as illustrated by the following case history.

2.7.4 Deep mixing process design in practice

On August 29, 2005, Hurricane Katrina passed southeast of New Orleans. The storm caused more than 50 breaches in drainage and navigational canal levees and precipitated one of the worst engineering disasters in the history of the United States of America. In response, construction and reconstruction of levees were planned; in several cases, with the help of the deep mixing technique. The LPV111 project is one of them. It consisted in the raising of an existing 8.5 kilometre levee, resting on a foundation of soft organic clay. The LPV111 project is part of the New Orleans East Back Levee, which is an essential component of the New Orleans Hurricane Protection System. For that project, the deep mixing method was selected to stabilize and support the burden of the new levee. As described in *Leoni and Bertero (2012)*, the QA/QC activities related to this project follow the previous schedule with the realisation of a Bench Scale Test programme (BST), a Field Scale Programme as Validation Test (VT) and a QC procedure during the production.

The main objective of the BST programme was to investigate the impacts of the type and the amount of binder and the influence of the water/binder ratio on the UCS of mixtures from the various types of soil that would be treated during production mixing.

According to *Leoni and Bertero (2012)*, the field tests should follow a comprehensive but flexible approach, allowing adjustments and modifications to the parameters initially foreseen as the operations proceed and new or more detailed information is acquired. For that purpose, at the LPV111 project, the Field Scale Programme or Validation Test (VT) was conceived to achieve the following objectives:
- verify and refine the type and content of binder, preliminarily determined through the BST, to attain the target mechanical characteristics of the treated soil,
- determine the most appropriate deep mixing operating parameters and equipment configuration,
- and develop QA/QC procedures for the deep mixing production stages.

In addition to laboratory mix test campaign, field trial installation is an essential step of the process design because the quality of the soil mix material, in terms of strength, stiffness, continuity and uniformity mainly depends on the execution process. Moreover, the laboratory test results are influenced by the test procedure, especially by the method of preparation (mixing time/energy, time between mixing and molding, molding procedure) and by the curing of the specimens (temperature, humidity, potential application of surcharge). Hence, in addition to the laboratory mix test programme, a field

387

test programme should systematically be performed prior to actual production mixing when the experience of the deep mixing contractor is limited for the encountered soil or field conditions.

According to *Leoni and Bertero (2012)*, during the production stage, over 500 soil mix samples were cored and tested for UCS assessment (related to the QA activities). All the results were tracked, analysed, and combined with the other available data from the BST and VT programmes on a day-by-day basis to fine tune the production parameters and the QA/QC procedures, and to determine whether corrective actions had to be taken.

All data obtained through the QA/QC programme at LPV 111 allowed the optimization of both installation time and cement consumption. Generally, the adjustments were the result of observations carried out in the field and through QA/QC testing during the production stages. Nonetheless, all changes, especially if entailing a reduction of the cement dosage, were validated before they were implemented for the production elements, as prescribed by the project requirements. This iteration gave an opportunity for a better design of the BST and VT stages as the project progressed. The most significant consequence of this process was primarily related to the cement consumption. The dosage of the binder was, in fact, reduced without affecting the overall quality of the ground improvement. As a conclusion, the consideration of the BST, VT, and QA/QC programmes allowed the optimization of the mixing parameters and equipment configuration to continuously improve economy and effectiveness as the project progressed.

Recent information concerning the QA/QC programme at LPV 111 can be found in *Bertero et al. (2012), Bertoni et al. (2015), Filz et al. (2015)* and *Bertero et al. (2015)*.

Concerning the added value of a field scale programme with regard to the laboratory mix test campaign, *Terashi and Kitazume (2011)* highlight the influence of the test procedure in laboratory, especially the method of preparation and the curing of the specimens. Methods differ from one country to another and even in the same region. These differences must be taken into account when comparisons are made. The sample preparation (mixing time/energy, time between mixing and moulding, moulding procedure) and the curing conditions (temperature, humidity, potential application of surcharge) differ from one procedure to another. Moreover, the quality of the soil mix material, in terms of strength, stiffness, continuity and uniformity depends on the execution process. Indeed, the way the soil is mixed during the process varies from one system to another. The result will be different if the soil stratification is preserved after the mix or if the soil is transported from depth to the surface during the process. Hence, field tests are essential in comparison with laboratory mix tests which can not entirely simulate the field conditions.

2.7.5 Quality Control by execution monitoring

Once the final mix design has been established (including the definition of the most convenient mix and the determination of the well-adapted execution parameters) with the previous described steps (lab and field programmes) or on the basis of the experience of the deep mixing contractor, the production of the final soil mix elements can start with the continuous monitoring and control of the construction process and the record of the monitored data. This is the phase of production with Quality Control.

According to Maswoswe (2001), the critical factor in the execution of soil mix walls is to maintain an auger withdrawal rate consistent with the grout flow rate. One way to control the success of the procedure and its efficiency is to estimate the cement factor or cement content (the cement mass per cubic metre of soil mix material) at different locations. The cement factor can be estimated considering the grout flow rate, the auger withdrawal rate and the assumed percentage of grout loss during the process.

Beyond the mechanical characterisation of the soil mix material (verified within the framework of the QA activities), the continuity and the overlapping of the soil mix wall elements (columns or panels) must be proved with regard to the execution tolerances. Locations and verticality of the soil mix elements should be controlled during execution.

The best way to ensure QC during execution is by monitoring, adjusting, recording and reporting the execution parameters. Current technology allows not only to monitor and record execution data but also to visualize the execution parameters during the production process, such as illustrated in Fig. 2.59 and 2.60.

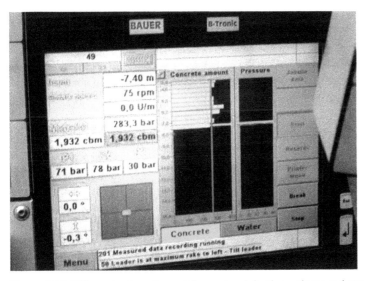

Figure 2.59. QC monitoring of grout flow rates, injection volume and pressure and penetration rate during execution of a CSM wall (photo with courtesy of Malcolm Drilling Company).

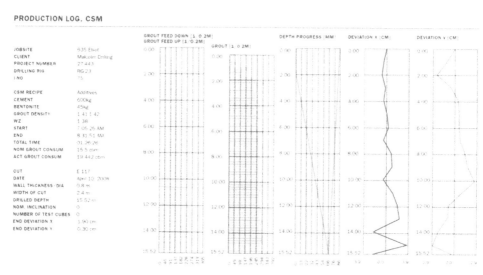

Figure 2.60. Typical production log (photo with courtesy of Malcolm Drilling Company).

For dry and wet mixing, the European standard *EN 14679* on deep mixing (2005) puts the emphasis on the continuous monitoring (or at least at a depth interval of 0.5 m) of the construction parameters as reported in Table 2.22. In addition, the material information should always be taken into account in the QC process. The binder handling and storage procedures should be described and available on demand from the deep mixing contractor. The slurry preparation procedures must be reported; notably the information with regard to the properties of the slurry (binder type, binder factor, water/binder ratio for wet method) and the holding time.

Finally, it is important to note that the recorded execution data must be summarised in a **report describing the production of each soil mix element for final control of the product.** According to the experience of *Bruce (2012)*, tabulated daily reports should contain a panel of information such as reported in Table 2.23.

Table 2.22. Construction parameters to be recorded according to the European standard EN 14679: 2005.

Dry mixing	Wet mixing
Air tank pressure	Slurry pressure; air pressure (if any)
Penetration and retrieval rate	Penetration and retrieval rate
Rotation speed (revs/min. during penetration and retrieval)	Rotation speed (revs/min. during penetration and retrieval)
Quantity of binder per meter of depth during penetration and retrieval	Quantity of slurry per meter of depth during penetration and retrieval

390

Table 2.23. Ideal content of the tabulated daily execution report, after Bruce (2012).

Specification about the mixing process (wet or dry mixing)
Type of mixing tool (single or multiple shaft, CSM, trenchmixing, mass stabilisation) and specificities related to the execution process
Rig or machine number
Name of the operator
Name of the construction site
Date and the time (start and finish) for each soil mix element
Soil mix element number and reference drawing number
Diameter or dimensions of the soil mix element
Top and bottom elevations of the soil mix element
Obstructions, interruptions, or any other difficulties arising during production and the resolution ways
Slurry mix design designation (reference to the design mix – should be related to the characteristics of the design slurry)
Slurry specific gravity measurements

2.7.6 Observational, laboratory and in-situ methods for quality assurance

Quality assurance is organised by the way of post-construction verification methods. For classical deep mixing projects, QA activities are mainly related to the assessment of
- the layout of the produced soil mix structure,
- the uniformity and homogeneity of the soil mix elements,
- the hydro-mechanical characteristics of the soil mix material.

Moreover, as underlined by *Terashi and Kitazume (2011)*, the acceptance criteria and verification procedures should also depend on the function of the deep soil mix element. The specifications of the project should define the QA activities in function of the application but also with regard to the temporary or permanent character of the construction.

Execution tolerances
The design and execution tolerances have to be first controlled with the help of the QC logs provided by the deep mixing contractors and on the basis of a field inspection.
The following criteria must be considered:
- the tolerance on the vertical positioning of each soil mix element including verification of the top and bottom elevations,
- the tolerance on the horizontal (in both directions) positioning of each soil mix element,
- the tolerance on the verticality of each soil mix element,
- verification of the overlap between neighbouring elements.

In the presence of cavities or large hard stones or in weak layers, large enlargements are unavoidable. As a consequence, tolerance on local enlargements must be indicated in the project specifications.

With the help of the monitored execution data, a direct overview of the underground construction can possibly be drawn, such as illustrated in the following case history.

As indicated in section 2.3.6 in the Part 2 of the handbook, *Benhamou and Mathieu (2012)* describe the use of the deep mixing method in the construction of two buildings on a soft alluvial soil in an area with a particularly high risk of earthquakes. To counter potential soil liquefaction and post-liquefaction damages, a new type of permanent foundation based on a Geomix caisson (36 × 40 m) was designed. In this particular case, the monitored verticality data was processed afterwards to construct the as-built 3D Geomix caisson structure, as illustrated in Fig. 2.61, so that verification of the continuity and overlap for each soil mix element was possible at any depth.

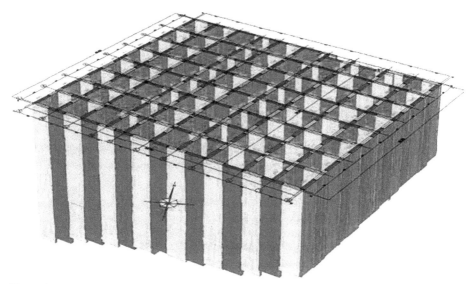

Figure 2.61. 3D view of the Geomix caisson structure, after Benhamou and Mathieu (2012).

Homogeneity and uniformity
For the control of the **homogeneity and uniformity** of the soil mix elements, several authors report the excavation of soil-cement columns or CSM panels. In the case history of *Benhamou and Mathieu (2012)*, partial excavation of the Geomix caisson arrangement was performed after completion of the works for the purpose of controlling the visual aspect of the produced soil mix material. Figure 2.62 illustrates the excavated Geomix caisson. In *Guimond-Barrett et al. (2012)*, three columns were excavated to verify the column geometry and to examine the homogeneity of the soil mix material (see Fig. 2.63).

Figure 2.62. Excavated Geomix caisson, after Benhamou and Mathieu (2012).

Figure 2.63. Excavated soil-cement columns 1 month after construction, after Guimond-Barrett et al. (2012).

As explained in *Holm (2000)*, uniformity and homogeneity of soil mix material can also be controlled computing the coefficient of variation of UCS test results on core samples or comparing field and laboratory UCS test results (calculation of q_{uf}/q_{ul}, where q_{uf} means UCS of core samples and q_{ul} UCS of laboratory mix samples).

Finally, within the framework of the BBRI 'Soil Mix' project (2009-2013), a methodology taking into account the amount of unmixed soft soil inclusions into the mix was developed (*Denies et al., 2012c*) and illustrated with case histories (*Denies et al., 2012b*). This methodology is based on the observation of core samples or excavated soil mix elements.

393

Hydro-mechanical characterisation of the soil mix material

In practice, the quality of the soil mix material is often controlled in a view of quality assurance with the help of UCS tests regardless of the soil mix application. Nevertheless, various tests can be performed to determine characteristics of soil mix material. The specifications of the project should specify the types and the number of required tests in function of the expected performances of the construction (its design functions, its lifetime…).

The most common testing method for the mechanical characterisation of the soil mix material is related to the **realisation of UCS tests on soil mix samples**. If the test result is directly influenced by the history of the sample, such as discussed in the Section 2.3.10 of the Part 2 of the handbook, differences also exist in practice with regard to the type of tested specimens. Indeed, UCS tests can be performed on slurry (water/binder mixture) samples directly collected in the batch mixing plan, on wet grab samples - from the surface or recovered at several depths - and finally on core samples (other sampling techniques are discussed in Section 8.4.2 in the first part of the handbook). For several authors and according to the experience acquired within the framework of the BBRI 'Soil Mix' project (2009-2013), cores can be considered more representative than wet grab samples. Besides for the US practice, Bruce (2012) has recently proposed that wet grab samples should not be used for acceptance.

Nevertheless, coring also presents challenges as explained in Section 2.3.10 of the second part of the handbook. Ideally samples from fresh material and cores should be collected, from test and final panels, in order to assess the strength and the homogeneity of the material, as well as to perform tests at different ages.

Gaib et al. (2012) reports interesting case history on a large slope stabilisation project performed with the help of the CSM technique. UCS tests were conducted on core samples at 7, 14 and 28 curing days. The seven day tests were used to provide an indication of any suspect columns, the fourteen day tests to allow comparison to the specification values, and the 28 day tests to ensure that the target strength is met, if there was concern regarding a 14 day result.

Density measurements of the slurry (water/binder) mixture and review of the automated batch plant records should be taken into account to verify the w/b ratio of the injected mixture. Similar measurements can also be done on the spoil produced during mixing to control the blending of the water/binder mixture with the soil. Finally the density of core sample should be carefully measured and correlated with UCS test results in such a way to understand the influences of the binder factor and the w/b ratio on the strength of the soil mix material.

In order to integrate the **modulus of elasticity** in the design, its value can be obtained from soil mix samples. As explained in the Section 6.3.7 (in the first part of the handbook), the modulus of elasticity should be deduced from measurements performed directly on the tested sample (with LVDTs, DEMEC mechanical strain gauges or electrical strain gauges) and not on the basis of the relative displacement of the plates.

The tensile strength of the soil mix material can be measured with the help of **direct or tensile splitting tests**.

For water retaining function, the **coefficient of permeability** of the soil mix material should be determined to verify the efficiency of the cut-off wall. Nevertheless, the real waterproof qualities of the wall will be directly related to the correct overlap and positioning of the soil mix elements. Sometimes measurements of the porosity of the soil mix material are alternately required even if it remains difficult to correlate both parameters, such as previously discussed in §2.3.9 in the second part of the handbook.

In parallel to laboratory works, mechanical characterisation with the help of field tests can be envisaged. Large reviews of methods are given in *Porbaha (2002)* and *Larsson (2005)*. The characterisation of the soil mix material can be performed with the help of penetration tests, loading tests, geophysical methods and non-destructive tests.

Penetration methods can be performed with the help of the following techniques in function of the site condition and with regard to the expected soil mix strength: static cone penetration test CPT (in but also between the element if necessary), standard penetration test SPT, rotary penetration test RPT (mainly used in Japan), pressuremeter test PMT, conventional column penetrometer (CCP) test, reverse column penetrometer (RCP) test, modified column penetrometer (MCP) test, column vane penetrometer (CVP) test, dynamic cone penetration (DCP) test and the static-dynamic penetration (SDP) test (*Porbaha, 2002*). It is still difficult to perform penetration test in soil mix element with a structural function (retaining or bearing function) with regard to the strength of the soil mix material after a few days of hardening.

Quasthoff (2012) recently refers to the conventional column penetration test which is the most widely spread testing method for lime-cement columns in Sweden. According to the European standard *EN 14679*, column penetration test is carried out using a probe that is pressed down into the centre of the soil-cement column at a speed of about 20 mm/s and with continuous registration of the penetration resistance. As illustrated in Fig. 2.64, the probe is equipped with two opposite vanes.

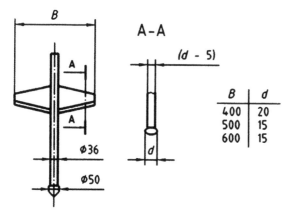

Figure 2.64. Vanes used in the conventional column penetration test, after EN 14679-2005.

The method can normally be used on columns with a maximum length of 8 m and with UCS smaller than 300 kPa. The undrained shear strength of the soil mix material can be deduced from the column penetration test. Quasthoff (2012) illustrates the use of this method with the help of a case history.

Nevertheless it can be noted that if these test methods are well-convenient for the characterisation of soil mix in the field of soil stabilisation, the suitability of the penetration tests must be questioned with regard to the strength of the soil mix material installed with the wet mixing method for the construction of an excavation. For that kind of application, in the Netherlands and in Belgium, the UCS of the soil mix material is generally greater than 300 kPa.

For the control of the bearing capacity and lateral response of soil mix elements, vertical and lateral static load tests can be performed. Plate load tests can be possibly envisaged when deep response is not required.

Concerning the use of geophysical methods (seismic tomography and electrical resistivity) and non-destructive tests (pile integrity testing, cross-hole sonic logging, parallel seismic testing) for the control of the soil mix elements scientific knowledge remains up to now limited. As a consequence, for soil mix characterisation the results of these methods should be regarded with caution. Ultrasonic tests performed on laboratory mix homogeneous samples with the same age present limited variability (*Hird and Chan, 2005*). The same measurement performed on *in-situ* cored samples can be considered as a good indicator of the homogeneity (or non-homogeneity) of the soil mix material (*Denies et al., 2012b*). Moreover, if the full-characterisation of the soil mix material cannot exclusively be obtained from the results of ultrasonic tests, once correlated with the UCS test results they give a direct trend of the long-term evolution of the material strength. Non-destructive tests can then be performed on preserved samples (see § 2.3.10 of Part 2 for storage conditions).

Finally, in case the acceptance criteria are not encountered associated remedial actions for non-compliance should be foreseen. The project specifications shall take into account the effect of all the tolerances on the implantation of the underground constructions and the possible related additional costs.

2.7.7 Quality Control of the soil mix construction by monitoring plan

For permanent and temporary earth/water retaining walls, the implementation of a monitoring plan allows the reduction of the risk for the neighbouring buildings and structures and for the workers in the excavation. Preventive or rescue measures (if necessary) can then be applied. Given that this topic goes beyond the scope of the present state-of-the art, guidance rules for the monitoring of an excavation can be found in the guidelines of the *CUR 223 (2010)*. The aspects related to the geotechnical instrumentation for monitoring field performance are as for them discussed in *Dunnicliff (1993)*.

All the aspects related to the QC/QA requirements for soil mix walls in Belgium and in the Netherlands are treated in detail in Chapter 8 of the first part of the handbook.

Finally, it can be noted that a lot of control procedures are also discussed in the *DFI Guidelines for soil mixing (2016)* with a full-Chapter dedicated to the topic of QC/QA and guide specifications in the Appendix. As for them, *Cali and Filz (2015)* provide an interesting critique of the job specifications for deep mixing works. In their paper, the authors give recommendations to improve the deep mixing specifications and to reach a better balance between the costs and the safety character of the construction.

Bibliography

ACI 318-08. 2008. American Concrete Institute. Building Code Requirements for Structural Concrete (ACI 318-08) and Commentary. An ACI Standard. Fourth printing January 2011

ALLU. 2010. Mass Stabilisation Manual, ALLU Finland Oy Al-Tabbaa, A., and King, S. D. 1998. Time effects of three contaminants on the durability and permeability of a solidified sand. Environmental technology, 19(4), 401-407

Al-Tabbaa, A., and Evans, C. 2003. Deep soil mixing in the UK: geoenvironmental research and recent applications. Land Contamination & Reclamation, 11, 1-14

Al-Tabbaa, A. 2005. State of practice report - stabilisation/ solidification of contaminated materials with wet deep soil mixing. Proceedings of the International Conference on Deep Mixing Best Practice and Recent Advances, Stockholm. Swedish Deep Stabilisation Research Center, Linköping, Sweden, 697-731

Al-Tabbaa, A., Barker, P. and Evans, C. W. 2009. Innovation in soil mix technology for remediation of contaminated land. International Symposium on Soil Mixing and Admixture Stabilisation, 19-21 may, Okinawa, Japan

Al-Tabbaa, A., Liska, M., McGall, R. and Critchlow, C. 2012. Soil Mix Technology for Integrated Remediation and Ground Improvement: Field Trials. Proceedings of the International symposium of ISSMGE - TC211. Recent research, advances & execution aspects of ground improvement works. N. Denies and N. Huybrechts (eds.). 31 May-1 June 2012, Brussels, Belgium

Al-Tabbaa, A., O'Connor, D. & Abunada, Z. 2014. Field trials for deep mixing in land re-mediation: execution and early age monitoring, QC and lessons learnt. DFI-EFFC International Conference on Piling and Deep Foundations, 21-23 May 2014, Stockholm, pp. 621-630

Al-Tabbaa, A., Jin, F., O'Connor, D. and Abunada, Z. 2015. Deep mixing field trials for land remediation: some 3-year results from the SMiRT project. DFI Deep mixing conference, San Francisco, pp. 521-530

Arnold, M. 2015. Seepage cut-off wall installation using Cutter Soil Mixing for Herbert Hoover dike rehabilitation. DFI Deep mixing conference, San Francisco, pp. 213-222

Aulrajah, A., Abdullah, A, Bo, M.W. & Bouazza, A. 2009. Ground improvement techniques for railway embankments, Ground Improvement, Vol. 162, issue 1, pp. 3-14

ASTM C 597 – 09. Standaardtestmethode voor drukgolfsnelheid door beton

ASTM D 559-89. 1994. Standard Test Methods for Wetting and Drying Compacted Soil-Cement Mixtures. Annual Book of ASTM Standards Vol. 04.08 of 1994

ASTM D 560-89. 1994. Standard Test Methods for Freezing and Thawing Compacted Soil-Cement Mixtures. Annual Book of ASTM Standards Vol. 04.08 of 1994

BBRI, 2007. Technische voorlichting 231. Herstellingen en bescherming van beton

BBRI, 2009 - WTCB Rapport nr. 12 : Richtlijnen voor de toepassing van de Eurocode 7 in België, Deel 1 : Het grondmechanische ontwerp in de uiterste grenstoe-stand van axiaal op druk belaste funderingspalen

BBRI "Soil Mix" project (2009-2013). IWT 080736 soil mix project: SOIL MIX in constructieve en permanente toepassingen – Karakterisatie van het materiaal en ontwikkeling van nieuwe mechanische wetmatigheden. For information: www.bbri.be

Barber, D. & Miller, D. 2015. Rocky Mountain Arsenal Section 36 Lime Basin Soil Remediation Project – seepage barrier. DFI Deep mixing conference, San Fransisco, pp. 333-340

Bellato, D., Dalle Coste, A., Gerressen, F.-W. and Simonini, P. 2012. Long-term performance of CSM walls in slightly overconsolidated clays. International symposium of ISSMGE - TC211. Recent research, advances & execution aspects of ground improvement works. 31 May-1 June 2012, Brussels, Belgium

Benhamou, L. and Mathieu, F. 2012. Geomix Caissons against liquefaction. International symposium of ISSMGE - TC211. Recent research, advances & execution aspects of ground improvement works. 31 May-1 June 2012, Brussels, Belgium

Bertero, A., Leoni, F.M., Filz, G., Nozu, M. & Druss, D. 2012. Bench-Scale Testing and QC/QA Testing for Deep Mixing at Levee LPV111, Proceedings of the 4th International Conference on Grouting and Deep Mixing, ASCE, Geotechnical Special Publication No. 228, pp. 694-705

Bertero, A., Morales, C., Leoni, F.M. & Filz, G. 2015. Three deep mixing projects: comparison between laboratory and field test results. DFI Deep mixing conference, San Francisco, pp. 699-713

Bertoni, M., Leoni, F.M., Schmutzler, W. and Filz, G. 2015. Soil mixing for the LPV-111 levee improvements, New Orleans: A case history. DFI Deep mixing conference, San Francisco, pp. 247-256

Besluit Bodemkwaliteit. 2014. http://wetten.overheid.nl/BWBR0022929/geldigheids-datum_11-11-2014#

BGGG. 2013. Handboek beschoeiingen, draft versie 13 juni 2013, www.bggg-gbms.be

BGGG. 2015a. Standaardprocedures voor geotechnisch onderzoek: algemene bepalingen – 8 januari 2015, www.bggg-gbms.be

BGGG. 2015b. Proceedings van de BGGG Thema-avond Beschoeiingen & Grondankers – Deel 2 Ontwerp, Brussel, 12 & 19 maart, www.bggg-gbms.be

Borel, S. 2007. Soil mixing innovations: Geomix, SpringSol and Trenchmix. Presentation at Joint BGA/ CFMS meeting, London

Borel, S. 2012. Soil Mixing Equipment. Latest Advances in Deep Mixing. Presentation of the Short Courses of the International Symposium on Ground Improvement IS-GI Brussels, 30 May 2012

BRL SIKB 1000. 2013. Beoordelingsrichtlijn. Monsterneming voor partijkeuringen. Stichting Infrastructuur Kwaliteitsborging Bodembeheer

BRL 2834. 2002. Nationale beoordelingsrichtlijn. Kathodische bescherming van wapeningstaal in beton

Brouwer, R, Veldhuizen, F. 2011. Lekdetectie bij bouwputten, toepassing van de elektrische potentiaalmethode. Geotechniek, pp. 60-65, januari 2011

Bruce, D. A., Bruce, M. E. C. and DiMillio, A. F. 1998. Deep mixing method: a global perspective. ASCE, Geotechnical special publication, n°81, pp. 1-26

Bruce, D. A., Bruce, M. E. C. & DiMillio, A. F. 1999. Dry Mix Methods: A brief overview of international practice. International Conference on Dry Mix Methods for Deep Soil Stabilization, 13-15 October, Stockholm, Sweden, A. A. Balkema, pp. 15-25

Bruce, M. E. C. 2012. QA/QC recommendations. FHWA Design Manual for Deep Mixing for Embankment and Foundation Support. DFI presentation

Cali, P. & Filz, G. 2015. Critique of specifications for deep mixing. DFI Deep mixing conference, San Francisco, pp. 1051-1060

Carnevale, F. Pampanin, P. and Bracci, L.A. 2014. Slurry walls executed with the cutting soil mixing technique: a case history. DFI-EFFC conference, Stockholm, pp. 735-753

CDIT. Coastal Development Institute of Technology. 2002. The Deep Mixing Method – Principle, Design and Construction. Edited by CDIT, Japan. A. A. Balkema Publishers/Lisse/ Abingdon/Exton (PA)/Tokyo

Chapman, G.A., Denny, R.J., Knowles, M. & Uren, J.G. 2012. Cutter Soil Mixed Columns for an LNG Export Tank Foundation. Grouting and Deep Mixing 2012. ASCE Geotechnical Special Publication no 228, Vol. 1, pp. 788-797

Chapman, G.A., Bouazza, A. & Russel, A.S. 2015. Stabilization trials for soft soils in a high sulphate environment. DFI-EFFC conference, Stockholm, pp. 611-620

Chu, J., Varaskin, S., Klotz, U. and Mengé, P. 2009. Construction Processes, Proceedings of the 17th International Conference on Soil Mechanics and Geotechnical Engineering, 5-9 October 2009, Alexandria, Egypt, M. Hamza et al. (Eds.), IOS Press, Amsterdam, Vol. 4, pp. 3006-3135

Cruz, R. B., Knop, A., Heineck, K. S. and Consoli, N. C. 2004. Encapsulation of a soil contaminated by hydrocarbons. Amélioration des sols en place. In Dhouib, Magnan et Mestat (ed.) Amélioration des sols en place. Paris : Presses de l'ENPC/LCPC

CUR-Aanbeveling 84. 2002. Cement-bentoniet wanden

CUR 166. 2012. Damwandconstructies, zie sbrcurnet.nl

CUR 189. 1997. Cement-bentoniet schermen. Stichting CUR, Gouda, zie sbrcurnet.nl

CUR Rapport 199. 2001. Handreiking toepassing no-recess technieken. Stichting CUR, Gouda, zie sbrcurnet.nl

CUR 223. 2010. Richtlijn meten en monitoren van bouwputten. Voor kwaliteits- en risico-management. Stichting CURNET, Gouda, zie sbrcurnet.nl

CUR 231. 2010. Handboek diepwanden. Ontwerp en uitvoering, zie sbrcurnet.nl

CUR 243. 2016. Durability of geosynthetics, zie sbrcurnet.nl

CUR 247. 2013. Risicogestuurd grondonderzoek, van planfase tot realisatie, zie sbrcurnet.nl

CUR 2001-10. 2001. Diepe grondstabilisatie in Nederland. Handleiding voor toepassing, ontwerp en uitvoering, zie sbrcurnet.nl

CUR Leidraad 1. Duurzaamheid van constructief beton met betrekking tot chloride-geïnitieerde wapeningscorrosie, zie sbrcurnet.nl

Deltares, 2005a. Duurzaamheidsonderzoek Mix In Place Kolommen, projectnummer CO-402811-0019, versie 01 Concept, GeoDelft-maart 2005

Deltares. 2005b. Inside 'Mixed in Place'. Uitwerking van een aantal witte vlekken, project-nummer CO-402812-0009, versie 02 Definitief, GeoDelft-november 2005

Denies, N., Huybrechts, N., De Cock, F., Lameire, B., Maertens, J. and Vervoort, A. 2012a. Soil Mix walls as retaining structures – Belgian practice. International symposium of ISSMGE - TC211. Recent research, advances & execution aspects of ground improvement works. 31 May-1 June 2012, Brussels, Belgium

Denies, N., Huybrechts, N., De Cock, F., Lameire, B., Vervoort, A., Van Lysebetten, G. and Maertens, J. 2012b. Soil Mix walls as retaining structures – mechanical characterization. International symposium of ISSMGE - TC211. Recent research, advances & execution aspects of ground improvement works. 31 May-1 June 2012, Brussels, Belgium

Denies, N., Huybrechts, N., De Cock, F., Lameire, B., Vervoort, A. and Maertens, J. 2012c. Mechanical characterization of deep soil mix material – procedure description. International symposium of ISSMGE - TC211. Recent research, advances & execution aspects of ground improvement works. 31 May-1 June 2012, Brussels, Belgium

Denies, N. and Van Lysebetten, G. 2012. General report Session 4 – Soil Mixing 2 – Deep Mixing. Proceedings of the International symposium of ISSMGE - TC211. Recent research, advances & execution aspects of ground improvement works. N. Denies and N. Huybrechts (eds.). 31 May-1 June 2012, Brussels, Belgium, Vol. I, pp. 87-124

Denies, N., Van Lysebetten, G., Huybrechts, N., De Cock, F., Lameire, B., Maertens, J. and Vervoort, A. 2013. Design of Deep Soil Mix Structures: considerations on the UCS characteristic value. Proceedings of the 18th International Conference on Soil Mechanics and Geotechnical Engineering, Paris, France, 2013, Vol. 3, pp. 2465-2468

Denies, N., Van Lysebetten, G., Huybrechts, N., De Cock, F., Lameire, B., Maertens, J. and Vervoort, A. 2014. Real-Scale Tests on Soil Mix Elements. Proceedings of the International Conference on Piling & Deep Foundations, Stockholm, Sweden, 21-23 May 2014, Eds. DFI & EFFC, pp. 647-656

Denies, N., Huybrechts, N., De Cock, F., Lameire, B., Maertens, J. and Vervoort, A. 2015. Large-Scale Bending Tests on Soil Mix Elements. ASCE Proceedings of the International Foundations Congress and Equipment Expo of San Antonio IFCEE 2015. Texas, March 17-21, 2015, pp. 2394-2409, doi: 10.1061/9780784479087.222

Denies, N. and Huybrechts, N. 2015. Deep Mixing Method: Equipment and Field of Applications. Ground Improvement Case Histories, Book chapter 11, Elsevier, pp. 311-350

DFI. 2016. Guidelines for Soil Mixing. Prepared by the Soil Mixing Committee of Deep Foundations Institute

DIN 4093. 2012. Design of ground improvement – Jet grouting, deep mixing or grouting. August 2012

DIN 4150-1. Erschütterungen im Bauwesen: Vorermittlung von Schwingungsgrößen

DIN 4150-2. Erschütterungen im Bauwesen: Einwirkungen auf Menschen in Gebäuden

DIN 4150-3. Einwirkungen auf bauliche Anlagen

DIN 18130-1. 1998. Laboratory tests for determining the coefficient of permeability of soil

Dunnicliff, J. 1993. Geotechnical instrumentation for monitoring field performance. A Wiley-Interscience Publication, John Wiley & Sons, Inc

De Jong, E., Gerritsen, R., te Boekhorst, C. & van der Velde, E. 2015. MIP cut-off walls create artificial polders in the Netherlands, DFI Deep mixing conference, San Francisco, pp. 319-332

EN 206-1:2000. Beton - Deel 1 : Specificatie, eigenschappen, vervaardiging en conformiteit

EN 1504-9:2008. Products and systems for the protection and repair of concrete structures - Definitions, requirements, quality control and evaluation of conformity – Part 9: General principles for the use of products and systems

EN 1538: 2010. Uitvoering van bijzonder geotechnisch werk - Diepwanden

EN 1990. 2002. Eurocode - Grondslagen van het constructief ontwerp

EN 1991-1-1. 2002. Eurocode 1 - Belastingen op constructies - Deel 1-1 : Algemene belastingen – Volumieke gewichten, eigen gewicht en opgelegde belastingen voor gebouwen

EN 1991-2. 2004. Eurocode 1: Belastingen op constructies - Deel 2: Verkeersbelasting op bruggen

EN 1992-1-1. 2004. Eurocode 2: Ontwerp en berekening van betonconstructies - Deel 1-1: Algemene regels en regels voor gebouwen

EN 1992-1-2. 2005. Eurocode 2: Ontwerp en berekening van betonconstructies - Deel 1-2: Algemene regels - Ontwerp en berekening van constructies bij brand

EN 1993-1-1. 2005. Eurocode 3: Ontwerp en berekening van staalconstructies - Deel 1-1: Algemene regels en regels voor gebouwen

EN 1993-1-2. 2005. Eurocode 3: Ontwerp en berekening van staalconstructies - Deel 1-2: Algemene regels - Ontwerp en berekening van constructies bij brand

EN 1993-5. 2007. Eurocode 3 - Ontwerp en berekening van staalconstructies - Deel 5: Palen en damwanden

EN 1994-1-1. 2005. Eurocode 4: Ontwerp en berekening van staal-betonconstructies - Deel 1-1: Algemene regels en regels voor gebouwen

EN 1994-1-2. 2005. Eurocode 4 - Ontwerp en berekening van staal-betonconstructies - Deel 1-2: Algemene regels - Ontwerp en berekening van constructies bij brand

EN 1997-1. 2004. Eurocode 7: Geotechnisch ontwerp - Deel 1: Algemene regels

EN 1997-2. 2007. Eurocode 7 - Geotechnisch ontwerp - Deel 2 : Grondonderzoek en beproeving

EN 12390-2 – 2009. Testing hardened concrete - Part 2: Making and curing specimens for strength tests

EN 12390-3. 2009. Testing hardened concrete - Part 3: Compressive strength of test specimens

EN 12390-6 – 2010. Testing hardened concrete - Part 6: Tensile splitting strength of test specimens

EN 12390-7. 2009. Testing concrete in structures – Part 1: Cored specimens - Taking, examining and testing in compression

EN 12504-1. 2009. Testing concrete in structures – Part 1: Cored specimens - Taking, examining and testing in compression

EN 12504-3 – 2005. Testing concrete in structures – Part 3: Determination of pull-out force

EN 14679. 2005. Execution of special geotechnical works - Deep mixing

Erkel, van B., Kerks, T., 2013. Waterdichtheid geolock slot, Kiwa N.V., September 2013

Essler, R. and Kitazume, M. 2008. Application of Ground Improvement: Deep Mixing. TC17 website: www.tc211.be

Eurosoilstab. 2002. Development of design and construction methods to stabilise soft organic soils. Design Guide Soft Soil Stabilisation. EC project BE 96-3177

FHWA-RD-99-167. 2001. An introduction to the Deep Soil Mixing Methods as used in geotechnical applications: verification and properties of treated soil. Prepared by Geosystems (D.A. Bruce) for US Department of Transportation, FHWA, p. 434

FHWA-HRT-13-046. 2013. Federal Highway Administration Design Manual: Deep Mixing for Embankment and Foundation Support. US Department of Transportation, Federal Highway Administration, October 2013, p. 244

FIB. International Federation for structural concrete. Bulletin n°34. 2006. Model Code for Service Life Design

Filz, G., Adams, T., Navin, M. and Templeton, A. E. 2012. Design of Deep Mixing for Support of Levees and Floodwalls. Grouting and Deep Mixing 2012. ASCE Geotechnical Special Publication no 228, Vol. 1, pp. 89-133

Filz, G., Bruce, D., Schmutzler, W., Nozu, M. & Reeb, A. 2015. Deep Mixing Core Data from LPV 111, DFI Deep mixing conference, San Francisco, pp. 543-552

Forsman, J., Lindroos, N. & Korkiala-Tanttu, L. 2015. Three mass stabilization phases in the West Harbor of Helsinski, Finland – Geotechnical and environmental properties of mass stabilized dredged sediments as construction material. DFI Deep mixing conference, San Francisco, pp. 671-680

Gaib, S., Wilson, B. and Lapointe, E. 2012. Design, Construction and Monitoring of a Test Section for the stabilization of an Active Slide Area utilizing Soil Mixed Shear Keys installed using Cutter Soil Mixing. International symposium of ISSMGE - TC211. Recent research, advances & execution aspects of ground improvement works. 31 May-1 June 2012, Brussels, Belgium

Ganne, P., Huybrechts, N., De Cock, F., Lameire, B. and Maertens, J. 2010. Soil mix walls as retaining structures – critical analysis of the material design parameters, International conference on geotechnical challenges in megacities, June 07-10, 2010, Moscow, Russia, pp. 991-998

Ganne, P., Denies, N., Huybrechts, N., Vervoort, A., Tavallali, A., Maertens, J, Lameire, B. and De Cock, F. 2011. Soil mix: influence of soil inclusions on structural behavior. Proceedings of the XV European conference on soil mechanics and geotechnical engineering, Sept. 12-15, 2011, Athens, Greece, pp. 977-982

Ganne, P., Denies, N., Huybrechts, N., Vervoort, A., Tavallali, A., Maertens, J., Lameire, B. and De Cock, F. 2012. Deep Soil Mix technology in Belgium: Effect of inclusions on design properties. Grouting and Deep Mixing 2012. ASCE Geotechnical Special Publication no 228, Vol. 1, pp. 357-366

Geo-Con, Inc. 1998. Promotional Information Geotech Tools: Geo-construction Information & Technology Selection Guidance for Geotechnical, Structural, & Pavement Engineers. 2013. http://geotechtools.org/ Copyright © 2010–2013 Iowa State University

Gerressen, F.-W. and Vohs, T. 2012. CSM-Cutter Soil Mixing – Worldwide experiences of a young soil mixing method in challenging soil conditions. International symposium of ISSMGE - TC211. Recent research, advances & execution aspects of ground improvement works. 31 May-1 June 2012, Brussels, Belgium

Gerritsen, R. 2013. Polderconstructie en dichtwanden Aquaduct Westelijke Invalsweg. Mixed-in-place (MIP) en folieschermen. Presentatie in het kader van de CUR commissie Soilmix-wanden: 28 november 2013

Grisolia, M., Kitazume, M., Leder, E., Marzano, I.P. and Morikawa, Y. (2012). Laboratory study on the applicability of molding procedures for the preparation of cement stabilised specimens. International symposium of ISSMGE - TC211. Recent research, advances & execution aspects of ground improvement works. 31 May-1 June 2012, Brussels, Belgium

Guimond-Barrett, A., Mosser, J.-F., Calon, N., Reiffsteck, P., Pantet, A. And Le Kouby, A. 2012. Deep mixing for reinforcement of railway platforms with a spreadable tool. International symposium of ISSMGE - TC211. Recent research, advances & execution aspects of ground improvement works. 31 May-1 June 2012, Brussels, Belgium

Guimond-Barrett, A. 2013. Influence of mixing and curing conditions on the characteristics and durability of soils stabilised by deep mixing. Ph D Thesis, Le Havre Université-IFST-TAR (September 2013)

Hall, B.E., Azizian, A., Naesgaard, E., Amini, A., Baez, J.I. and Preece, M.L. 2015. Evergreen line rapid transit project: analysis and design of deep mix panels to mitigate seismic lateral spreading, DFI Deep mixing conference, San Francisco, pp. 67-78

Helson, O. 2017. Comportement mécanique et physique des bétons de sol : étude expérimentale paramétrique, predictive, et de durabilité. PhD Thesis, Université de Cergy-Pontoise. [in French]

Hird, C. C. and Chan, C. M. 2005. Correlation of shear wave velocity with unconfined compressive strength of cement-stabilised clay. Proceedings of the International Conference on Deep Mixing: Best Practice and Recent Advances, May 23-25, 2005, Stockholm, Sweden, pp. 79-85

Holm, G. 2000. Deep Mixing. ASCE, Geotechnical special publication, N°112, pp. 105-122

Holm, G., Macsik, J., Makusa, G., Kennedy, H. & Rogbeck, Y. 2015. Stabilization of materials – upgrading of soils and sediments to use in geo-constructions. DFI Deep mixing conference, San Francisco, pp. 405-414

Horpibulsuk, S., Phojan, W., Suddeepong, A., Chinkulkijniwat, A. and Liu, M. D. 2012. Strength development in blended cement admixed saline clay. Applied Clay Science, 55, 44-52

ISO 834. 1999. Fire-resistance tests -- Elements of building construction -- Part 1: General requirements

ISO 1920-10. 2010. Testing of concrete -- Part 10: Determination of static modulus of elasticity in compression Kitazume, M. and Terashi, M. 2013. The Deep Mixing Method. CRC Press/Balkema. Taylor and Francis Group, London, UK

Iwasa, D., Lopez, R., Lindquist, E. & Bussiere, J. 2015. Deep soil mixing ground improvement for the U.S. Federal Courthouse in downtown Los Angeles, California. DFI Deep mixing conference, San Francisco, pp. 415-425

Kitazume and Terashi. 2013. The Deep Mixing Mehotd. CRC Press, Taylor and Francis Group

Kleist, F. 1999. Die Systemdurchlässigkeit von Schmalwänden, Ein Beitrag zur Herstellung von Schmalwänden und zur Prognose der Systemdurchlässigkeit, Lehrstuhls und der Versuchsanstalt für Wasserbau und Wasserwirtschaft der Technischen Universität München

Kondoh, M., Miyatake, H., Ohbayashi, J., Shinkawa, N. & Yagiura, Y. 2015. Case histories of arch action low improvement ratio cement column method (ALiCC Method). DFI Deep mixing conference, San Francisco, pp. 1005-1014

Kynett, M., Chowdhury, K., Millet, R., Perlea, M., Walberg, F. 2015. Improvements to commonly used cut-off wall specifications. DFI Deep mixing conference, San Francisco, pp. 1027-1040

Lambert, S., Rocher-Lacoste, F. and Le Kouby, A. 2012. Soil-cement columns, an alternative soil improvement method. International symposium of ISSMGE - TC211. Recent research, advances & execution aspects of ground improvement works. 31 May-1 June 2012, Brussels, Belgium

Lantinga, C. 2012. Wat is de toepasbaarheid van de Cuttersoilmixwand in de Nederlandse bodem. Afstudeeronderzoek. Avans Hogeschool te 's-Hertogenbosh, The Netherlands

Larsson, S.M. 2005. State of practice report - Execution, monitoring and quality control, International Conference on Deep Mixing, pp. 732-785

Leach, C. 2014. Use of CSM to support a heritage building. DFI-EFFC conference, Stockholm, pp. 685-691

Leoni, F. M. and Bertero, A. 2012. Soil mixing in highly organic materials: the experience of LPV111, New Orleans, Louisiana (USA). International symposium of ISSMGE - TC211. Recent research, advances & execution aspects of ground improvement works. 31 May-1 June 2012, Brussels, Belgium

Li, M., Williams, R.R., Filz, G.M. and Wilson, B.W. 2015. Load-deformation based design of deep mixing for supporting MSE walls at KITIMAT lng, BC, DFI Deep mixing conference, San Francisco, pp. 101-110

Maghsoudloo and Can. 2014. A case study of wet soil mixing for bearing capacity improvement in Turkey. DFI-EFFC conference, Stockholm, pp. 693-701

Martel, D.W. and George T. 2015. Dam wall toe stabilisation using mass soil mixing techniques. DFI Deep mixing conference, San Francisco, pp. 181-191

Maswoswe, J. J. G. 2001. QA/QC for CA/T Deep Soil-Cement. ASCE, Geotechnical special publication, N°113, pp. 610-624

Mathieu, F., Mosser, J.-F., Utter, N. and Darson-Balleur, S. 2012. Deep Soil Mixing with Geomix Method: Influence of Dispersion in UCS Values on Design Calculations. Grouting and Deep Mixing 2012. ASCE Geotechnical Special Publication no 228, Vol. 1, pp. 334-342

Matsui, H., Ishii, H. & Horikoshi, K. 2013. Hybrid application of deep mixing columns combined with walls as soft ground improvement method under embankment. 18th International Conference on Soil Mechanics and Geotechnical Engineering, 2-6 September, Paris, Vol. 3, pp. 2545-2548

McGinn, A.J. and O'Rourke, T.D. 2003. Performance of Deep Mixing Methods at Fort Point Channel. Report prepared for Bechtel/Parsons Brinckerhoff, the Massachusetts Turnpike Authority, and the Federal Highway Administration at Cornell University, Ithaca, NY (also discussed in Rutherford et al. 2005)

McGuire, M., Templeton, E. & Filz, G. 2012. Stability Analyses of a Floodwall with Deep-Mixed Ground Improvement at Orleans Avenue Canal, New Orleans. International Symposium of ISSMGE - TC211, 31 May-1 June 2012, Brussels, Vol. III, pp. 199-209

NBN B 15-203. 1990. Concrete testing – statical module of elasticity with compression

NBN B 15-215. 1989. Concrete testing – Absorption of water by immersion

NBN EN 1997-1 ANB : 2014. Eurocode 7 : Geotechnisch ontwerp - Deel 1 : Algemene regels - Nationale bijlage

NEN 5740. 2009. Bodem - Landbodem – Strategie voor het uitvoeren van verkennend bodemonderzoek - Onderzoek naar de milieuhygiënische kwaliteit van bodem en grond

NEN 8005:2008. Nederlandse invulling van NEN-EN 206-1: Beton - Deel 1: Specificatie, eigenschappen, vervaardiging en conformiteit

NEN 9997-1+C1. Geotechnisch ontwerp van constructies – Deel 1: Algemene regels, ICS 91.080.01; 93.020, april 2012

NEN-EN 1997-2. Eurocode 7: Geotechnisch ontwerp - Deel 2: Grondonderzoek en beproeving - ICS 91.080.01; 93.020, juli 2010

NF P94-282 – Mars 2009. Calcul géotechnique - Ouvrages de soutènement – Écrans

Niina, A., Satoh, S., Babasaki, R., Tsutsumi, I. and Kawasaki, T. 1977. Study on DMM using cement hardening agent (Part 1). Proc. Of the 12th Japan National Conference on Soil Mechanics and Foundation engineering: 1325-1328 (in Japanese)

Nijland, T.G., van der Zon, W.H., Pachen, H.M.A. en van Hille, T. 2005. Duurzaamheid van grondstabilisaties, Cement, januari 2005

Ortigao, A., Falk, E., Felix, M. & Koehler, T. 2015. Deep soil mixing trials at Porto Alegre Airport, Brazil. DFI Deep mixing conference, San Francisco, pp. 997-1004

Packard, R.G. 1962. Alternate Methods for Measuring Freeze-Thaw and Wet-Dry Resistance of Soil-Cement Mixtures. Highway Research Board Bulletin No. 353

Packard, R.G. and Chapman, G.A. 1963. Developments in Durability Testing of Soil-Cement Mixtures. Highway Research Record, No. 36

Paul, D.B. 2014. Army Corps of Engineers seepage control cutoffs for dam and levees engineering manual (EM). DFI-EFFC conference, Stockholm, pp. 763-785

Pauw, A. 1960. Static modulus of elasticity of concrete as affected by density. Journal of the American Concrete Institute, Vol. 32, N°6, pp. 679-687

Pearlman, S.L. and Himick, D. E. 1993. Anchored Excavation Support using SMW. Deep Foundation Institute, 18th Annual Conference, Pittsburgh, PA, pp. 101-120

Perera, A. S. R., Al-Tabbaa, A., Reid, J. M., and Johnson, D. 2005. State of practice report UK stabilisation/solidification treatment and remediation, Part V: Long-term performance and environmental impact. Proceedings of the International Conference on Stabilisation/ Solidification Treatment and Remediation, April, Cambridge, UK, 437-457

Peixoto, A., Sousa, E. & Gomes, P. 2012. Solutions for soil foundation improvement of an industrial building using Cutter Soil Mixing technology at Fréjus, France. International Symposium of ISSMGE - TC211, 31 May-1 June 2012, Brussels, Vol. III, pp. 243-250

Pinto, A., Tomásio, R., Pita, X., Godinho, P. & Peixoto, A. 2012. Ground Improvement Solutions using CSM Technology. International Symposium of ISSMGE - TC211, 31 May-1 June 2012, Brussels, Vol. III, pp. 271-284

Porbaha, A. 1998. State of the art in deep mixing technology: part I. Basic concepts and overview. Ground Improvement, Vol. 2, pp. 81-92

Porbaha, A., Tanaka H. and Kobayashi M. 1998. State of the art in deep mixing technology, part II. Applications. Ground Improvement Journal, Vol. 3, pp. 125-139

Porbaha, A. Shibuya, S. and Kishida, T. 2000. State of the art in deep mixing technology. Part III: geomaterial characterization. Ground Improvement, Vol. 3, pp. 91-110

Porbaha, A. 2000. State of the art in deep mixing technology. Part IV: design considerations. Ground improvement. Vol. 3, pp. 111-125

Porbaha, A. 2002. State of the Art in quality assessment of deep mixing technology. Ground Improvement (2002) 6, No. 3, pp. 95-120

Pye, N., O'Brien, A., Essler, R. & Adams, D. 2012. Deep dry soil mixing to stabilize a live railway embankment across Thrandeston Bog. Grouting and Deep Mixing 2012. ASCE Geotechnical Special Publication no 228, Vol. 1, pp. 543-553

Quality Services Testing B.V., Verwachte levensduur PEHD folie en profiel Geolock Systeem van Cofra B.V., referentie Q ST-13046004.01, datum 06-09-2013

Quasthoff, P. 2012. State of the art in "Dry Soil Mixing" – Basics and case study. International symposium of ISSMGE - TC211. Recent research, advances & execution aspects of ground improvement works. 31 May-1 June 2012, Brussels, Belgium

Rijkswaterstaat. 2013. ROK Richtlijnen Ontwerp Kunstwerken. Rijkswaterstaat, Dienst-infrastructuur

Rutherford, C., Biscontin, G. and Briaud, J.-L. 2005. Design manual for excavation support using deep mixing technology. Texas A&M University. March 31, 2005

Sasa, K. Kurumada, Y. and Watanabe, M. 2015. Mixing proportion design and construction management for application of cement deep mixing method to construction project on controlled final landfill site in Japan. DFI Deep mixing conference, San Francisco, pp. 193-202

Shihata, S.A. and Baghdadi, Z.A. 2001. Long-term strength and durability of soil cement. Journal of materials in civil engineering, May-June 2001, pp. 161-165

Shrestha, R., Griffin, R., Cali, P., Kafle, S., Watanabe, M. & Yang, D. 2015. Deep Mixing for Levee Repair at Hurricane Protection Project, P-17A, Louisiana, USA, DFI Deep mixing conference, San Francisco, pp. 257-266

Sondermann, W. 2014. Deep mixing technologies for lateral support systems. DFI-EFFC conference, Stockholm, pp. 787-796

Tagushi, H., Asasuma, T. Tanaka, Y, Ootsuka, Y. & Tokunaga, S. 2015. Application of the cement deep mixing method for shallow water areas using the CDM-FLOAT technique. DFI Deep mixing conference, San Francisco, pp. 943-950

Takahashi, H., Morikawa, Y. Taguchi, H., Tokunaga, S. and Maruyama, K. 2014. Efficacy and execution of floating lattice-shaped cement treatment against liquefaction. DFI-EFFC conference, Stockholm, pp. 713-722

Taki, O. and Yang, D.S. 1991. Soil–Cement Mixed Wall Technique. ASCE Geotechnical Special Publication no 27, pp. 298–309

Terashi, M., Okumura, T. and Mitsumoto, T. 1977. Fundamentals properties of lime treated soil (1st report). Report of the Harbour Research Institute, 16(1), pp. 3-28 (in Japanese)

Terashi M. 1997. Theme lecture: Deep mixing method – Brief state of the art. Proceedings of the 14th International Conference of Soil Mechanics and Foundation Engineering, Hambourg, 6-12 September 1997. A. A. Balkema/Rotterdam/Brookfield/1999. Vol. 4, pp. 2475- 2478

Terashi, M. 2002. Long-term strength gain vs. deterioration of soils treated by lime and cement, Proceedings of the Tokyo Workshop 2002 on Deep Mixing, pp. 39-57

Terashi, M. 2003. The State of Practice in Deep Mixing Methods. Grouting and Ground Treatment (GSP 120), 3rd International Specialty Conference on Grouting and Ground Treatment New Orleans, Louisiana, USA, pp. 25-49

Terashi, M. and Kitazume, M. 2011. QA/QC for deepmixing ground: current practice and future research needs. Ground improvement, Vol. 164, Issue GI3, pp. 161-177

Terashi, M. & Kitazume. M. 2015. Deep mixing – Four decades of experience, research and development. DFI Deep mixing conference, San Francisco, pp. 781-800

Tokunaga, S. Kitazume, M., Morikawa, Y., Takahashi, H., Nagatsu, T., Honda, N., Onishi, T., Asanuma, T., Kubo, S. & Hogashi, S. 2015. Performance of cement deep mixing method in 2011 Tohoku earthquake. DFI Deep mixing conference, San Francisco, pp.1071-1080

Topolnicki, M. 2004. In situ soil mixing. In M. P. Moseley & K. Kirsch (Eds.), Ground improvement, 2nd ed., Spon Press

Topolnicki, M. 2009. Design and execution practice of wet Soil Mixing in Poland. International Symposium on Deep Mixing & Admixture Stabilization, Okinawa, 19-21 May, 2009, pp. 195-202

Topolnicki, M. and Pandrea, P. 2012. Design of in-situ soil mixing. International symposium of ISSMGE - TC211. Recent research, advances & execution aspects of ground improvement works. 31 May-1 June 2012, Brussels, Belgium

Topolnicki, M. and Trunk, U. 2006. Tiefreichende bodenstabilisierung. Einsatz im Verkehrswegebau für Baugrundverbesserung und Gründungen. Tieàau, Vol. 118, N°6, pp. 319 – 330 (in German)

Topolnicki, M. & Slotys, G. 2012. Novel application of wet deep soil mixing for foundation of modern wind turbines. Grouting and Deep Mixing 2012. ASCE Geotechnical Special Publication no 228, Vol. 1, pp. 533-542

Topolnicki, M. 2015. Geotechnical design and performance of road and railway viaducts supported on DSM columns – a summary of practice. DFI Deep mixing conference, San Francisco, pp. 131-150

Towhata, I. 2008. Geotechnical Earthquake Enginee-ring. Springer

Van Calster,P., De Cock, F., De Vos, M., Maertens, J. & Van Alboom, G. 2009. Richtlijnen Bemalingen. Document beschikbaar op www.bggg-gbms.be

Van Lysebetten G., Vervoort A., Denies, N., Huybrechts, N., Maertens, J., De Cock, F. and Lameire B. 2013. Numerical modeling of fracturing in soil mix material. Proceedigns of the International Conference on Instal-lation Effects in Geotechnical Engineering. March 24 – 27, 2013. Rotterdam. The Netherlands

Varaksin, S., Hamidi, B., Huybrechts, N. & Denies, N. 2016. Ground Improvement vs. Pile Foundations/ Keynote in the Proceedings of the ISSMGE - ETC 3 International Symposium on Design of Piles in Europe. Leuven, Belgium, 28 & 29 April 2016, Vol. I. pp. 157-205

Vervoort, A., Tavallali, A., Van Lysebetten, G., Maertens, J., Denies, N., Huybrechts, N., De Cock, F. and Lameire, B. 2012. Mechanical characterization of large scale soil mix samples and the analysis of the influence of soil inclusions. International symposium of ISSMGE -TC211. Recent research, advances & execution aspects of ground improvement works. 31 May-1 June 2012, Brussels, Belgium

Wilk, C. 2015. Soil mixing treatment of marginal soil for railway construction and maintenance. DFI Deep mixing conference, San Francisco, pp. 1015-1025

Yamashita, K., Wakai, S. & Hamada, J. 2013. Large-scale piled raft with grid-form deep mixing walls on soft ground. 18th International Conference on Soil Mechanics and Geotechnical Engineering, 2-6 September, Paris, Vol. 3, pp. 2637-2640

Yan, W., Friesen, S.J., Driller, M.W. 2015. Seismic remediation of Perris dam. DFI Deep mixing conference, San Francisco, pp. 21-34

Yang, D.S., Notaro, A.P., Wang, Y.D. and Watanabe, G. 2015. Application of deep soil mixing walls in the Warren Avenue Grade Separation project, Fremont, CA. DFI Deep mixing conference, San Francisco, pp. 171-180

Zitny, B. & Deklavs, T. 2015. Mass Soil Mixing for Wind Tower Turbine Foundations. ASCE Proceedings of the International Foundations Congress and Equipment Expo of San Antonio IFCEE 2015. Texas, March 17-21, 2015, pp. 1764-1772, doi: 10.1061/978078447908

Milton Keynes UK
Ingram Content Group UK Ltd.
UKHW050448071024
449327UK00014B/286